THE CITY

THE GOVERNMENT AND MISGOVERNMENT OF LONDON

POLITICS OF THE CITY

THE GOVERNMENT AND
MISGOVERNMENT OF LONDON

WILLIAM A. ROBSON

Routledge
Taylor & Francis Group

LONDON AND NEW YORK

First published in 1939

This edition published in 2007
Routledge
2 Park Square, Milton Park, Abingdon, Oxon, OX14 4RN

Simultaneously published in the USA and Canada by Routledge
605 Third Avenue, New York, NY 10017

Routledge is an imprint of Taylor & Francis Group, an informa business

First issued in paperback 2013

The publishers have made every effort to contact authors and copyright
holders of the works reprinted in the *The City* series. This has not been
possible in every case, however, and we would welcome correspondence
from those individuals or organisations we have been unable to trace.

These reprints are taken from original copies of each book. In many cases
the condition of these originals is not perfect. The publisher has gone to
great lengths to ensure the quality of these reprints, but wishes to point out
that certain characteristics of the original copies will, of necessity, be
apparent in reprints thereof.

British Library Cataloguing in Publication Data
A CIP catalogue record for this book
is available from the British Library

The Government and Misgovernment of London

ISBN13: 978-0-415-41826-3 hbk (volume)
ISBN13: 978-0-415-41930-7 (subset)
ISBN13: 978-0-415-41318-3 (set)
ISBN13: 978-0-415-86476-3 pbk (volume)

Routledge Library Editions: The City

THE GOVERNMENT
AND MISGOVERNMENT
OF LONDON

by

WILLIAM A. ROBSON
*Professor of Public Administration in the
University of London*

LONDON
GEORGE ALLEN & UNWIN LTD
MUSEUM STREET

FIRST PUBLISHED IN 1939
SECOND EDITION . 1948

TO
LORD SIMON OF WYTHENSHAWE

PRINTED IN GREAT BRITAIN BY
JARROLD AND SONS, LIMITED, THE EMPIRE PRESS, NORWICH

London, thou art of townes A per se,
Soveraign of cities, semeliest in sight,
Of high renoun, riches and royaltie;
Of lordis, barons, and many goodly knyght;
Of most delectable lusty ladies bright;
Of famous prelatis, in habitis clericall;
Of merchauntis full of substaunce and of myght:
London, thou art the flour of Cities all.

        .   .   .   .   .

Gemme of all joy, jasper of jocunditie,
Most myghty carbuncle of vertue and valour;
Strong Troy in vigour and in strenuytie;
Of royall cities rose and geraflour;
Empresse of townes, exalt in honour,
In beawtie beryng the crone imperiall;
Swete paradise precellyng in pleasure:
London, thou art the flour of Cities all.

Above all ryvers thy Ryver hath renowne,
Whose beryall stremys, pleasaunt and preclare,
Under thy lusty wallys renneth down,
Where many a swanne doth swymme with wyngis fare;
Where many a barge doth saile, and row with are,
Where many a ship doth rest with toppe-royall,
O towne of townes, patrone and not compare,
London, thou art the floure of cities all.

WILLIAM DUNBAR

"The point to which in every Kingdom a native looks
with pride, and a foreigner with curiosity, is undoubtedly
its Metropolis. Other cities may be the especial deposi-
tories of learning, of science, of the arts, of manufactures,

or of commerce, but the foreigner expects to find these all more or less represented in the chief city of the Kingdom; and no enlightened native considers his acquaintance with the country complete till he has visited her Capital.''

First Report of the Commissioners appointed to consider the most effectual means of Improving the Metropolis. B.P.P. Reports from Commissioners (1844), XV, p. 3.

PREFACE TO THE SECOND EDITION

In preparing a second edition for the press I have corrected a few errors which inadvertently appeared in the first edition and to which readers kindly drew my attention. My thanks are due to them for doing so.

I have brought the book up to date in regard to local government finance (pages 458-9), election returns (pages 171-2, 350-1) and civil aerodromes (page 412). Apart from these and a few other matters I have left the factual basis of the work untouched, and have made no attempt to substitute later figures. To do so would have been a lengthy and indeed impossible task, for in regard to many aspects of London Government no very recent information is available. In any event it would have been a waste of time, for the general picture of London Government which I drew has not been substantially modified by changes which have occurred since 1939.

The most important new developments in regard to the Metropolis consist, first, of the publication of the Barlow Commission's Report and the changes in government policy and public opinion which it produced; and, in the second place, of the creation of the City, London County, and Greater London Plans. Chapter XIII of Part III, which I wrote when the Barlow Commission was still sitting, has been revised and enlarged to enable me to discuss the Reports of the Commission. I have added an Epilogue, which reviews all the main events which have occurred since 1939 affecting the Metropolis, both in regard to planning and local government. By this means I have been able to bring the book up to date without introducing innumerable amendments of minor details which in themselves would be of slight value.

I am indebted to Mr. J. R. Howard Roberts, Clerk of the London County Council, Mr. A. Wood, the Comptroller of the Council, and other officers at London County Hall for kindly helping me with information. I wish to thank Mr.

Geoffrey Sutton, Managing Editor of Todd Reference Books, Ltd., for permission to reproduce part of an article on the Greater London Plan which I contributed to *Planning and Reconstruction*, 1946.

WILLIAM A. ROBSON

THE LONDON SCHOOL OF ECONOMICS
AND POLITICAL SCIENCE

July 1947

PREFACE TO THE FIRST EDITION

I hope no one will imagine that this book deals with a subject of merely local concern. The proper government of London, in the wide sense in which the term is used here, is a question of national importance. A dim realisation is beginning to dawn on the public that the size, growth, and development of the metropolis have nation-wide effects which are felt ultimately in remote quarters of Great Britain; and many people have recently become aware of the sinister military significance of the immense concentration of population, wealth and power in London. But so far there is little or no understanding of the connection between these matters and the general administration and planning of London Government.

There is a tendency for the government of the principal metropolitan cities to be either ignored by social scientists, or else treated as part of the general subject of local government. It is obvious, however, that giant conurbations such as London are in a category by themselves and demand separate and individual treatment.

Such treatment has so far been either lacking or else inadequate in depth and extent, in view of the magnitude and complexity of the subject. I have not so far discovered any satisfactory studies of Paris, New York or Berlin. London has been dealt with in two competent handbooks by men who have taken a leading part in her affairs—I refer to *London and its Government*, by Sir Percy Harris, M.P., L.C.C., and *How Greater London is Governed*, by the Rt. Hon. Herbert Morrison, M.P., L.C.C. But these are both simple treatises explaining little more than the anatomy of local government in London. They throw only a sidelong glance at the problems which seem to me of central and urgent importance.

My own approach is broadly as follows. I regard the metropolis as a vast and difficult problem in municipal administration, different both quantitatively and qualitatively from anything we have previously known. In order to discover how it should be dealt with it is necessary to study the working of the present machinery with care and in detail. But the present organisation cannot be understood without a knowledge of the manner and causes of its growth. Hence, the book is divided into three parts dealing respectively with the past, the present and the future.

As regards the historical part, I have not found it necessary to go back further than 1835, except in regard to a few matters relating to the ancient City. Thus, Part I covers in outline roughly the last 100 years. In the course of preparing it, I was struck by the immense mass of Parliamentary papers of all kinds referring to the metropolis, such as the reports of Royal Commissions, Select Committees, Special Commissioners, and Departmental Committees, which are available to the student of London government in the 19th and 20th centuries. Up to the present this material has been almost unworked. It is of great interest and would well repay the attention of modern social and political historians. The dearth of studies in the history of London government contrasts sharply with the plethora of books dealing with the social life, manners and customs of London in the past.

My historical outline may have a special interest at the present time in view of the fact that the 50th anniversary of the London County Council is about to be celebrated. The publication of this book will, indeed, coincide with that event. I hope that the later parts of the book will serve to deter those who take part in the celebrations, no less than the general public, from indulging in any excessive satisfaction at the present state of affairs. The celebrations will serve a valuable purpose only if they assist in making the need for reform more widely felt.

The present study is in no sense a *pièce d'occasion* though it happens to come out at a moment when municipal affairs in London are likely to be engaging an unusually large amount of attention. I hope that it will be read not only by councillors and officers, students and candidates for the service, but also by ordinary citizens, voters and ratepayers throughout the country.

I have to acknowledge a great deal of courtesy in giving information and kindness in discussing points and projects. My thanks are due in particular to Sir Arthur Robinson,

G.C.B., G.B.E., former Permanent Secretary to the Ministry of Health, Lord Snell, former Chairman of the London County Council; Alderman Ewart G. Culpin, the present Chairman of the Council, Alderman Mrs. Barbara Drake, L.C.C., Alderman Mrs. H. Dalton, L.C.C., Mr. Charles Robertson, L.C.C., Chairman of the Education Committee, Mr. Charles Latham, L.C.C., Chairman of the Finance Committee; Sir George Gater, Clerk to the London County Council; Mr. J. R. Howard Roberts, Solicitor to the London County Council; Sir William Prescott, C.B.E., Chairman of the Metropolitan Water Board; Mr. Frank Pick, Vice-Chairman of the London Passenger Transport Board; Mr. Leslie Bowker, City Remembrancer, Mr. C. W. Radcliffe, Clerk of the Middlesex County Council, Mr. Parker Morris, Town Clerk of Westminster, Mr. P. H. Harrold, Town Clerk of Hampstead, Dr. R. H. Tee, Town Clerk of Hackney, Mr. A. E. Lauder, formerly Town Clerk of Southgate, Mr. J. C. Dawes and Mr. F. N. Bath, of the Ministry of Health, Mr. S. H. Wood of the Board of Education, Mr. Philip Allen, of the Home Office, Mr. H. Claughton, Clerk to the Court of the University of London, Mr. E. M. Rich, Education Officer to the London County Council, Mr. Graeme Haldane, Mr. R. Hardy Syms, Mr. Thomas Sharp, Mr. A. V. Huson, Accountant to the Metropolitan Water Board, and Mr. A. G. Huson, B.Sc.(Econ.). This list is by no means exhaustive—there are many others to whom I am indebted. I would emphasise, however, that none of these persons above mentioned should be regarded as in any sense committed to my conclusions or proposals. Indeed, in several instances I am certain they would meet with definite disagreement.

I am specially indebted to Miss Emmeline Cohen, B.A., who acted as my research assistant for some months on the historical part of the enquiry. Her interest and ability in mastering the intricacies of official documents was remarkable, and so, too, was her capacity for keeping the broader aspects of the study in view. I have also to thank Professor Luther Gulick, the Director, and Miss Sarah Greer, the librarian, of the Institute of Public Administration in New York. During my visit to the United States in 1938 Miss Greer was indefatigable in assisting me with information and documents relating to the City Government of New York. Thanks to the kindness of Mayor la Guardia, Mr. Joseph D. McGoldrick, the Comptroller to the New York City Council, and Mr. A. A. Berle, Jr., then Chairman of the City Planning Commission, I had an opportunity of learning something about the City Government of New York under the new Charter. I hope that some of my friends and colleagues across the Atlantic who are concerned with the special problems of metropolitan cities will find this study useful to them.

My thanks are also due to the staff of the British Library of Political and Economic Science for many helpful services in tracing and obtaining obscure documents. Miss V. Martin, of the Statistics Department of the London School of Economics and Political Science, kindly assisted me in computing several of the tables contained in the text.

An acknowledgment to a more impersonal source of help is due to recent volumes of *London Statistics*. My obligations to this work are evident from the many footnote references; but I should like to pay a special tribute to the unique value of the work. I know of no other official publication containing the statistics of a particular locality which can compare with it in lucidity, accuracy, and comprehensiveness. The London County Council has kindly permitted me to reproduce some of the maps published in recent volumes. A similar permission was granted by the Editor of *The Economist* in regard to the maps and diagrams on pages 176–8 and 197–8. The Index was prepared by Mr. C. Fuller, B.A.

WILLIAM A. ROBSON

THE LONDON SCHOOL OF ECONOMICS
AND POLITICAL SCIENCE

December 1938

CONTENTS

PART I. THE PAST (since 1835)

PART II. THE PRESENT

PLATES

FOLDING MAPS

DIAGRAMS AND MAPS

Part I

THE PAST
(since 1835)

CHAPTER I

A REVOLUTION MISSED

It is usually assumed that to undergo a revolution is a misfortune to be avoided at all costs. But there are circumstances in which to forgo a revolution is a much greater disaster. The failure of London to pass through the municipal revolution of 1835 was a circumstance of this kind. The Municipal Corporations Act of that year transformed the local government of the provincial towns, and laid the basis of municipal democracy as it now exists. The incalculable benefits which followed in the train of this great event have recently been celebrated throughout the country.[1] But London, to which the Royal Commission on Municipal Corporations devoted the whole of its second report, was not included in the Act of 1835. From this calamity the metropolis has never recovered.

The exclusion of London was in no way due to the excellence of the municipal administration which at that time existed in the capital. A hundred years ago the government of the metropolis was divided between the City Corporation and the city companies, 7 boards of commissioners for sewers, nearly 100 paving, lighting and cleansing boards, about 172 vestries of one kind or another (including select vestries, open vestries, and those appointed under various local and general Acts), boards of guardians established under the Poor Law Act, 1834, the commissioners of highways and bridges, turnpike trusts, the commissioners of police and of woods and forests, the commissioners of courts of requests, grand juries, inquest juries, leet and annoyance juries, the Middlesex bench of magistrates and various other bodies such as the salaried police magistrates.[2] Even to the unfastidious eye of the early Victorian Londoner such a collection of authorities was scarcely reassuring.

[1] Cf. *A Century of Municipal Progress*, edited by W. I. Jennings, H. J. Laski and W. A. Robson (published by Allen and Unwin for the National Association of Local Government Officers) (1935), second impression.

[2] W. E. Hickson: *The Local Government of the Metropolis*. Reprinted from *The London and Westminster Review*, April 1836, p. 7; Report from Select Committee on Metropolis Sewers, B.P.P., 1834, vol. xv, p. 197.

The multitudinous commissioners of sewers and the boards for paving, lighting and cleansing the highways were generally irresponsible, frequently extravagant and often corrupt. The membership of these boards was often absurdly large—the Westminster sewers trust had no less than 250 commissioners, while St. Pancras was afflicted with 21 paving and lighting boards on which sat 900 commissioners, many of whom lived outside the parish or abroad. Most of the boards were heavily in debt; and in St. Pancras there were several parts of the parish where there was no lighting or paving beyond what each householder was able to provide for himself.[1]

The City Corporation, the oldest and most important of the local governing bodies, was primarily concerned with maintaining its ancient privileges and exclusive rights against the potential encroachments of the rapidly growing town outside its narrow boundaries. "A common councilman," wrote a contemporary observer, "whether Whig, Radical, or Tory (the exceptions are remarkably few) has not the least notion that he exists to forward any common object in which the inhabitants of the whole metropolis are interested, but will honestly state that he considers it his duty to maintain the exclusive rights and privileges of the City of London regardless of any other consideration."[2] For example, in 1836 horned cattle were still being driven through the principal streets of the town, to the terror and danger of pedestrians, in order to maintain the City Corporation's cattle market at Smithfield.[3] In the opinion of the editor of *The Westminster Review*, the machinery of local government in the metropolis was about the worst that could be devised.[4]

The Royal Commission on Municipal Corporations, whose outspoken report was largely responsible for the Municipal Corporations Act, 1835, were under no illusion as to the need for improvement in London. The second report of the Royal

---

[1] Hickson: *op. cit.*, p. 21.

[2] *Op. cit.*, p. 8. The writer was William Edward Hickson (1803–70), editor and proprietor of *The Westminster Review*. He was associated with Nassau Senior on the Royal Commission of 1837 to enquire into the condition of unemployed handloom weavers. He was a pioneer of national education.

[3] *Ibid.*, p. 9.   [4] *Ibid.*, p. 7.

Commission, which appeared in 1837, was devoted entirely to the metropolis.

The Royal Commission grasped without hesitation the fundamental need of the capital for a unified administration covering the whole of its effective area. "We do not find any argument," they remarked, "on which the course pursued with regard to other towns could be justified which does not apply by the same force to London, unless the magnitude of the change in this case should be considered as converting that which would otherwise be only a practical difficulty into an obligation of principle." They declared there were no circumstances to justify the separation of a small area within the municipal boundary from the remainder of the town. It will hardly be suggested, they remarked, with a grim irony which only future generations could fully appreciate, that in order to remove the appearance of singularity arising from the distinction between the city and the rest of London, independent communities should be formed in the other quarters of the town. Such a plan, they said, in getting rid of an anomaly would tend to multiply and perpetuate an evil.[1] London must have a single and unified system of local government; but whether it should resemble the newly-established municipal corporations in the provincial boroughs, or whether it should come under the jurisdiction of the central government, was a question which the Commissioners asked but did not answer.[2] They felt that special considerations must necessarily arise in regard to London which do not apply elsewhere, from the fact of its being the seat of the legislature and of the supreme executive power of the state. The only real point for decision, they reported, is how far such duties as paving, sewage, lighting and police for the whole metropolis "could be placed in the hands of a Metropolitan Municipality, or how far they should be entrusted to the Officers of Your Majesty's Government."[3]

The possibility of central government control was by no means an unlikely prospect a hundred years ago. The practice existed, and continued for many years, of appointing Select

[1] Royal Commission on Municipal Corporations (1837), Second Report, p. 4.
[2] *Ibid.*, p. 4.   [3] *Ibid.*, p. 5.

Committees of Parliament to consider plans for the improvement of the metropolis. The appointment of these committees was due to the absence of responsible organs of municipal administration, and although the method was quite unsuited to the task, some of these committees, and the commissions which succeeded them, were not unaware of the large issues involved and might eventually have developed into permanent administrative bodies. There was a general feeling that the London of 1837 was not equal to its position as capital of the British Empire. The Select Committee of 1838, for example, took occasion to regret that the "magnificent and judicious" plans of Wren and Evelyn for rebuilding London after the Great Fire should have been set aside by "the perverse self-interest of the citizens of London."[1] The Commissioners for Improving the Metropolis of 1844, referring once more to those self-same plans, remarked that their fate demonstrates "the difficulty of effecting great and systematic changes in such a metropolis. They were thwarted by public bodies and individual interests."[2]

It is ironical to reflect that the politicians who were thus regretting the stupidity and selfishness of the past were incapable of taking any steps to safeguard the future. Yet so it was. The Government of the day signified its intention in 1836 of introducing a Bill for the reform of the Corporation of the City of London;[3] but the days and weeks and months and years passed, and nothing was done. The project was dropped and no legislative measure of importance was enacted until 1855.

The vital importance of making the areas of local government authorities correspond with the prevailing social and economic realities was fully understood by intelligent students of the

[1] Second Report from the Select Committee on Metropolis Improvements (1838), B.P.P., 1838, vol. xvi, p. vi.

[2] First Report of the Commissioners appointed to consider Means of Improving the Metropolis, B.P.P., 1844, vol. xv, p. 4.

[3] Hickson: *op. cit.* See also *The Corporation of London and Municipal Reform*, reprinted from *The Westminster Review* (1843), p. 9.

Lord John Russell gave notice of an intention to introduce a Bill to reform the Corporation, if it would not reform itself. In consequence a few feeble half-measures were taken which in no sense affected the essential situation.

early Victorian age. The Royal Commission on Municipal Corporations pointed out in their first report that in most important towns the suburbs extended far beyond the limits of the corporate authority, and they instanced such places as Bristol, Rochester, Carlisle and Hull. The result was that in a very large number of boroughs to which the Act of 1835 was applied the boundaries of the town were substantially enlarged.[1] In London, the Commission reported, expansion had taken place to such a degree that the word suburb could no longer be applied with its usual signification to describe "the vast extent of uninterrupted Town which forms the Metropolis of the British Empire."[2] The Commission not only emphasised the small proportion of the metropolis governed by the City Corporation, but declared that there was no justification for the distinction between that particular district and the rest except that in fact it had been for long so distinguished.[3] *The Westminster Review* printed in 1836 an article by the editor pointing out that London is one and indivisible; that the inhabitants of the metropolis have a variety of interests in common which cannot be provided for by legislating for one district without reference to the surrounding neighbourhood. Much of the business of local administration could not be divided up among different districts of London without detriment to the public interest, including such functions as the lighting and paving of the streets, traffic regulation, public health, river navigation and the construction and management of sewers.[4]

It is interesting to recall for a moment the physical appearance of London in these early decades of the 19th century.[5] When the first census was taken in 1801, Chelsea was a solitary suburban retreat, Paddington and Westbourne were rural hamlets, and Kensington scarcely more than "the old court

---

[1] Report of the Royal Commission to enquire into the Corporation of the City of London, B.P.P., 1854, vol. xxvi, p. xi.

[2] Royal Commission on Municipal Corporations (1837), Second Report, B.P.P., vol. xxv, p. 2.

[3] *Ibid.*, p. 4.　　　　　　　　　　　　　　[4] Hickson: *op. cit.*, p. 5.

[5] The best concise account of London between 1837–97 is to be found in Sir Laurence Gomme: *London in the Reign of Victoria.* Chapters I and II describe the London of 1837.

suburb."[1] "St. John's Wood was still a wood through which one or two roads were timidly making their way; a sparkling rivulet ran through the meadows of Kilburn; Haverstock Hill had a few villas; Hampstead was a village with a few large houses and fine gardens; Kentish Town had a sprinkling of houses; behind Old St. Pancras church was Mr. Agar's farm; Islington was almost a single street; Bow was a little village; Stratford did not exist; West Ham . . . was an open waste; Hackney was a pleasant suburb; Tottenham was a rural village whither the Londoners repaired on summer evenings."[2] Sir Walter Besant, the historian of London, wrote in 1909 that there were men still living who could remember the open fields of South Kensington, the market gardens of Bayswater, the fields and woods of Highgate, Hornsey and Tottenham.[3]

A hundred years ago the land between High Street, Kensington, and Earl's Court consisted largely of market gardens, and so it remained until the early '60's. The fields of Belgravia were in course of being built over, but the process was not complete until the middle of the 19th century. The line of buildings along Edgware Road extended as far as the canal on the east side of the road; but on the west side, only up to the junction of Harrow Road.[4] The London and Birmingham Railway was nearing completion in 1837. It ran into London almost entirely through open fields as far as Camden Town; and even between there and the terminus at Euston there were only patches of urban development. The entire north and north-east portions of London beyond a radius of less than two miles from the Thames were purely rural in character.[5] As late as 1868, when the ceremonial opening of Swiss Cottage Station took place, "the whole company spent a few minutes admiring the beautiful scenery."[6]

London has in the past grown largely by the process of agglomerating villages. It was said between 1821 and 1831 that such places as Camberwell, Bethnal Green, Stoke Newington, Highbury, Chelsea, Knightsbridge and even Ken-

[1] Report on Greater London Drainage, by Taylor, Humphreys and Frank, H.M.S.O. (1935), p. 13.
[2] Sir Walter Besant: London in the Nineteenth Century (1909), p. 5.
[3] Ibid., p. 27.   [4] G. A. Sekon: Locomotion in Victorian London, p. 3.
[5] Ibid., pp. 5–6.   Ibid., p. 162.

sington were making haste to join London. The process has continued up to the present day, when places three times the distance from the centre of the city, such as Bromley, Downham, Dagenham, Romford, Enfield, Barnet, Edgware, Southall, Kingston and Croydon, and even the more distant towns of Watford and Caterham, are becoming indissolubly connected with the vast urban conglomerate known as London.[1]

[1] Harold P. Clunn: *The Face of London* (1935), p. 3.

CHAPTER II

THE CITY CORPORATION

In the heart of the metropolis lies what is loosely called the square mile of the City, an area which formerly contained both residences and places of work but is now given over almost entirely to business premises. It contains the whole mass of London's financial institutions, such as the Bank of England, the Stock Exchange, the head offices of the great banks, Lloyd's, the Royal Exchange, and the issuing houses. Indeed, the term "the City" is today frequently used to connote those interests in the same way as "Wall Street" is regarded as synonymous with the large financial interests in New York. But this identification of the City with the world of finance is quite unimportant in connection with local government.

For nearly a thousand years the government of this central core has lain in the hands of the Corporation of the City of London. The long line of charters relating to the City[1] commences with one granted by William the Conqueror in 1070; and even Magna Charta contained an express provision that the City of London should have all its ancient liberties and customs.[2] For many centuries the City played a part of great importance in the nation's history. Its efforts were especially redoubtable in the struggle for popular rights against autocratic monarchs, until the Stuart period, when it capitulated in an ignoble manner to royal absolutism. The very power and prestige which the City had acquired through the long years were decisive influences in enabling the Corporation to resist reform in 1835. Yet even in 1835 the City was chiefly living on its past; for the decline of the City Corporation as a living force had set in with the opening of the 18th century.[3] The centre of gravity had already moved to Westminster.

[1] There are no less than 120 of them.
[2] Royal Commission on the City Corporation of London, B.P.P., 1854, vol. xxvi, p. xxiii; Alexander Pulling: *The City of London Corporation Enquiry* (pamphlet, 1854), p. 8.
[3] Percy Harris: *London and its Government* (1933), p. 7.

The boundaries of the City were fixed at an early date in its history and there is no record of their ever having been enlarged to correspond with the growing size of the town.[1] Even in 1837, although the City did not contain the Houses of Parliament, the Government offices, the law courts, the Royal palace or "the residences of the higher or more opulent classes,"[2] the Royal Commission on Municipal Corporations observed that much of the importance of the City was due to its being the daily resort of great numbers of people who slept outside its boundaries.[3]

The Corporation of the City of London has a structure of peculiar character; and this, like the area under its jurisdiction, has come down through the centuries substantially unchanged, with one exception shortly to be mentioned. It consists of three separate chambers: the Court of Common Hall, the Court of Common Council and the Court of Aldermen.

The full title of the first of these chambers is "an assembly of the Mayor, aldermen and liverymen of the several Companies of the City of London, in Common Hall assembled." It consists of the Lord Mayor, not less than four aldermen and such liverymen of the City Companies[4] as are of one year's standing, free of the City and have duly paid their livery fines. In former times the Common Hall had many important functions, but it has now lost all active control over municipal affairs.[5] It still meets, however, once a year to elect from the aldermen two men who have filled the office of sheriff; and from these the aldermen choose the Lord Mayor. In practice, the two senior aldermen are generally selected.[6] In addition the Common Hall elects the sheriffs, the chamberlain or treasurer, the bridgemaster, and the city auditors. Thus, not only the Lord Mayor, the principal magistrate of the City for ceremonial purposes, but also some of its leading administrative

[1] Royal Commission on the City Corporation of London, B.P.P., 1854 vol. xxvi, p. xii.
[2] Royal Commission on Municipal Corporations (1837), Second Report, p. 3.
[3] *Ibid.*   [4] For a description of the City Companies, see *post* pp. 35–41.
[5] J. B. Firth: *Municipal London* (1876), p. 42.
[6] Percy Harris: *op. cit.*, pp. 17, 30. The institution of Common Hall was drastically criticised by the Royal Commission on Municipal Corporations (1835).

officers, are the nominees of less than 10,000 liverymen, the majority of whom have no special knowledge of or interest in municipal questions.

The Court of Common Council is composed of the Lord Mayor, aldermen and 206 common councilmen. These latter are elected annually at the wardmotes by about 26,000 electors exercising a restricted franchise. There has not been a contest for some years.[1] Prior to 1867 the right of electing members of the Common Council was confined to freemen of the City, but in that year a statute was passed extending the vote to all £10 ratepayers. The voters thenceforth also included all the Parliamentary electors for the City.[2] This is the only constitutional change of note which has directly affected the City Corporation during the past century.

The Common Council is the principal governing body of the Corporation. It appoints the judicial officers of the Mayor's and City of London Court other than the Recorder,[3] the City solicitor, Town Clerk, Remembrancer, sword bearer and other administrative and ceremonial officers. It controls the property belonging to the Corporation and determines the policy by which the various services are administered.[4]

We come next to the Court of Aldermen, composed of 25 aldermen sitting with the Lord Mayor. The aldermen are elected separately by the wards and hold office for life. They are subject to approval by their own court as fit and proper persons, but are otherwise irremovable.[5] They also appoint to various valuable offices such as the Recorder, the Clerk to the Lord Mayor, the Steward of Southwark, the Clerk to the

[1] In December 1935 there was no contest. In 1934 there was one contested election in Billingsgate ward where 10 candidates offered themselves for 8 vacancies. Previously there was no contest since 1927. *London Statistics*, 1934–6, vol. 39, p. 23.

[2] 12 & 13 Vic., c. xciv; 30 & 31 Vic., c. i (both local Acts). George Whale: *Greater London and its Government* (1888), pp. 23, 91.

[3] The Court referred to is an amalgamation of the Mayor's Court, which has the unique privilege for a municipal Court of an unlimited jurisdiction, and the City of London Court, which deals with small claims. The Recorder sits as a judge on both of the Mayor's and City of London Court, and of the Central Criminal Court at the Old Bailey. In addition there are the Common Serjeant and two judges of the City of London Court.

[4] Hart and Hart: *An Introduction to the Law of Local Government and Administration* (1934), p. 226. [5] Harris: *op. cit.*, p. 18.

Guildhall magistrates and the district surveyors. The Court of
Aldermen controls the City police. It also claims the right of
drawing upon the City funds to an indefinite extent independently of any other authority, but does not in practice do so
without the approval of the Court of Common Council.[1] Each
alderman is *ex officio* a justice of the peace, and when sitting
alone possesses in that capacity the powers exercised elsewhere
by two or more justices sitting in petty sessions.

The qualifications required in law by common councilmen
and aldermen are much more stringent than those which obtain
in other towns. A common councilman must be either a
freeman of the City and a £10 occupier on the Parliamentary
register; or he must be a freeman householder "paying scot
and bearing lot." A City alderman must be a freeman approved
and admitted by the Aldermanic Court. Thus the aldermen
and the common councillors are all required to be free of the
City,[2] and in practice it would be very difficult if not impossible
to secure election without belonging to at least one of the
City Companies. But apart from these formal requirements
there are economic conditions to be satisfied, summed up
in traditional saying that an alderman must be worth
£40,000.

The City Corporation has for long rested on a system of
strictly limited political representation. The Royal Commission
on the Amalgamation of the City and the County of London
remarked in 1894 that the absence of any general legislation
affecting the City since 1835 might to some extent be attributed
to the fact that the City was, unlike most old corporations, a
popularly elected body. This was a highly misleading statement,
but there was enough truth in it to explain in part the complex
trend of events. There was just sufficient representation of the
inhabitants of the square mile to make it possible for the
Corporation to continue to exist during the era of democratic
reform.[3]

---

[1] *The Corporation of London and Municipal Reform* (1843), pp. 5, 7;
Harris: *op. cit.*, p. 18.

[2] As to the methods of becoming a freeman, see Herbert Morrison: *How
Greater London is Governed*, p. 14.

[3] The Royal Commission on the Corporation of the City of London,
1854, B.P.P., vol. xxvi, p. xii, pointed out that "although the constitution

The City Corporation was frequently accused of jobbery, corruption and extravagance from at least the early decades of the 18th century.[1] Precise information has usually been lacking as to the truth or falsity of these charges, since there is no independent public audit of the Corporation's accounts nor scrutiny of its activities, but the low standards which were common among the unreformed boroughs prior to 1835 doubtless prevailed in the City and may have persisted there longer than elsewhere for the very reason that reform was avoided. The immense wealth of the City's "privy purse" derived from sources other than rates, and the complete absence of any external control, must undoubtedly have provided temptation on a scale unknown elsewhere. The irresponsibility of the aldermen, who are elected for life and are answerable neither to their constituents nor to public opinion, has invited criticism.[2] The practice of paying pensions to the widows and dependants of late aldermen, of which complaint was made by *The Westminster Review* so long ago as 1843, still continues.[3] Enormous sums have been spent from time immemorial by the City on entertainment of one sort or another. A very serious charge of the improper use of its funds for political purposes was disclosed by a public enquiry in 1884.[4] Three years later a Select Committee of the House of Commons found that a mass of petitions presented to Parliament in support of the Coal and Wine Duties Continuation Bill 1887 were "wholly or in great part forgeries" concocted with money provided by the City Corporation. The duties enured to the benefit of the Corporation and the petitions were obtained by an agent

of the Corporation of London has secured to the inhabitants of the City a fair representation of their interests, that representation is confined to the persons dwelling within the narrow area of the City proper, and does not extend to the population of the larger portion of the metropolis."

[1] See, for an early example, *City Corruption and Mal-Administration Displayed; occasioned by the Ill-Management of the Publick Money in General*, by a Citizen, second edition (1738), pamphlet.

[2] *The Corporation of London and Municipal Reform* (1843), p. 5 (reprinted from *The Westminster Review*).

[3] *Ibid.*, p. 4. The article showed that nearly £2,500 a year was being paid in pensions granted by the Court of Aldermen, of which £1,304 was going to the widows, daughters or other relatives of deceased aldermen.

Post, pp. 78–81.

PLATE I. A general view of the City to-day. The rectangular building left centre is the Bank of England, with the Mansion House (the Lord Mayor's official residence) slightly below to the right. The disorderly jumble of buildings in narrow streets may be compared with Wren's plan for the rebuilding of London after the Great Fire (see Plate VII)

(Aerofilms)

instructed by the City Solicitor, who was severely censured by the Select Committee.[1]

The City Corporation possesses extensive municipal powers both old and new. It is responsible for the upkeep of four important cross-river bridges.[2] It is the local authority for street improvements, the protection of infant life, the storage of petroleum and explosives, the testing of gas, the regulation of shop hours, the inspection of weights and measures, the provision of reformatory schools, and many other matters.[3] It is the market authority, and under an ancient charter claims a monopoly over markets within an area of 7 miles. It has the right of representation on the managing bodies of a whole series of institutions, such as the United Westminster Schools, the Royal hospitals, the Thames and Lee Conservancies and Sir John Soane's Museum.[4] It is the sanitary authority for the entire Port of London, most of which lies outside its boundaries. In regard to the acquisition of open spaces, the Corporation is similarly not confined to its own area. Out of 6,703 acres of parks and open spaces owned and maintained by the City Corporation, only 3 acres lie within the City.[5]

During the 19th century much irritation and obstruction to trade was caused by a number of burdensome tolls and taxes exacted by the Corporation in the exercise of its ancient privileges. Among these was the metage of fruit, corn, oysters, salt and other commodities, which was carried out by fraternities of porters specially appointed for the purpose, and which in effect levied taxation on persons outside the City's boundaries. No brewer or other person could employ his own servants to measure or carry the barley needed for his business; no merchant could use his own porters to unload wagons; no

---

[1] Special Report from the Select Committee on Public Petitions, 1887, B.P.P., vol. xi, pp. iii, viii; J. F. B. Firth: *The Reform of London Government*, p. 18.

[2] London Bridge, Southwark Bridge, Blackfriars Bridge and Tower Bridge.

[3] Royal Commission on the Amalgamation of the City and County of London (1894), p. 18.   [4] *Ibid.*, p. 21.

[5] The open spaces outside the City are as follows: Epping Forest (5,560 acres), Burnham Beeches (492 acres), Coulsdon commons (392 acres), West Ham Park (77 acres), Highgate Woods (69 acres), Spring Park (51 acres), Queens Park (30 acres) and West Wickham Common (25 acres).

one could send a parcel from one part of the City to another unless it was delivered by a ticket-porter. No man could engage in any of this work without joining the appropriate fraternity or brotherhood and paying the fine and quarterages. Rents and tonnage dues were levied both on land and on water, including street tolls on carts and wagons, river tolls, and tolls for standing places in the public markets.[1] The duties on coal and corn were especially heavy. Some of these exactions were of doubtful legal validity; and in one important case the claim to impose it was successfully challenged in the Courts of Law.[2] From 1872 onwards the proceeds of some of these duties were earmarked to defray expenditure on particular services such as the cost of acquiring open spaces. Parliament finally abolished the whole system by the Metage on Grain (Port of London) Act, 1872, which did not come into effect until 1902.[3]

After the indignities which the City Fathers suffered during the Stuart period, and in particular after 1683, when they were deprived by the King of many of their most prized civic rights, the City Corporation allowed itself to fall back municipally "without an ideal, without even a message from its great past to tell the people who were building up a new London what the old London had accomplished."[4]

[1] *The Corporation of London and Municipal Reform* (1843), reprinted from *The Westminster Review*, p. 22; William Carpenter: *The Corporation of London, as it is, and as it should be* (1847), p. 85.

[2] The Royal Commission on the City Corporation of London stated in 1854 that many of the old charters are very obscure, and that the City claims privileges and rights which are highly doubtful or even "destitute of legal foundation" as resting on charters which have ceased to be in force. Litigation to challenge the right to measure all grain by the City meters was initiated by Messrs. Combe, Delafield & Co. The City at first appeared in support of their alleged right to compel involuntary metage but later abandoned the claim. B.P.P., 1854, vol. xxvi, p. xxiii.

[3] 35 & 36 Vic., c. c. This was a private Act.

[4] Sir George Laurence Gomme: *The Governance of London*, p. 393.

CHAPTER III

THE CITY COMPANIES

We have already noticed the part played by the Court of
Common Hall in the governance of the City Corporation.
This makes it necessary to say something about the Livery
Companies whose members constitute Common Hall.

The circumstances in which the Livery Companies originated
are not precisely known, but it is believed that they are
descended from the merchant guilds which flourished in the
Norman and Plantagenet eras. These guilds exercised in early
times extensive powers of regulating the various trades and
manufactures in which their members were engaged; but by
the opening of the Tudor period these vocational activities
had become of minor importance or had disappeared entirely.[1]
Their jurisdiction over economic life fell away in the 15th and
16th centuries; and in the 17th century the powers of search
and monopoly contained in their charters were in several cases
declared illegal by the courts, with the result that they no
longer possessed coercive power.[2]

When the London guilds ceased to be trade societies in any
genuine sense of the term they fell back on their non-economic
functions and became voluntary associations of well-to-do
citizens belonging to diverse occupations bound together for
social and philanthropic purposes, a feature which they shared
with most vocational bodies in medieval times. Nevertheless,
traces of their economic functions still persist, sometimes
reinforced by statute, and in a few instances are of some
importance. Thus, the Fishmongers' Company examines and
condemns fish at Billingsgate Market and has been empowered

---

[1] This was partly due to the refusal of the Companies to admit the
labourers who flocked to the towns on the break up of feudalism, which
made it impossible for the guilds to continue to claim to be representative
craft bodies, and partly to the admission of members by patrimony or
hereditary right, which had a similar effect.
  L. T. Dibdin: *The Livery Companies of London* (1886), pp. 45–7. Royal
Commission on the Livery Companies of London (1884) Report, p. 15.
  [2] L. T. Dibdin: *op. cit.*, p. 49.

by modern legislation to prosecute offenders offering under-sized fish for sale.[1] The Goldsmiths' Company assays and marks gold and silver plate and prosecutes persons selling plate bearing a counterfeit mark or selling plate requiring to be marked which is below standard.[2] The Vintners' Company confers upon its members the right to sell wine throughout England without a licence and employs porters who unload wines arriving at the London docks.[3] The Gunmakers' Company tests and stamps firearms and prosecutes offenders under the Gun Barrel Act, 1868. The Scriveners' Company conducts the examination for admission to the office of notary, and can prosecute persons practising without a licence. The Stationers' Company maintains a register of all publications under the Copyright Act and also publishes almanacs. A few trifling trade activities are performed by some of the other Companies.

In general, however, for two or three hundred years the fundamental purposes for which the societies have continued in existence have been those of entertainment and benevolence.[4] The promotion of social intercourse was of great importance while the wealthy bankers, merchants and shipowners who traded in the City also lived there and used the Companies' halls as dining and discussion clubs; but this phase had passed away by the opening of the 19th century. The merchants had already moved their houses to other parts of London.[5]

The Livery Companies have always been connected with the City Corporation. In early times the Companies were actually made electoral units for electing the Court of Aldermen and the Court of Common Council, in place of the wards into which the City area is divided. This formal acknowledgment of their civic supremacy lasted only a short time and came to an end in the 14th century.[6] But the substance of power remained with the Companies. For five hundred years the

---

[1] Fisheries Act, 40 & 41 Vic., c. 42.

[2] This is the origin of the so-called "Hall-mark," 12 George II, c. 26. Under the Coinage Act, 1870, the annual trial of the pyx takes place at Goldsmiths' Hall.

[3] Royal Commission on the Livery Companies of London (1884) Report, pp. 19–20.   [4] Ibid., p. 19.

[5] P. H. Ditchfield: The City Companies of London and their Good Works (1904),   6.   [6] L. T. Dibdin: op. cit., p. 49.

freedom of the City was reserved to members of the Livery Companies, and no one could exercise municipal rights within the City unless he was a freeman of a Company. This state of affairs continued until 1835, when the restriction was removed by the City Corporation.[1] Up to that date freedom of a Livery Company and freedom of the City were convertible terms.

The freemen of the City who are also liverymen continue to enjoy the exclusive right of comprising Common Hall, which meets annually to select the candidates for the Mayoralty and to appoint various officers.[2] They also possess the Parliamentary franchise.

One result of the condition that members of Common Hall should have the livery of a Company has been to impose an extremely high property qualification of an informal kind on the right to participate in City affairs, whereby, it has been naïvely observed, "unfit persons have been excluded."[3] For although in former times the freedom of a Livery Company was apparently not difficult to acquire and could be obtained by tradesmen and working craftsmen, in modern times membership of the livery has been confined to the well-to-do classes, and in most cases the fees are quite beyond the pocket of anyone without substantial means, quite apart from the difficulty of securing nomination. Hence, Common Hall, like the Court of Aldermen and the Mayoralty, is open to only a small section of the population, and is therefore undemocratic.

The London Livery Companies are 78 in number, of which 12 "great" companies are always distinguished on account of their superior wealth and prestige.[4] The total membership is divided into three grades: the simple freemen, the livery, and the court or governing body. The typical court consists of the master, wardens and assistants, who are chosen by co-option. The livery is usually selected by the court from among the freemen.[5] The freemen are recruited by patrimony, apprentice-

[1] L. T. Dibdin: *op. cit.*, pp. 68–9; J. F. B. Firth: *Municipal London*, pp. 46–7.   [2] *Ante*, p. 29.   [3] L. T. Dibdin: *loc. cit.*
[4] For details of the income, corporate and trust, of the twelve great Companies in 1879–80, see Report of the Royal Commission, p. 26; and for the present figures, *Whitaker's Almanack*.
[5] *Ibid.*, pp. 34–5. The members of the courts total about 1,500. *London's Heritage in the City Guilds*: Fabian Tract, No. 31.

ship or invitation. The Companies are entirely autonomous and the respective courts have the sole conduct of their affairs, including the management of property, the disposal of income and the appointment of salaried officers. The proceedings of the courts are usually secret and are not available even to the liverymen and the freemen. Accounts are not published.[1]

The London Companies are the possessors of great wealth, much of which has come to them through legacies from members. They are among the largest ground landlords in the metropolis and own real property far outside the City limits. They have acquired land by purchase and by legacy, and the enormous rise in the value of land and house property in the centre of London during the past century has largely augmented their funds. Their income in 1879–80 was stated by a Royal Commission to be between £750,000 and £800,000 and the capital value of their property to be £15 millions rising to £20 millions in the following 25 years.[2] It is probable that their income today is about £1,500,000 and the value of their property at least £30 millions.

The Royal Commission on the Livery Companies, 1884, found that about £200,000 a year was trust income to be expended on specified charitable objects, while the remainder (called corporate income) was subject to no legal control. But about £150,000 of the corporate income was in 1880 being devoted to benevolent purposes. Of the remainder, fees to members of the governing bodies absorbed £40,000 a year, salaries to officers £60,000, maintenance charges (such as rates, taxes and upkeep of halls and buildings) £75,000, and entertainments £100,000; the balance of £125,000 went to savings, redemption of debt, etc.[3] The greatly enhanced income of the Companies today is presumably allocated in substantially similar proportions.

A great deal has been said and written both by way of

---

[1] Royal Commission on the Livery Companies of London (1884) Report, p. 23.

[2] *Ibid.*, p. 26. The contributions of members amounted in 1897–80 to about £15,000 to £20,000 a year, or between 1/50th and 1/37th of their total income. *Ibid.*, p. 28.

[3] *Ibid.*, p. 36. According to L. T. Dibdin (*op. cit.*, pp. 115–16) the figure for entertainments should have been £75,000.

criticism of the Livery Companies and in support of them.[1] It is indisputable that the benevolent work of the Companies has been of great public advantage. Such excellent schools as St. Paul's, Tonbridge, Merchant Taylors', and Oundle, which are entirely due to the efforts of the Companies, have made a substantial contribution to middle-class education. Technological institutes like the City and Guilds College initiated a new standard in scientific training from which the whole country has benefited. The People's Palace in Mile End has helped to brighten the lives of the poorer inhabitants in the East End of London. Scholarships, Exhibitions and Chairs have been endowed at Oxford and Cambridge, and the University of London is today benefiting largely from the generosity of the Companies in the rebuilding of its headquarters. Some excellent hospitals and convalescent homes have also been established or assisted. On the other hand, a strong supporter of the Companies admitted fifty years ago that "the enormous expenditure" on entertainment does not commend itself to modern taste and feeling. It is the one point, he said, which had tended to lower the Companies in public opinion, and to make people somewhat careless as to their fate.[2]

As this study is concerned solely with the local government of London, it is no part of my task to evaluate the usefulness or otherwise of the Livery Companies as regards their social or philanthropic activities. Hence it is unnecessary to discuss the recommendations of the Royal Commission which enquired into these matters in 1884.[3]

At the same time, it is impossible to understand the development of London government in the 19th century without taking into consideration the immensely important part played by the Livery Companies in enabling the City Cor-

[1] Cf. J. F. B. Firth: *Municipal London* (1876), pp. 53, 57, 71, 97, 101; William Herbert: *The History of the Twelve Great Livery Companies of London*, 2 vols. (1836); P. H. Ditchfield: *The City Companies of London and their Good Works* (1904), *The Story of the City Companies* (1926); L. B. S.: *The City Livery Companies and their Corporate Property* (1885); George Unwin: *The Guilds and Companies of London* (1908); L. T. Dibdin: *The Livery Companies of London* (1886); Reports of the Royal Commission on the Livery Companies of London (1884); *London's Heritage in the City Guilds* (1898), Fabian Tract, No. 31.
[2] L. T. Dibdin: *op. cit.*, pp. 115–16.   [3] Report, pp. 42–4.

poration to withstand reform. The formal connection between the Companies and the Corporation through the Court of Common Hall and the right of members of the Companies to become freemen of the City has been less powerful than the intangible but pervasive influence exerted by the Livery Companies in rallying support in defence of the City. The identity of interest between the institutions goes beyond the constitutional fact that the Livery Companies are essential to the City Corporation in its present form, since without them there could be no election of the Lord Mayor and Sheriffs or the appointment of several of the principal officers. It is rooted fundamentally in the well-founded belief that the destinies of the Companies and the Corporation hang together. If a reforming finger were once permitted to touch the archaic constitution and ancient privileges of the Guildhall—the name is significant—no liveryman could henceforth eat his dinner in peace. The converse must be equally true, since practically everyone of importance in the City Corporation, from the Lord Mayor and the aldermen to the Common Councilmen and principal officers, are all members of one or more Livery Companies. Membership of one or more Companies is indeed a distinct advantage if not indispensable to a candidate seeking election or appointment to the Guildhall.[1]

Hence, the continuous influence which the Livery Companies have exerted since 1835 in support of the City Corporation has been not merely in order to preserve intact the ancient form of municipal government but also because their own existence was felt to depend on successful resistance to change at the Guildhall. Their influence has been not the less potent merely because it was exercised in subtle and informal ways. Institutions of great antiquity, with an unbroken history going back for centuries, with many quaint and picturesque traditions, inhabiting splendid halls filled with artistic treasures, and dispensing £100,000 a year on entertainment, can find many delightful ways to win the hearts of important guests and disarm potential criticism. The Livery Companies, wrote one of their critics, "are careful to be on the best possible terms with the powers that be. Officers of the Government and

[1] Sir Percy Harris: *London and its Government*, p. 17.

dignitaries of the Law are regularly invited to partake of their 'noble hospitality.' No one who has partaken of those gorgeous banquets can find it in his heart to treat his hosts harshly."[1] Many charges have been brought against the Companies relating to the conduct of their affairs and the expenditure of their funds, but "the perfected skill of four hundred years' continuous dining" has never been questioned.

The City Corporation in its turn has also a sound understanding of the uses of lavish hospitality. The Lord Mayor's annual banquet, attended by the Prime Minister and his colleagues and many other notables, the yearly dinner to the High Court judges, the magnificent entertainments to Royal guests, can scarcely have failed to inspire in the minds of many important visitors a feeling of affection towards the ancient Corporation which might otherwise be lacking. It is not necessary to suggest that these festive occasions were deliberately devised to curry favour in high places in order to appreciate the subtle result of their cumulative effect over a very long span of years.

[1] J. F. B. Firth: *Municipal London*, p. 103.

CHAPTER IV

THE GROWTH OF POPULATION

The position of the City in relation to the rest of the metropolis, and the general problem of administrative areas in London, can best be understood in the light of some acquaintance with the growth and distribution of population. I shall therefore give a brief account of the more significant changes which have occurred since the beginning of the 19th century.

A census taken by Captain John Graunt in 1631 gave the population of the City as 130,178; and by 1661 he estimated it had increased to 179,000. The population of the area within the Bills of Mortality he reckoned to be 460,000.[1] A great increase took place towards the close of the 17th century, despite the visitation of the plague in 1665 and the great fire in the following year. In 1683 Sir William Petty gave the figure of 696,000 for the area of the Bills of Mortality including Westminster and Southwark.[2] Gregory King in 1694 gave a reduced figure of only 530,000. From about this time, however, London superseded Paris as the largest city in Europe.

The first official census was made in 1801. The results of the first four enumerations were as follows:

TABLE I

	1801	1811	1821	1831
The City	127,621	119,113	123,888	122,412
Southwark	67,448	72,119	85,905	91,501
Inns of Court	1,907	1,796	1,546	1,271
Westminster	153,272	162,085	182,085	201,842
Outer parishes	514,597	654,433	832,270	1,054,915
Total for metropolis	864,845	1,009,546	1,225,694	1,471,941

[1] J. R. Taylor, Sir George Humphreys and T. Peirson Frank: Report on Greater London Drainage (1935), H.M.S.O., p. 13.
[2] William Petty: Essay on the Growth of London in *Essays on Political Arithmetic.*

The Royal Commission on Municipal Corporations duly noted in 1837 that the population of the City had decreased by over 4 per cent while the population of the metropolis as a whole had increased by more than 70 per cent since the opening of the century.[1]

Before considering the subsequent decades it should be pointed out that the registration districts for census purposes did not remain unchanged. The territory comprised within the Bills of Mortality in the year 1801 was only 21,587 acres. After that it was enlarged to 30,002 acres, until 1841, when the limits were further extended to 44,850 acres. Further additions brought the area to 75,334 in 1851, at which figure it remained with slight variations until 1891, when the administrative county, which is a few hundred acres smaller, was taken as the basis.[2] In Table II the area for 1841 has been enlarged so as to enable a comparison to be made.

The figures given in Table II overleaf show the growth of population in the administrative county from just under two millions in 1841 to the peak figure of four and a half millions in 1901, followed by a slow but steadily increasing decline to 4,396,821 in 1931.[3]

The coming of the railways led to a great exodus from the ancient city from 1861, and the movement was helped later in the century by the development of tramways and other forms of street transport. From 1861 the resident population in the City halved itself every 20 years.[4] The flight from the ancient stronghold is shown in Table III on page 45.

But this was by no means the most important change which occurred during the century. Of greater moment was the shift of population within the administrative county. The figures, in Table IV (page 46) taken from Professor Bowley's analysis,[5]

[1] Royal Commission on Municipal Corporations (Second Report) (1837), B.P.P., vol. xxv, p. 3. These figures do not agree with those given in Table III, but the discrepancies are not important.

[2] H. Price-Williams: *The Population of London 1801–1881*, vol. xlviii, *Statistical Society Journal*, pp. 349–50.

[3] Royal Commission on London Government, 1923, Minutes of evidence, p. 153, Appendix III; A. F. Weber: *The Growth of Cities in the 19th Century* (1899), pp. 47–8.

[4] J. F. P. Thornhill: *Greater London*, p. 60.

[5] *New Survey of London Life and Labour*, vol. i, chap. ii, Table I, pp. 72–3.

TABLE II

	Population of London[1]	Inter-censal Change	Percentage Inter-censal Change	Population of England and Wales	London's Percentage of Total Population	Acreage of London	London's Population per Acre
1841	1,948,417	+ 293,423	+ 17·73	15,914,148	12·24	75,334	25·9
1851	2,362,236	+ 413,819	+ 21·24	17,927,609	13·18	75,334	31·6
1861	2,803,989	+ 441,753	+ 18·70	20,066,224	13·97	75,334	37·2
1871	3,254,260	+ 450,271	+ 16·06	22,712,266	14·33	75,334	43·2
1881	3,816,483	+ 562,223	+ 17·28	25,974,439	14·69	75,334	50·7
1891	4,211,743	+ 395,260	+ 10·36	29,002,525	14·52	75,442	55·8
1901	4,536,063	+ 324,320	+ 7·70	32,527,843	13·94	74,672	60·8
1911	4,521,685	− 14,378	− 0·32	36,070,492	12·54	74,816	59·9
1921	4,483,249	− 38,436	− 0·85	37,886,699	11·83	74,850	59·9
1931	4,396,821	− 86,428	− 1·53	39,947,931	11·00	74,850	58·7

[1] From 1841 to 1881 the "London" referred to is the total of the 29 registration districts, which includes the City of London, listed on page 398, vol. xlviii, *Royal Statistical Society's Journal* (article by Price-Williams). Since 1881 census figures for the administrative County of London are given. From 1841 to 1881 London population figures have been adjusted on the basis of an area of 75,334 acres.

show the remarkable growth of the outer rings and the slowing down and eventual decline of population in the registration county as a whole at the close of the century.

TABLE III

RATIO OF POPULATION OF LONDON: "WITHIN" AND "WITHOUT WALLS"

	a	*b*	*b − a*	$\frac{a}{b-a}$%
	Population of the City of London	Population of Administrative County of London	Population of Administrative County (exclusive of City)	Ratio of City to remainder of Administrative County
1801	128,833	958,863	830,030	15·5
1811	121,124	1,138,815	1,017,691	11·9
1821	125,065	1,378,947	1,253,882	10·0
1831	123,608	1,654,994	1,531,386	8·1
1841	124,717	1,948,417	1,823,700	6·8
1851	129,128	2,362,236	2,233,108	5·8
1861	113,387	2,803,989	2,690,602	3·8
1871	75,983	3,254,260	3,178,277	2·4
1881	51,439	3,816,483	3,765,044	1·4
1891	38,320	4,211,743	4,173,423	0·9
1901	26,897	4,536,063	4,509,166	0·6
1911	19,657	4,521,685	4,502,028	0·4
1921	13,706	4,483,249	4,469,543	0·3
1931	10,996	4,396,821	4,385,825	0·25

Before 1881 Price-Williams' figures of the 29 registration districts are used.

Broadly speaking, while the central area consisting of the inner ring of boroughs grew in absolute numbers and in density up to 1861, it did not keep pace with the outer districts within

TABLE IV

SUMMARY OF POPULATION CHANGES IN GREATER LONDON, 1851-1921
(in thousands)

	Area (Acres)	1851	1881	1891	1901	1911	1921
(a) London (i.e. present county) ..	74,800	2,363	3,830	4,228	4,536	4,522	4,485
(b) Outside to 10 miles	130,300	164	634	1,055	1,593	2,138	2,328
(c) 10-12 miles	72,900	57	147	187	270	359	397
(d) 12-15 miles	153,800	68	110	132	173	221	251
Total	431,800	2,652	4,721	5,602	6,572	7,240	7,461

	1921 as percentage of 1891	1921 as percentage of 1911
(a) .. : : : :	106	99
(b) .. : : : :	221	109
(c) .. : : : :	212	111
(d) .. : : : :	190	114
Total ..	133	103

the county area; its proportion to the population of London as a whole therefore fell steadily throughout the century.[1] The census of 1911 revealed that the population of the administrative county was shifting towards the boundary, and that much of it was overflowing into the areas immediately outside,[2] while some of it was migrating to places situated up to 30 miles from London.

The shifting of population as between the inner and outer zones of greater London during the years 1861–91 can be seen from Table V on page 48, which also gives the position in the outer ring of areas beyond the county boundary.

The transference of growth from near the county boundary to a more distant zone could easily be demonstrated by the population figures for particular areas within the metropolis. The population of the Finsbury, Holborn and Westminster areas had diminished before 1891 and has decreased still more since then. The thirty years between 1891–1921 saw the decrease spreading to a further ring containing such places as Bermondsey, Shoreditch, Bethnal Green, St. Pancras, St. Marylebone and Chelsea. Small increases were registered between 1911–21 in a few of the more remote metropolitan boroughs such as Wandsworth (6 per cent), Lewisham (8 per cent) and Hammersmith (7 per cent); but the spectacular instances of population growth took place beyond the frontiers of the administrative county. Essex and Middlesex, which Londoners had already begun to colonise some time before 1891, became vast dormitory areas in the following twenty years.[3] Surrey also felt the impact from 1911 onwards of London's teeming millions migrating in search of rural surroundings. Southgate, to take an example from the northern slopes, grew from 39,000 in 1921 to 55,000 in 1931; and other significant increases during the same decade were 160 per cent in Hornchurch, 84 per cent in Romford, 132 per cent in Chingford, 879 per cent in Dagenham (all in Essex), 796 per cent in Kingsbury,

[1] A. F. Weber: *The Growth of Cities in the 19th Century* (1899), p. 463.

[2] London Traffic Branch, Board of Trade. Fourth Annual Report, 1911, p. 7.

[3] A. L. Bowley in *New Survey of London Life and Labour* (1930), vol. i, p. 60.

TABLE V

POPULATION OF VARIOUS "LONDON" AREAS AT EACH CENSUS, 1801–1931 [1]
(Approximate figures are indicated below by italic type)
Inter-censal Increase or Decrease

Decennium	NUMBER				
	"Central Area"[2]	Rest of London	Administrative County	Extra London	Greater London
1801–11	+ *109,000*	+ *71,000*	+ 180,045	+ 29,210	+ 209,255
1811–21	+ *162,000*	+ *78,000*	+ 240,188	+ 32,264	+ 272,452
1821–31	+ *165,000*	+ *111,000*	+ 276,039	+ 31,182	+ 307,221
1831–41	+ *164,000*	+ *130,000*	+ 293,695	+ 38,077	+ 331,772
1841–51	+ *177,000*	+ *237,000*	+ 414,064	+ 31,527	+ 445,591
1851–61	+ *102,000*	+ *343,000*	+ 445,153	+ 96,632	+ 541,785
1861–71	+ *5,000*	+ *448,000*	+ 452,902	+ 210,019	+ 662,921
1871–81	− *18,000*	+ *587,000*	+ 568,901	+ 312,119	+ 881,020
1881–91	− 69,333	+ 466,990	+ 397,657	+ 469,488	+ 867,145
1891–01	− 50,125	+ 358,438	+ 308,313	+ 639,283	+ 947,596
1901–11	− 136,123	+ 121,541	− 14,582	+ 684,538	+ 669,956
1911–21	− 126,170	+ 89,008	− 37,162	+ 266,005	+ 228,843
1921–31	− 101,172	+ 13,652	− 87,520	+ 811,261	+ 723,741

Decennium	PER CENT				
	"Central Area"[2]	Rest of London	Administrative County	Extra London	Greater London
1801–11	+ *13·9*	+ *40·3*	+ 18·8	+ 18·8	+ 18·8
1811–21	+ *18·2*	+ *31·6*	+ 21·1	+ 17·5	+ 20·6
1821–31	+ *15·7*	+ *34·2*	+ 20·0	+ 14·4	+ 19·2
1831–41	+ *13·5*	+ *29·8*	+ 17·7	+ 15·4	+ 17·4
1841–51	+ *12·8*	+ *41·9*	+ 21·2	+ 11·0	+ 19·9
1851–61	+ 6·5	+ *42·7*	+ 18·8	+ 30·4	+ 20·2
1861–71	+ 0·3	+ *39·1*	+ 16·1	+ 50·7	+ 20·6
1871–81	− *1·1*	+ *36·8*	+ 17·4	+ 50·0	+ 22·7
1881–91	− 4·2	+ 21·4	+ 10·4	+ 50·1	+ 18·2
1891–01	− 3·2	+ 13·5	+ 7·3	+ 45·5	+ 16·8
1901–11	− 8·9	+ 4·0	− 0·3	+ 33·5	+ 10·2
1911–21	− 9·1	+ 2·8	− 0·8	+ 9·7	+ 3·2
1921–31	− 8·0	+ 0·4	− 2·0	+ 27·1	+ 9·7

[1] Taken from *London Statistics*, vol. xxxix, 1934–6, p. 27.

[2] The central area includes the City of London, the City of Westminster, and the metropolitan boroughs of St. Marylebone, St. Pancras (south of Euston Road), Holborn, Finsbury, Shoreditch, Bethnal Green, Stepney, Southwark, Bermondsey, and Lambeth (north of Kennington Lane), an area of 12,268 acres (about 19 square miles).

135 per cent in Merton and Morden, 105 per cent in Carshalton.[1]

Three distinct movements emerge from this short survey of population trends during the last hundred years. First in point of time was the exodus from the ancient city. The second was the preponderating growth of the outer fringe of boroughs within the administrative county. This was succeeded by the third process, which consisted of an enormous growth in the population of the dormitory areas lying outside the boundaries of the administrative county. Thus, the expansion of London by the method of agglomerating villages and country towns situated outside the built-up area has been assisted, if not effected, by the colonisation of these and neighbouring places by Londoners migrating for residential purposes farther and farther away from the centre.

These three movements can perhaps be more accurately described as three phases of a single movement: namely, the tendency of the better-off Londoner to live as far away as possible from the congested centre of the town where he earns his living. The distance he can go is obviously determined by the speed and cost of transport. As transportation has improved the distance has increased, and will no doubt continue to do so.

A great city such as London is centrifugal at night and centripetal during the day. I shall refer later to the huge population tide which ebbs and flows in the metropolis on every working day, and discuss the important problems of local government to which this phenomenon gives rise. In these few pages I have done no more than trace the steps by which the largest urban aggregate in the world has come into existence; and to show the vital changes which have accompanied its evolution. It is worth noting that at no point in the modern history of London has the machinery of local government been adjusted so as to take the facts of population into account.

[1] Report on Greater London Drainage (1935), p. 4.

4

CHAPTER V

THE ESTABLISHMENT OF THE POLICE FORCE

The one function of London government which received serious and drastic treatment during the first half of the 19th century was the police. The protection of property was regarded by the ruling classes in early Victorian London as more important than the protection of health, the preservation of amenities, the promotion of education, the development of the highways, the maintenance of public utilities, or, indeed, any other aspect of local government whatever. Hence it came about that while precious years were allowed to slip away without any general reorganisation of the local government of the metropolis being attempted until the reforming movement of 1835 had spent itself and great opportunities had been irrevocably lost, Parliament intervened with a firm hand at an early stage to sweep away the archaic collection of Bow Street runners and other quaint devices which served as a substitute for an effective police force.

Crimes of violence had long been rampant in the capital, culminating in occasional outbreaks of disorder and riot.[1] Burglary and robbery in the central area led to seven Parliamentary enquiries into the police system between 1770 and 1829,[2] the year when Sir Robert Peel introduced his Metropolitan Police Bill. The essential cause of the measure is explicitly stated in the preamble, which declares that "offences against property have of late increased in and near the Metropolis: and the local establishments of the nightly watch and nightly police have been found inadequate to the prevention and detection of crime, by reason of the frequent unfitness of the individuals employed, the insufficiency of their number, the limited sphere of their authority, and their want of connection and co-operation with each other."

[1] As on the occasions of the Gordon Riots in 1780 and the funeral of Queen Charlotte in 1820.

[2] H. Finer: "The Police and Public Safety," pp. 275–7 in *A Century of Municipal Progress*, edited by H. J. Laski, W. I. Jennings and W. A. Robson (1935).

The Act of 1829 established a unified organisation controlled by two salaried justices of the peace who in 1839 became respectively Commissioner and Assistant Commissioner.[1] They operated from a new police office set up in Westminster. The area of their jurisdiction after the latter date extended to a radius of 15 miles from Charing Cross. This Metropolitan Police District seems considerable even today; a hundred years ago it was a colossal territory for administering a police force.

Despite these thoroughgoing changes, the policing of the square mile of the City was left in the hands of the City Corporation. The Royal Commission on Municipal Corporations 1835 had reported strongly against leaving such an enclave,[2] but the Government of the day, like all its successors, did not dare to lay a hand on the rights and privileges of the ancient stronghold. Hence, the City Corporation was permitted to have a separate police force with its own Commissioner appointed by a committee composed of the Lord Mayor, the Aldermen and 29 Common Councilmen.[3] The City Corporation, with its vast wealth, was for long able to manage without the 50 per cent grant-in-aid from Parliamentary money which other local authorities had for many years received as the price of inspection and supervision by the Home Office; but in 1919 it could no longer afford to do so. Since then it has received a grant of one-half its approved expenditure after deducting the amount which would be realised by a 4d. rate in the £. A Royal Commission pointed out in 1894 that the control of the

[1] In 1856 the force was placed under the charge of one Commissioner and two Assistant Commissioners acting under the direction of the Home Secretary. In 1933 the executive officers became one Commissioner and five Assistant Commissioners appointed by the Crown. *Cf.* Hart and Hart: *An Introduction to the Law of Local Government and Administration*, p. 229.

[2] "The present system of City Police is burthensome, without possessing vigilance or unity; and we doubt whether any exertions of a civic authority could give the requisite vigour and efficiency to a Police under their direction, while their operations must be confined to the limits of the City. . . . There may be room for doubt which is the proper authority to whom the superintendence of the Police of the Metropolis should be entrusted; we apprehend there can be none that the authority, whatever it be, should be supreme and undivided throughout the whole district." Royal Commission on Municipal Corporations, Second Report (1837), p. 15.

[3] The appointment is subject to the Home Secretary's approval. 2 & 3 Vic., c. cxciv.

City Corporation over the Commissioner of the City Police is much less than that exercised by the Watch Committee of an ordinary municipality over their chief constable. This same Royal Commission estimated that the fusion of the City and the Metropolitan forces would result in a saving to the rate-payers of £50,000 a year—a figure which would probably be doubled by now.

The Metropolitan Police was completely removed from the domain of local government by the legislation of 1829–39. The Home Secretary acquired control over the police forces of the metropolis, exercising his functions through the Commissioner and his assistants. No change was effected when in course of time the Metropolitan Board of Works and later the London County Council came in turn to be established. One reason of overwhelming cogency ruled out the possibility of transfer to either of these local authorities: namely, the fact that the Metropolitan Police District was a large and comprehensive area, while the administrative county comprised merely a town within a town. "The prime necessity a hundred years ago," writes a contemporary exponent of London police administration, "was the introduction of unity, order and efficiency into the confusion and helplessness of the numerous petty police jurisdictions in the Metropolitan area."[1] Other solutions than the one adopted by Peel were possible: for example, the parishes could have been grouped in divisions and placed under divisional justices of the peace; or the Bow Street magistrates could have been given a general power of supervision. But neither of these expedients would have achieved the important principle of separating police and judicial functions. When the separation was effected and the new police office established in 1829 there were no local authorities worthy of serious consideration to challenge Home Office control of the police. Hence the new police office, like its predecessors, came under the jurisdiction of the Home Secretary. There was little, if any, political doctrine involved in the decision.[2] It was chiefly a matter of expedience.

The very effectiveness of this centralised solution of the only

---

[1] J. F. Moylan: *Scotland Yard and the Metropolitan Police*, pp. 60–1.
[2] *Ibid.*

problem which was regarded as urgent helped to postpone the general reform of metropolitan government by relieving the pressure at a vital spot. And when reform eventually came in 1855, it was far weaker than it would have been if the question of police administration had remained to be dealt with in terms of local government.

CHAPTER VI

THE METROPOLIS MANAGEMENT ACT, 1855

Prior to 1855 there was no administrative machinery of any kind responsible for the local government of the metropolis as a whole. All that existed, outside the narrow limits of the City, were about three hundred parochial boards operating under as many separate Acts of Parliament, and comprising more than 10,000 members. Some districts were entirely lacking in any sort of administrative control whatever.[1]

It is scarcely an exaggeration to say that before 1855 a condition of utter chaos prevailed. "In many places, and notably in districts which had become densely peopled with the poorest of the poor, there was not even a pretence of management, no public or quasi-public body existing at all for any sanitary purposes. And where there were such bodies their administration was usually a mere mockery of local government, the only reality of which was its entire freedom from control and its consequent inefficiency and extravagance. Miscellaneous bodies of paving commissioners, lighting commissioners, turnpike boards, directors of the poor, etc., were scattered at random over the town, with no regulations for their guidance, no attempt at uniformity of administration, no bond of union, no security for the proper performance of their functions. In most cases these bodies were entirely self-selected, and even where in theory the ratepayers were the electors, the process of election was conducted in a hole-and-corner fashion, and was utterly corrupt."[2] Outside the square mile of the antiquated

---

[1] J. B. Firth: *The Reform of London Government*, p. 40.

[2] A. Bassett Hopkins: *The Boroughs of the Metropolis*, pp. 3–4. There were also the vestries in some of the London parishes. The open vestry was nominally supposed to be the "town meeting." It assembled each year to elect the people's churchwarden and to fix the church-rate. Everyone might attend and voice his opinions. The chief paid official was the vestry clerk who was elected for life. There were also surveyors of highways and constables—unpaid offices to which any householder might be appointed. There was much corruption but in the close vestries matters were much worse. Sir Percy Harris: *London and its Government*, pp. 25–30.

City Corporation, London was a mere geographical expression.[1] There was not even a body to protect the rights of Londoners against the encroachments of railway companies desiring to run their lines across public commons, or to oppose other interests which threatened the welfare of the people.[2]

The metropolis was a veritable jungle of areas and authorities and a nightmare of inefficiency. In the parish of St. Pancras alone there were no less than 21 distinct and independent bodies of commissioners responsible for paving, cleansing and lighting the streets. These bodies operated under 35 local and private Acts. Only 4 of them were even in theory elected by the rate-payers, the remainder being self-elected or appointed by the proprietors.[3] Oxford Street was divided up between four separate parishes.[4] Some of the districts of the various commissioners and boards were a little over 300 yards long, others were 450 yards, and a few 1,500 yards. Some commissioners were self-elected or co-opted, and sat for life. In St. George the Martyr there were 6 different paving boards; in St. George in the East there were 5. St. Mary Newington had 2 paving and 4 lighting boards. In the Strand Union there were 7 different paving boards with separate staffs, and among the surveyors were a tailor and a law stationer. Within 1,336 yards of Northumberland House were 9 paving boards.[5]

In such circumstances it was useless for the General Board of Health to point out in 1850 that unless the elementary task of keeping the streets clean was carried out generally and simultaneously, much of the dirt in one district was carried by the traffic into another. The Board reported that insufficient use was made of sweeping machines, the more expensive method of hand-cleaning being employed by the contractors who flourished under the system. The Board of Health explicitly stated that they had received much evidence to show that the local boards were to a great extent in the hands of contractors who indirectly procured the return of friendly tradesmen as

[1] Sidney Webb: *The Reform of London* (pamphlet), p. 4.
[2] A. Emil Davies: *The Story of the London County Council*, p. 19.
[3] A. Bassett Hopkins: *op. cit.*, pp. 4–5.
[4] Report of the Metropolitan Board of Works (1857), p. 12.
[5] *The Government of London (The Westminster Review)* (1876), vol. 49, n.s., p. 107.

members. Against this result the ratepayers were unable to protect themselves.[1]

In view of the almost incredible mismanagement which existed, it is scarcely surprising that even so conservative a defender of ancient privileges as Toulmin Smith should declare, in 1852, that "the present condition of this huge metropolis exhibits the most extraordinary anomaly in England. Abounding in wealth and in intelligence, by far the greater part of it is yet absolutely without any municipal government whatever."[2]

The following year the Government appointed a Royal Commission consisting of three members, with Henry Labouchere as chairman, to enquire into the Corporation of the City of London. The Royal Commission of 1853 criticised in some detail the structure and practices of the City Corporation, and put forward a number of suggestions for its improvement.[3] But its recommendations left the City in substance untouched.

The commissioners proposed to apply the Municipal Corporations Act, 1835, to the Corporation, but they advised against extending the City's boundaries to cover the entire metropolis, on the remarkable ground that to do so would not only alter the whole character of the City Corporation but would "defeat the main purpose of municipal institutions."[4] It might have been thought that the main purpose of municipal institutions is to secure the good government of the areas to which they ought to be related; and that in order to achieve

[1] Report of the General Board of Health on Metropolitan Water (1850), pp. 242–3.

[2] J. Toulmin Smith: *The Metropolis and its Municipal Administration*, p. 6.

[3] Among the proposals were the following: A new charter for the City, the election of the Lord Mayor by the Common Council, the election of aldermen for a three-year term of office, the appointment of stipendiary magistrates, the abolition of the Court of Aldermen, an extension of the franchise, election by Common Hall to be abolished, election of sheriffs by the Common Council, abolition of the exclusive privileges of fellowship porters, application of the Municipal Reform Act in part to the City, the incorporation of the City with the Metropolitan Police, the transfer to the City of the Thames Conservancy, and the consolidation of accounts. *Cf. The Government of London (The Westminster Review)* (1876), vol. 49, n.s., p. 104.

[4] Royal Commission to enquire into the Corporation of the City of London, B.P.P., 1854, vol. xxvi, p. xiv.

this end an alteration in the character of the City Corporation would have been highly desirable. But the Commission thought otherwise, and they accordingly recommended that a municipal council should be set up in each of the 7 Parliamentary boroughs lying within the metropolitan district.[1]

This feeble report, which entirely overlooked the need for a unified administration in regard to a number of large-scale functions, had no other result than to defeat the movement for enlarging the City boundaries and transforming the Corporation into a great metropolitan municipality. The positive proposals of the Commission were ignored in the changes introduced by the Metropolis Management Act, 1855.

This statute, which was a private member's Bill, took as its area the metropolis as defined by the Registrar-General for the purposes of the Bills of Mortality. The territory in question contained 74,029 acres, a population of 2,803,034 persons living in 360,237 houses, and an assessable value of £12,450,416.[2] There were situated within it 99 parishes. In 23 of the largest parishes (such as Marylebone, St. Pancras, Islington, Shoreditch and St. George's, Hanover Square) a vestry was established elected by the ratepayers, while 59 smaller parishes were grouped into 15 districts administered by district boards. In each district the ratepayers of the constituent parishes were to elect members of the parish vestry, and the united vestries then elected the district board. The vestrymen were elected by householders rated for the poor rate at not less than £40 a year, except in some of the poorer places where the proportion of assessments at this figure did not exceed one-sixth of the whole number, in which case the qualifying minimum was reduced to £25. The number of vestrymen depended on the number of rated households;[3] the minimum was 18 and the

[1] Royal Commission to enquire into the Corporation of the City of London, B.P.P., 1854, vol. xxvi, p. xxxv. The boroughs and their populations were Finsbury (323,772), Marylebone (370,957), Tower Hamlets (539,111), Westminster (241,611), Lambeth (251,345), Southwark (172,863), and Greenwich (105,784).

[2] The figures relate to 1861.

[3] Parishes of less than 1,000 inhabitants had 18 vestrymen, those over 1,000 had 24, those with 2,000 or more had 36, with a further 12 vestrymen for every thousand inhabitants. The vestries also included a small proportion of ex officio members, numbering about 90 in all. Prior to 1894 the

maximum 120 members. The total number of members serving on all the vestries and district boards was about 3,100.[1] One-third of the members of each body retired annually.

These directly elected vestries and indirectly elected boards became the authorities for administering in their localities a series of functions shortly to be described. They were also made responsible for electing a central organ called the Metropolitan Board of Works. This consisted of 45 members. Three of them were elected by the Common Council of the City Corporation, two by each of the 6 largest vestries, one by each of 17 other vestries and the remainder by the district boards. One-third of the whole number retired each year.

The limited objectives which Parliament had in mind in enacting the Metropolis Management Act, 1855, are clearly indicated in the preamble to the statute which declares that the Act is passed in order to make provision "for the better management of the metropolis in respect of the sewerage and drainage and the paving, cleansing, lighting and improvements thereof." This narrow outlook prevails throughout the statute.

The vestries and district boards were entrusted with the management of local sewage and drainage; with paving, lighting, watering, cleansing and improving their parishes; and with all other duties and powers relating to the sanitary affairs of their localities. They were expressly empowered to regulate underground vaults and cellars; to restrain the occupation of underground dwellings; to remove projections and obstructions from houses; to control the erection of hoardings; to regulate the scavenging of houses; to object before the justices to licences for slaughterhouses being granted. An extensive Nuisances Removal Act[2] had recently been passed for the whole country, including London, and the new vestries and district boards were made the authorities for administering the Act. Each vestry or board was to appoint a Medical Officer of Health and Inspectors of Nuisances.

The area assigned to the Metropolitan Board of Works was

vicar was *ex officio* chairman. For details, see Sir G. Laurence Gomme: *London in the Reign of Victoria*, p. 198. The qualification for a vestryman was that he must be an inhabitant householder rated at £40 a year.

[1] J. F. B. Firth: *The Reform of London Government* (1888), p. 42.

[2] 18 & 19 Vic., c. 121.

identical with that which had come under the jurisdiction of the Commissioners of Sewers from 1848 onwards. The boundary was utterly unscientific even in the conditions then prevailing. The territory included, for example, a sparsely populated district such as Fulham, which was growing slowly, while leaving outside a populous and rapidly expanding area like West Ham.

The Metropolitan Board of Works superseded the Commissioners of Sewers and was explicitly charged with the task of designing and constructing a system of sewers "for preventing all or any part of the sewage within the metropolis from flowing or passing into the river Thames in or near the metropolis."[1] It took over from the Metropolitan Commissioners of Sewers and the City Commissioners of Sewers all drainage works previously vested in them. No local vestry could henceforth install a new sewer in its parish without its approval. The Board was further empowered to make, widen or improve highways to facilitate communication between different parts of London, and to name and number the streets. It was given important powers under the Metropolitan Building Act, 1855,[2] in regard to such matters as the issuing of general building rules, the appointment of a superintending architect of metropolitan buildings, the regulation of the thickness of walls, the appointment, suspension and (with the consent of the Home Secretary) dismissal of the district surveyors who were primarily responsible for enforcing the provisions of the Act. The approval of building plans also came within the jurisdiction of the Board. The Board could make by-laws in respect of a number of matters, to be enforced by the vestries and district boards.

The original functions of the Metropolitan Board of Works were enlarged by subsequent legislation as the Board gradually

[1] Section 135.
[2] See also the Metropolitan Building Act, 1844, which introduced elaborate provisions for regulating the construction, maintenance and repair of buildings. This Act required surveyors to be appointed for each district by the City Corporation in the City and the justices in quarter sessions in the counties. The surveyors were to be superintended by two official referees. There was also to be a Registrar of Metropolitan Buildings to be appointed by the Commissioners for Works and Buildings to report to them any matter appearing contrary to law.

came to be recognised as the most suitable authority for administering new services in London. Thus, the fire brigade was placed in its hands in 1866, the inspection of the gas supply in 1860, and in the ensuing years the Board received powers relating to the formation and maintenance of parks and open spaces, the construction of tramways, slum clearance, the superintendence of measures for the suppression of cattle plague, the right to protect the public interest before Parliamentary Committees considering private Bills,[1] and the collection of a rate to defray the cost of relieving the casual poor.[2] By 1876 some 80 statutes had been passed explaining, amplifying or extending the Metropolis Management Act, 1855, by conferring further powers on the Metropolitan Board.[3]

It was, of course, a great advance on the previous state of affairs to have a central body responsible even to a limited extent for the whole metropolis. But the defects of the legislation were as conspicuous as its advantages. The most glaring mistake was to make the Metropolitan Board of Works an indirectly elected body. In many instances its members were indirectly elected by indirectly elected district boards. In the second place, the Act of 1855 did nothing to supersede the pre-existing parochial units: it merely grouped some of them into districts. Nor did it touch the Poor Law Unions. It left intact the jurisdiction of the Lords Lieutenant of Middlesex, Tower Hamlets, Surrey, Kent and the other counties, and made no attempt to transfer the administrative functions exercised by the justices of the peace of Middlesex and the other counties. It permitted the Corporation of the City of Westminster to continue to be appointed by the Dean and Chapter of Westminster. It avoided any measure of reform of the City Corporation beyond vesting the City drains in the Board and enabling it to run main drains through the square mile. The Metropolitan Board of Works was not even given

---

[1] A Select Committee of the House of Commons proposed in 1861 that the Metropolitan Board of Works should act in this capacity; and the standing orders of both Houses of Parliament were amended accordingly.
[2] *Cf.* Select Committee on Metropolitan Local Government: First Report, 1866, B.P.P., vol. xiii.
[3] *The Government of London (The Westminster Review)* (1876), vol. 49, n.s., p. 95.

power to levy a rate in the City.[1] The Board had nothing to do with the drainage of the City.[2] "For the last 8 years," a spokesman of the City Corporation informed a Select Committee in 1871, "they have never been allowed in any Act of Parliament to have jurisdiction in the City, except where we consented in one case, and that was the Fire Brigade Act."[3]

A third fundamental defect was the absence of any provision enabling the Board to compel negligent or recalcitrant local authorities to carry out their statutory duties or to enforce the by-laws of the Board. The vestries and district boards might and did with absolute impunity neglect to carry out the imperative directions which Parliament had imposed on them, with disastrous consequences to the health and welfare not merely of a particular parish but of the whole London community.[4]

Nor could the Board require any degree of economy or uniformity of action to be observed by the parochial authorities, whose expenditure in course of time amounted to about £3 millions a year.[5] The Board could not even appoint a medical officer of health or sanitary inspectors to advise it on matters affecting the health of London as a whole. Another great weakness in the Act was the lack of any supervision by the central government over the work of the Board or the vestries beyond that implied by the requirement that the Metropolitan Board of Works should render an annual report to the Secretary of State.

In view of these obvious defects there is ample justification for saying that the setting up of the Metropolitan Board of Works in 1856 was "a lamentably half-hearted attempt" at reform. It still denied democratic control over local government in London; it kept in existence in a slightly changed form the old vestries whose corruption and incompetence were notorious; and it left the City untouched.[6]

---

[1] The City made a contribution, according to its rental, to the expenses of the Board "particularly as to sewers." The Board had no power to enforce a rate in case of default of payment. Evidence of William Corrie for the City of London: Select Committee on Metropolis Water (No. 2) Bill, 1871.
[2] *Ibid.*, p. 206.   [3] *Ibid.*
[4] Henry Jephson: *The Sanitary Evolution of London* (1907), p. 85.
[5] J. F. B. Firth: *The Reform of London Government*, p. 46.
[6] A. G. Gardiner: *John Benn and the Reform Movement*, p. 86.

CHAPTER VII

THE METROPOLITAN BOARD OF WORKS
AND THE VESTRIES

In view of the inherent defects in the legislation of 1855 and the feeble intentions of Parliament, the surprising thing about the Metropolitan Board of Works is not that it was a corrupt and almost irresponsible body, but that in spite of these evils it accomplished as much as it did in the thirty years of its existence.

By far the most important task to which the Board set its hand was the construction of a main drainage system for the metropolis. Within ten years this was virtually complete; 82 miles of main intercepting sewers had been constructed and the sewage was being conveyed through them to outfalls several miles away from the town. This enabled the open sewers to be filled in, and the frightful danger to health which they constituted to be removed. The cost of the main drainage work was reported by the Metropolitan Board in 1881 to be £5¾ millions.[1]

Another important achievement was the embanking of the north side of the Thames. The Metropolitan Board of Works was not responsible for the plan of the embankment, which was authorised by statute in 1852; but it carried out the construction of it at a cost of £2 millions. It freed from vexatious and obstructive tolls 10 bridges crossing the Thames, at a cost of £1,376,000; and established a steam ferry service at Woolwich. It created a municipal fire brigade for London after powers had been conferred in 1866. It acquired and maintained more than 30 parks containing about 2,600 acres, including such notable open spaces as Hampstead Heath, Clapham Common, Finsbury Park, Wormwood Scrubs and Southwark Park. Above all, it pursued a vigorous policy in regard to the construction of new highways, and was responsible for making several important thoroughfares, such as Victoria

[1] Henry Jephson: *The Sanitary Evolution of London*, pp. 158, 281; see also the Report of the Metropolitan Board of Works for 1881.

Street, Queen Victoria Street, Shaftesbury Avenue, Clerken-
well Road and Charing Cross Road, at a cost of about
£2 millions. During the latter half of the 19th century London
underwent a process of structural improvement far more
extensive than anything which has occurred in the 20th century.[1] Londoners of the mid- and late-Victorian eras were
accustomed to witness the execution of large and ambitious
projects to an extent which far exceeds the expectations of their
Edwardian and Georgian descendants. A large proportion of the
£28 millions which the Metropolitan Board of Works spent on
capital outlay was absorbed by major structural improvements.[2]

The Metropolitan Board of Works showed during its short
lifetime that public bodies which are corrupt are not always
necessarily apathetic, especially in regard to the more spectacular feats of municipal construction. Indeed, ambitious
projects of public works are sometimes embarked upon by such
bodies for the double reason that they offer good opportunities
for corruptly lining the pockets of their members and officers,
while at the same time impressing the general public with a
feeling that it is getting something for its money. It was,
however, the exposure of corruption incidental to the execution
of some of its most striking highway schemes which brought
the Metropolitan Board of Works to an inglorious end.

The particular circumstances of the scandals are of no
special interest at this distance of time. A full account of them
is on record in the official reports for those who are curious
about the details. The worst cases concerned the letting of the
extremely valuable sites which became available after land had
been cleared for the building of Shaftesbury Avenue and other
streets. The sites for the Piccadilly Restaurant and the London
Pavilion were let by private negotiation in circumstances which
aroused suspicion and several of the vestries passed resolutions
condemning the Metropolitan Board of Works for not inviting

[1] The City Corporation was also active within its area. Thus, Gresham
Street was an early Victorian improvement, Cannon Street was widened
and extended in 1860, and the Holborn Viaduct constructed in 1869. G. A.
Sekon: *Locomotion in Victorian London*, pp. 10–11. A good account of street
improvements in the 19th century is given in Sir G. Laurence Gomme:
London in the Reign of Victoria, Chapter VIII.

[2] H. Haward: *The London County Council from Within*, p. 109.

public competition. After an unsatisfactory "enquiry" by the Board itself a Royal Commission was appointed in 1888.

The Commission found that a number of grossly improper transactions had taken place. Two practising architects named Saunders and Fowler who had served on the Board of Works for 25 and 20 years respectively, and were also members of the Building Act Committee of the Board, and of its Theatres sub-committee, were most seriously involved in the scandal. They were found to have secured the withdrawal of theatre licences and the rejection of building plans, in order to get themselves employed on the preparation of fresh plans which were duly passed by the Board without debate. These members were implicated in a whole series of corrupt transactions relating to such widely spread projects as the London Pavilion, the Colonial Institute site, the building of the Avenue Theatre, the Northumberland Avenue sites, the fireproofing of the Albert Palace, and the Albany Road estate in Camberwell.[1]

Nor was it only the members of the Board who emerged from this enquiry with damaged reputations. Members of the staff were also implicated. The chief offenders among the officers were Goddard, the Chief Assistant Surveyor and Valuer, and Robertson, the Assistant Surveyor, who were proved to have systematically received "improper monies" in large sums as bribes for allocating sites and valuable privileges belonging to the Metropolitan Board of Works to unscrupulous speculators anxious to pervert the powers of the Board to their own purposes.[2] "The abuse of their position which characterised the official careers of both Goddard and Robinson was not confined to these departmental heads. It extended also to their subordinates."[3]

The Metropolitan Board of Works became increasingly corrupt as it grew older; and in the end it overreached itself. The appointment of the Royal Commission into these unsavoury affairs sounded its death-knell.[4]

The real causes for the rottenness of the Board lay deeper

[1] Royal Commissioner appointed to inquire into certain matters connected with the working of the Metropolitan Board of Works: B.P.P. (1888), Interim Report, vol. lvi, pp. 22, 32.
[2] *Ibid.*, pp. 6–9.   [3] *Ibid.*
[4] J. F. B. Firth: *The Reform of London Government*, p. 39.

PLATE II. The City, showing St. Paul's. The deep cleft running down the right-hand side is Queen Victoria Street, one of several large highway improvements carried out by the Metropolitan Board of Works (1855–88)
(*Aerofilms*)

than the rapacity of officials and contractors. The fundamental defect of the Metropolitan Board of Works was that it completely failed to awaken any civic spirit in the minds of London's inhabitants. The proceedings of the Board evoked neither interest nor enthusiasm. At no point in its existence did there emerge anything analogous to the vigorous municipal life which by this time was flourishing in the provincial towns. Its very offices in Spring Gardens were described by Lord Rosebery as "cavernous and tavernous."[1]

There is nothing surprising in this fact when we recall the ill-considered principles on which the Metropolis Management Act, 1855, was based. The conception of an indirectly elected authority precluded the possibility of responsibility on the part of the central body and of watchfulness or interest on the part of the public. It resulted in the election of the members being shrouded in mystery and secrecy so far as the ordinary man in the street was concerned. The very name of the Metropolitan Board of Works was ugly and repellent—a symbol of the narrow and uninspiring conception of London government which Parliament entertained throughout the century.

The multiplicity of authorities in the metropolitan area was another factor which encouraged apathy in the London citizen. We have already seen that the pre-existing bodies were permitted to continue in existence after 1855.[2] But even after this date new organs were brought into being. The Metropolitan Asylums Board was established in 1867 to deal with poor law patients, and later came to provide infectious disease hospitals and other classes of hospital accommodation for non-pauper residents, so that the Metropolitan Board of Works was not the only central metropolitan authority even in the sanitary field.[3] In 1870 the London School Board was set up under the Education Act to administer elementary schools. In 1872 the Port of London sanitary authority was created and the City Corporation constituted the responsible body. How could Londoners fail to be confused and distracted by all these diverse administrative organs?

[1] A. G. Gardiner: *Sir John Benn and the Progressive Movement*, p. 91.
[2] *Ante*, p. 60.
[3] *Cf.* Ministry of Health's 16th Annual Report, 1934–5, p. 46.

In 1867 a Select Committee appointed to enquire into the local government of the capital had put forward a scheme of reform which might have considerably improved the situation. Its proposals included the direct election by the ratepayers of a proportion of the members of the Metropolitan Board of Works; the inclusion of some justices of the peace to represent property owners; the representation of the various districts on the Board to be determined according to their rateable value, population and area; the revision of existing divisions in order to eliminate the small and inconvenient areas. The name of the Board was to be changed to "The Municipal Council of London"; it was to have the dignity of a president nominated by the members and approved by the Crown. The powers of the new Council were to include not only all those conferred on the Metropolitan Board but also the supervision of gas and water supplies, the protection of public interests regarding railway and other undertakings, and "such other functions as Parliament may from time to time appoint to be exercised by the chief authority for the general management of the affairs of the Metropolis."[1] The Committee emphasised that greater authority would attach to the deliberations of the principal London body if its members were directly elected by the ratepayers.[2]

This intelligent report unfortunately fell on deaf ears so far as the reform of the Metropolitan Board of Works was concerned. Its chief influence appears to have been to ensure that the London School Board was a directly elected authority. The general reform of the central administrative body in London was delayed for more than twenty years after the Select Committee had reported; and even then it was taken in hand only in consequence of the particular scandals to which reference has already been made rather than from a desire to remedy the serious defects in the legislation of 1855.

We may turn now to the vestries and district boards established in conjunction with the central Board of Works.

There is almost unanimous agreement among those who

---

[1] Select Committee on Local Government of the Metropolis, B.P.P., 1867, vol. xii. Second Report, p. vi.
[2] *Ibid.*, p. vi.

have enquired into the subject that the minor authorities created in 1855 failed miserably to fulfil with credit or efficiency the tasks which Parliament had assigned to them. "During the last ten years," complained Sir William Fraser, M.P., in 1866, "the condition of the streets has been gradually deteriorating; it is scarcely possible to conceive anything worse than the present state."[1] It would, he continued, be impossible to imagine anything more sordid than the state of the principal London thoroughfares. Vegetable and animal refuse was left to rot for weeks; "all is allowed to remain, and decay, and then, when his favourite daughter dies of scarlet fever, or leaves her bed deaf, or a cripple for life, let him remember that this might have been prevented, had the lanes of London been like those of a civilized community."[2]

Between 1856 and 1870 the vestries and district boards spent nearly £6½ millions on paving, lighting and improvement works. But, complained a writer in *The Westminster Review* in 1876, "on the most important point of sanitary action they have spent eighteenpence only per head of population, and for improvements under 2s. 9d. per head during 20 years of their authority, whilst Newington has spent £13,000 on its Town Hall, and Shoreditch £30,000. It is regrettable to find so little interest taken in the working of important Acts, like those regulating baths and washhouses, public libraries, common lodging houses, those enabling mortuaries to be established, crossing sweepers to be employed, disinfecting houses to be established, nuisances removed . . . and that a want of thought is shown in elementary action as to otherwise essential advantages."[3] The working of the system, declared J. F. B. Firth in 1888, has been accompanied by an incredible apathy. There are many vestries where a contested election is the exception rather than the rule; and the very date of an election is not known to the vast majority of the inhabitants. A Parliamentary return showed that in 1885 less than 1 in 30 of the electors used their votes—less than 4 per cent.[4] "The duties neglected

[1] Sir William Fraser: *London Self-Governed*, p. 9.   [2] *Ibid.*, p. 21.
[3] *The Government of London* (*The Westminster Review*) (1876), vol. 49, n.s., p. 110.
[4] J. F. B. Firth: *The Reform of London Government*, pp. 44-5.

by these vestries and district boards," wrote Mr. Sidney Webb in 1892, "are more important than those they attempt to perform."[1]

The work which had been entrusted to the vestries in connection with the sanitary condition of houses was very far-reaching and might have been of great importance. They had power to condemn and to close insanitary dwellings; to acquire and demolish condemned houses; to enforce stringent rules for premises let in lodgings or tenements. They could acquire land and provide public lodging or tenement houses for the poor. They could establish and maintain public libraries, baths and wash-houses, mortuaries, open spaces and other amenities.

The vestries and district boards had far greater legal powers than the central board so far as the sanitary state of dwellings was concerned. Most of these powers they neglected, particularly those connected with a pressing need of the day—the improvement of conditions in the tenement houses of the metropolis.

Various causes can be ascribed to the failure of the vestries. There is no doubt that jobbery and corruption by vested interests were the principal deterrents to effective action. "Vested rights in filth and dirt" were strongly represented on the vestries and district boards. "So long as vestrymen own little properties," observed a witness before a Select Committee in 1882, "and so long as their relations and friends do the same thing, and they are all mixed up in a friendly association, you can never get the prevention of the continuance of unhealthy tenements carried through."[2]

A feature of the Act of 1855 which ultimately became of great importance was the obligation of each vestry or district board to appoint a medical officer of health, with one or more inspectors of nuisances to assist him. The medical officer was required by law to inspect and report from time to time on the sanitary condition of the parish; to enquire into the existence of diseases increasing the death rate; to indicate the causes likely to originate or maintain those diseases and to

[1] Sidney Webb: *The London Programme*, pp. 17 et seq.
[2] H. Jephson: *The Sanitary Evolution of London*, pp. 267-8.

recommend measures to remedy them. They were also entrusted with various other important health duties.

The far-reaching effect which the establishment of public health departments under qualified men had upon the entire government of London were probably not foreseen either inside or outside Parliament at the time the legislation was passed. Nevertheless, when the appointments were once made there was for the first time a body of some 40 professional medical men who drew attention year after year in their annual reports to the appalling conditions of disease and filth existing in the metropolis.[1] Many of them brought both public spirit and energy to their work, and although their reports seldom managed to circulate outside the walls of the vestry hall, in course of time this mass attack on the insanitary condition of London produced an effect despite the apathy, indifference and jobbery of the vestry members.[2]

The lot of a medical officer in a London vestry was no easy one. Some of the local authorities put pressure on their medical officers in order to restrain their zeal. Thus, the medical officer for St. George-the-Martyr, Southwark, resigned in disgust because he was not allowed to carry out the duties of his office.[3] The vestry of St. James, Westminster, when they discovered that their medical officer took his work very seriously, reduced his salary from £200 to £150 a year in order to discourage him.[4] Dr. W. Farr, of the Registrar-General's office, told a Select Committee in 1866 that he believed in certain London districts "the medical officer of health was under all sorts of restraints. If he is active, they look upon him with disfavour, and he is in great danger of dismissal."[5]

Another method of frustrating the efforts of the medical officers was to refuse them an adequate staff of inspectors of nuisances for the purpose of inspecting the district. In Bethnal Green a single inspector was appointed in 1861 to deal with a population of 108,000 persons living in nearly 15,000 houses. Shoreditch, with a population of 129,000 persons living in

---

[1] A very careful and detailed study of these reports is to be found in H. Jephson: *The Sanitary Evolution of London* (1907), *passim*.
[2] *Ibid.*, p. 92.   [3] *Ibid.*, p. 189.   [4] *Loc. cit.*
[5] Select Committee on Metropolitan Local Government, 1866, Second Report, B.P.P. vol. xiii. Minutes of Evidence, p. 27, Q. 2250.

17,000 houses, had also only one inspector.[1] A similar situation existed in St. George's, Hanover Square, in Bermondsey, in Paddington and many other districts. The medical officer for St. Pancras tendered his resignation in 1875 on the ground that while he was responsible for the sanitary condition of the parish, he was denied the assistance which he considered necessary to conduct outdoor inspection of houses visited with contagious diseases or habitually in an unsatisfactory condition. "I feel," he said, "that the severe condemnation which a house-to-house visitation of the poorer parts of the parish has received from a majority of the sanitary committee must of necessity hopelessly weaken my authority with the sanitary inspectors, and render nugatory my efforts to carry out the Sanitary Acts."[2]

In his careful history of the public health movement in London, Jephson has expressed the opinion that the greater part of the sanitary progress which was made during the period of vestry rule was directly due to the unceasing labour, the courageous efforts and the insistence of many of these officers.[3] As we have seen, they had to face the frequent disapproval, obstruction and discouragement of their own councils. In regard to several matters, however, the vestries were powerless even if they had desired to take effective action. Thus, the water supply was in the hands of a number of commercial water companies; and no matter how great the danger to the health of the district from an impure or insufficient supply, the vestry had no remedy whatever in its hands.[4]

The Local Government Board had come into existence in 1871; but so far as London was concerned, its supervision was a matter of trifling importance during the lifetime of the Metropolitan Board of Works.[5] The administration of the capital city was not notably improved by the establishment of a central department responsible for local government, at any rate during the existence of the Metropolitan Board of Works and the vestries.

[1] Jephson: *op. cit.*, p. 188.   [2] *Ibid.*, p. 273.
[3] *Ibid.*, p. 399. See also p. 92.
[4] See the complaint of the medical officer for St. George-in-the-East in 1856. Jephson, p. 106.
[5] The water supply, so far as it was subject to any control at all by the central government, came under the jurisdiction of the Board of Trade.

CHAPTER VIII

THE CITY'S OPPOSITION TO REFORM

Before considering the abolition of the Metropolitan Board of Works and its supersession in 1889 by the London County Council we must examine the attitude of the City Corporation to the reform of London government, for without knowing something of that attitude it is impossible to understand the legislation of 1888.

During the 19th century there were frequent enquiries by Royal Commissions, Select Committees and Departmental Committees as to the best method of improving the local government of the metropolis. The most important of these were the Royal Commissions of 1835, 1853 and 1894; and the Select Committees of the House of Commons of 1861, 1866 and 1867. There were also many official enquiries into particular services in the metropolis such as drainage, transport, and water supply. The reports of some of these enquiries have already been mentioned; and reference will be made to others in due course. All through the century a stream of books, pamphlets and articles poured out on the subject of the reform of London government.[1]

Numerous Bills were introduced to improve the chaotic conditions which prevailed in the capital. Sir George Cornewall Lewis introduced a Bill in 1860,[2] and Mr. Ayrton sponsored a measure to reform and unify London government from 1860 onwards.[3] John Stuart Mill introduced a Bill in 1867 to set up 11 municipal corporations in the metropolis to supersede the vestries.[4] In 1869 and 1870 Charles Buxton introduced a

---

[1] The bibliography is, of course, far too large to be cited. The following are a few examples, and the footnotes to Part I of this work refer to many others:—William Carpenter: *The Corporation of London, as it is, and as it should be* (1847); J. Toulmin Smith: *The Metropolis and its Municipal Administration* (1852); Alexander Pulling: *The City of London Corporation Inquiry* (1854); Sir William Fraser: *London Self-governed* (1866); George Horton: *The Municipal Government of the Metropolis* (1866); Sidney Webb: *The London Programme* (1891).

[2] George Whale: *Greater London and its Government* (1888), p. 23.

[3] Percy Harris: *London and its Government*, p. 36.

[4] J. F. B. Firth: *The Reform of London Government*, Chapter II.

somewhat similar measure and also another Bill establishing a new central authority for the metropolis. The Corporation of London Bill, as it was called, set up a Municipal Council consisting of a Lord Mayor, Aldermen and Councillors. The Councillors were to comprise 32 Common Councilmen from the City, 135 representatives from 9 new boroughs, and members of the Metropolitan Board of Works. A third Bill was also before Parliament at this time for the purpose of constituting the metropolitan area a single county. These Bills were all withdrawn when the Home Secretary declared that the Government intended to deal with the matter. They were, however, reintroduced in 1870 and after second reading were referred to a Select Committee.

In 1875 Lord Elcho introduced a Bill to extend the City Corporation so as to give it jurisdiction over the whole of London. The new organ was to absorb the functions of the City, the Metropolitan Board of Works and the vestries. For purposes of representation there were to be 10 boroughs. This Bill received strong support from *The Times*, which said in a principal leader[1] that the first condition for a solution of the London problem was the creation of a municipality for the entire town. "A reform of the Municipal Government of London has confessedly been so long needed, and it has been so long left unperformed, that we are glad at last to be assured that it will be taken vigorously in hand in the coming Parliamentary session." "Lord Elcho's Bill," continued *The Times*, "will of course have for its firm aim to get rid of the old evil of separate and conflicting municipal jurisdictions. . . ." The Bill seems to meet very completely the difficulties of the situation and the problem of adjusting the different interests; and at the same time "to pay all due regard to the very stubborn claims of the vested interests of the old City Corporation." The article concluded with a hope that all Londoners would agree heartily in putting an end to the anomalies and inconvenience of the existing artificial divisions of power within the single area of the metropolis "and that the whole of London may thus be united into one great Municipality."[2] *The Times* also drew attention to the more economical administration

[1] *The Times*, 7th October, 1874, p. 9.   [2] *Ibid.*

which would be brought about by the proposed change, since one staff of officers acting in concert would replace many staffs serving separate authorities.

But the Thunderer thundered in vain. The Bill was withdrawn before its second reading. Five years later another private member's Bill was introduced with the backing of J. F. B. Firth, Thorold Rogers, and others. The 1880 Bill, like its forerunner, set up a central representative body to control local administration throughout London, with more or less the same powers as those set out in Lord Elcho's Bill. These later Bills tended to give larger powers to the central authority than the earlier ones.

This Bill shared the fate of its predecessors. But it was obvious that the existing situation could not continue indefinitely, and the volume of discontent was growing. In 1875 Sir William Fraser moved a resolution in the Commons to the effect that the condition of the metropolis as regards lighting, paving and cleansing called for legislation. In the course of his speech he asked the House whether the metropolis should continue "in its present state of sordid anarchy" in regard to these important matters.[1] The motion was withdrawn, but in 1878 the House of Commons passed a resolution declaring the present state of London government to be unsatisfactory and requiring reform; that the whole metropolis should be united under one directly elected authority commanding general confidence; and that these reforms should be undertaken by the Government without delay.[2] Parliament had so far failed to deal with the problem of London government with either courage or foresight; but these straws in the wind showed that even the complaisance and indifference of the legislature had limits.

In 1884 the first Government Bill ever framed to deal with the matter was introduced by the Home Secretary, Sir William Harcourt.[3] The London Government Bill sought to establish a unified system of local administration by transforming the City Corporation into an organ for the whole metropolis, to

[1] Hansard: 1875, vol. 222, col. 190.
[2] Percy Harris: *London and its Government*, p. 38.
[3] The Metropolis Management Act, 1855, was a private member's Bill.

be operated in association with district councils to which it would delegate powers.[1] The City boundaries were to be extended to cover the whole area; the local districts would elect the Common Council, which would in turn elect the Lord Mayor. The district councils would be directly elected; but they would possess no powers other than those conferred by the Central Council. The latter body would have a majority of elected members (159), although 44 members were to be elected by the old Common Council. The whole membership of the Metropolitan Board of Works was to be retained on the new authority. This was an obvious attempt to conciliate both the City Corporation and the Metropolitan Board of Works.[2]

The Harcourt Bill was a genuine measure of unification. The new municipality of London was to possess all the powers of the City Corporation, of the Metropolitan Board of Works, of the vestries and district boards, of the commissioners of paving, baths and wash-houses, etc.; and of the justices of the peace as regards their administrative functions.[3] The district councils were to become, in effect, committees of the central body rather than independent organs. The area chosen was the area of the Metropolitan Board of Works; but it was generally understood that this was a preliminary step and that the territory defined in 1855 would in due course be enlarged.[4] The Select Committee of 1867 had emphatically recommended that the metropolis should be constituted a county unto itself.[5]

The Harcourt Bill had a number of defects; but it cannot be doubted that it would have removed once and for all the worst evils which had afflicted London since the early years of the 19th century: namely, the inefficiency of parochial bodies

[1] A. G. Gardiner: *Sir John Benn and the Progressive Movement*, p. 87.
[2] Harris: *op. cit.*, p. 39. [3] A. G. Gardiner: *op. cit.*, p. 204.
[4] J. F. B. Firth: *The Reform of London Government* (1888), pp. 26–7.
[5] "Had the metropolis been situate within an entire county, there can be little doubt that it would long since have enjoyed the same advantages as other large cities or boroughs, and would have had within its own limits a complete organisation for all purposes of local government and administration. Your Committee, therefore, consider the first step towards an improved and efficient system of local government is to correct this anomaly and to constitute the metropolis a county of itself, according to the principles recognised in the government of this country from the earliest times." Select Committee on Local Government and Local Taxation of the Metropolis, B.P.P., 1867, vol. xii, p. v.

and the independence of the City. London would today be a far finer and better-governed city if Harcourt's London Government Bill had been passed into law.

The drafting of the Bill was preceded by a severe struggle in the Cabinet on the subject of police control. Harcourt had been much concerned during his tenure of office with the outbreaks of violence in Ireland; and he took the view that the Home Secretary should be a "Chief of Police." He was therefore unwilling to hand over the metropolitan police to the new council. Gladstone, Joseph Chamberlain and Dilke protested in vain at this discrimination between London and the provinces.[1]

These disputes were, however, resolved by the time the Bill was introduced, and Gladstone as Prime Minister spoke strongly in support of the measure. "The local government of London," he told the House of Commons, "is, or if it is not, it certainly ought to be, the crown of all our local and municipal institutions. The principle of unity (of London) has already been established, under the pressure of necessity, as a matter which could not be resisted. It has been established in the Metropolitan Board of Works. . . . There can be no doubt that we have established a principle of unity and that we have found it satisfactory. . . . London, large as it is, is a natural unit—united by common features, united by common approximation, by common neighbourhood, by common dangers—depending upon common supplies, having common wants and common conveniences. . . . Unity of government in the metropolis is the only method on which we can proceed for producing municipal reform."[2]

The Harcourt Bill never even reached the committee stage. It was introduced into an almost deserted House, and was then put into cold storage for many weeks. Nearly three months elapsed before there was an opportunity for a general debate on its principles. By this time it was too late in the session for the subject to be properly discussed, while on the other hand the City Corporation had made the fullest use of the long interval to marshal up every ounce of opposition—genuine or spurious—which could be mustered against the measure. After

[1] A. G. Gardiner: *op. cit.*, pp. 87–8.
[2] Hansard: 1884, vol. 290, cols. 541, 547, 552, 555.

three days' debate the Government obtained a majority of 71 on a motion for the adjournment. An optimist could regard this as a decision by the House in favour of a Bill which had to be abandoned owing to shortage of time.[1] The measure had in fact been killed.

The Government fell in 1885 and was replaced by a Conservative administration under Salisbury. The main reason for the defeat of the Harcourt Bill lay in the bitter opposition to its provisions manifested by the City Corporation, although the immediate cause was shortage of Parliamentary time. If the Corporation had been willing to merge its own interests in the common welfare of the whole capital, there is little doubt that the same or a similar measure would have been enacted after the fall of Gladstone. But the City was adamant.

The record of the City Corporation in delaying, obstructing or defeating legislation aimed at the reform of London government between 1835 and 1880 was one of unbroken success from their point of view. They defeated the intention of Lord John Russell to introduce a Bill in 1837. They managed to exclude the square mile of the City from the Police Bill of 1839 defining the Metropolitan Police area, and thus secured a separate police force for the City. They defeated Lord Grey's City Reform Bill in 1856 and procured the withdrawal of the Government Bill of 1858. They used all their influence against measures introduced in 1859, 1863, 1867, 1868, 1869, 1870 and 1875. According to J. F. B. Firth, who had inside knowledge of the subject, several of these Bills were defeated by the City's exclusive agency.[2]

In 1884 an unusually determined effort was called for if the *status quo* was to be preserved; and the City rose—or perhaps one should say descended—to the occasion. We are not accurately informed as to the normal practices of the Guildhall in combating legislation before or after 1884; but we do know what occurred on that occasion.

The Court of Common Council appointed in 1882 a Special

---

[1] Annual Register, 1884, pp. 180-1; J. F. B. Firth: *The Reform of London Government* (1888), p. 26.
[2] J. F. B. Firth: *A Practical Scheme of London Municipal Reform* (1881), pamphlet published by London Municipal Reform League.

Committee to consider the announcement in the Queen's Speech referring to the reform of the Corporation of London and to do whatever they might deem expedient in the matter. This Special Committee spent nearly £20,000 out of the City's cash between 1883 and 1885 for the purpose of influencing Parliament by misrepresenting the state of public opinion. No less than £14,139 was spent in 1884,[1] the year of greatest danger from the City's point of view. A great deal of this money was undoubtedly spent on bribery and corruption of the crudest kind; and every conceivable trick was employed to make it appear that large sections of the public were opposed to the reforms contained in the Bill.[2]

The Bill was being supported by a sincere and honourable organisation known as the Municipal Reform League, run mainly by J. F. B. Firth, a public-spirited and able barrister who later became a Member of Parliament and deputy chairman of the first London County Council. The Special Committee of the City Corporation paid speakers and persons to attend the meetings of the League and to oppose the views expressed there;[3] it hired bullies to disturb and break up those and similar meetings; it promoted a bogus organisation called the Metropolitan Ratepayers' Protection Association to oppose the Bill[4] and suppressed by violence any manifestations of dissent shown at their meetings. Under the management of a journalist acting for the Corporation "this practice assumed proportions which could scarcely have been consistent with the public safety."[5] It forged counterfeit tickets of admission to meetings of the Municipal Reform League in order to pack the audiences.[6] It paid men to organise sham conferences, deputations and public meetings, and spent large sums on obtaining the insertion in the press of bogus resolutions, concocted reports of meetings, and the abuse of those in favour of reform, in order to create a hostile atmosphere against the

[1] Select Committee on London Corporation (charges of Malversation), B.P.P., 1887, Reports, vol. x.
[2] J. F. B. Firth: *The Reform of London Government* (1888), p. 17.
[3] *Ibid.*
[4] Select Committee on London Corporation (charges of Malversation), p. xi.
[5] *Ibid.*   [6] Firth: *loc. cit.*

legislation. It promoted sham charter movements in various parts of London in order to lead people to believe that there was a desire in the metropolitan districts for incorporation. As soon as the London Government Bill was dropped these charter movements quickly faded away.[1]

The extent and character of the Corporation's activities in wrecking the Harcourt Bill were considered so scandalous that Parliament appointed a Select Committee to enquire whether malversation of funds had taken place. The Committee investigated the facts at length and reported that a conclusive judgment on the question of malversation could only be made by a court of law—a decision which might well have been made before the Committee was appointed. All that the Committee could say on that point was that malversation was not established by the evidence given before them.[2]

The real gravamen of the charge against the Corporation was, however, that it had been guilty of dishonest political manipulation; and this the Committee found to be clearly proved. It condemned the extravagant and excessive expenditure on advertisements[3] and the absence of proper supervision by the Corporation over its agents. It declared that much of the money which passed through their hands was used for improper and indefensible purposes.[4] It denounced the subsidisation by a public body of so-called political associations such as the Metropolitan Ratepayers Association. It found that the Corporation had improperly used part of its funds for the purpose of misleading Parliament by the appearance of an active and organised public opinion hostile to the reforms.[5]

The Report of the Select Committee was a severe blow to the honour and prestige of the City; but it had nevertheless gained the day so far as the practical issue was concerned. The privileges and possessions of the Corporation were left intact, its jurisdiction remained untrammelled, its narrow outlook untouched by a wider vision. If a measure similar to the Harcourt Bill had been reintroduced soon after the Select Committee had reported, the City Corporation could scarcely have maintained the struggle after such a damaging exposure

---

[1] Select Committee, p. xii.
[2] Ibid. p. xiii.
[3] Ibid., p. xiv.
[4] Ibid.
[5] Ibid.

of its methods. But the tide was running against the Liberal reformers and time was on the side of those who resisted change. Hence, when the next important step was taken in 1888, the ancient rights of the City were not challenged.

There are two important points to be noted in connection with the City's opposition to reform. One is that vested municipal interests can be, and often are, as obstructive to political changes and as indifferent to social welfare as private financial interests. Local authorities have nothing to learn from business men in the matter of putting their own interests before those of the public welfare when the question of survival is at stake. The case of the City Corporation of London provides an outstanding example of this, but there are plenty of other instances.

Second, the City Corporation would not have been able successfully to withstand the many attempts to transform it into a municipal corporation administering the whole metropolis unless Parliament had been imbued with a peculiar animus against London. The City may have been selfish in refusing to broaden its basis; but the resistance of the Guildhall touched a responsive chord in the hearts of Members of Parliament. A deep-seated reluctance to create a great elected municipal council for the capital city of the British Empire, embodying all the prestige and power which such a body should properly possess, has undoubtedly been entertained by successive Parliaments. The motives of jealousy and fear of a potential rival may have played an unconscious but decisive part in enabling the opponents of reform in the City to obtain that ready defeat of legislative intentions which it is otherwise so difficult to explain.[1]

[1] A. G. Gardiner: *Sir John Benn and the Progressive Movement*, p. 86.

CHAPTER IX

THE LONDON COUNTY COUNCIL

When the reform of London government was eventually brought about, it arrived by the back-stairs method characteristic of almost all the changes in the metropolis which occurred during the past century. The Local Government Act, 1888, which established the London County Council, was primarily a measure for instituting elected councils in the counties, where local administration was still in the hands of the justices of the peace appointed by the central government. The Public Health Act of 1875 set up urban and rural sanitary authorities within the county districts; but county government as such was left unchanged. No breath of democracy had so far touched the face of the ancient shires. In the process of providing the rural counties with elected councils, Mr. Ritchie, the President of the Local Government Board, took the opportunity of creating London an administrative county and treating it in similar manner to the others.

The statute provided that the area of the Metropolitan Board of Works—in which the London County Council was shortly to be established—should be an administrative county called the administrative county of London, and that such portions of this new county as formed parts of Middlesex, Surrey and Kent should be severed from those counties and form a separate county for all non-administrative purposes under the name of "The County of London."[1]

The point about this distinction is that for certain administrative purposes the Act created the administrative county which includes the City, and for which the London County Council is elected;[2] while for so-called "non-administrative" purposes[3] the City of London remains a separate county in its

[1] Section 40, Local Government Act, 1888. *Cf.* Herbert Morrison: *How Greater London is Governed*, p. 32.
[2] Report of the Royal Commission on the Amalgamation of the City and County of London, 1894, B.P.P. vol. xvii, p. 15.
[3] By "non-administrative" in this connection is meant such functions as those performed by quarter sessions, the justices of the peace, the coroner, sheriffs and so forth.

own right. Thus, for non-administrative purposes there are
two counties, while for so-called administrative purposes there
is only one—in theory. But in practice the City is engaged in
administering many or most of the services which are provided
within its area.

The London County Council consists of 2 councillors elected
for three years by the electors in each of the 60 Parliamentary
divisions into which London is divided, together with 4 members elected by the City electors, making a total of 124. To
this must be added 20 aldermen elected for six-year terms of
office by the elected councillors of the London County Council.

The Local Government Act, 1888, did not affect either the
constitution or the functions of the Metropolitan Asylums
Board which was providing hospital and other institutional
accommodation for the sick poor; the School Board for London,
which was responsible for public education; the Thames and
Lee Conservancy Boards; the boards of guardians engaged in
administering the poor law; the several burial boards; the
Metropolitan and City police forces; or the various commissioners of baths, wash-houses and public libraries which
existed in various parts of London. Even the vestries and
district boards were left essentially untouched: the only ways
in which they were directly affected were that they were no
longer required to bear the cost of repairing main roads, and
the London County Council could also pay half the salaries of
their medical officers of health and bear part of the cost of
maintaining minor streets and footpaths.

It can be seen, therefore, that the Local Government Act,
1888, did not even pretend to solve the problem of London
government. It merely superimposed on top of the existing
confused structure a new type of organ designed primarily for
large, sparsely populated rural areas. It gave county government
to a metropolis which was above all else a town. This, it has
been truly said,[1] is the root of all the struggles and troubles
which have subsequently disfigured the scene. In one important
respect the legislation of 1888 made the proper organisation of
government in London more difficult to obtain in that it
established powerful elected councils in the home counties

---

[1] *Cf.* A. G. Gardiner: *Sir John Benn and the Progressive Movement*, p. 90.

6

bordering the metropolis, and thereby strengthened their potential resistance to any widening of boundaries. This problem of the neighbouring counties is one of the most intractable questions requiring to be dealt with at the present time.

The London County Council obtained very few new powers under the Local Government Act, 1888. Nearly all the functions were transferred from the Metropolitan Board of Works, which it supplanted, together with a number of duties transferred from the former county authorities or justices out of sessions, and from the highway authorities. The principal new powers acquired in 1888 were the right to oppose Bills in Parliament, to appoint a staff of medical officers, and to contribute to the maintenance or enlargement of highways even although not main roads.

The most important matters in which the London County Council was given jurisdiction over the whole administrative county, including the City, were main drainage, main roads, tramways, the fire brigade, the embankment of the river and flood prevention, commons, parks and open spaces (other than those managed by the City Corporation, the vestries or the Crown), the issuing of by-laws relating to overhead wires, the naming and numbering of streets, the formation of streets and control over the frontage lines, together with numerous minor duties.

The London County Council was thus from its birth charged with administering within the square mile of the City a number of services which the City Corporation, if it had the status of a county borough in a provincial town, would itself perform. To these original services many others have been added by subsequent legislation, including education, public assistance and the provision of hospitals. On the other hand, the City Corporation not only retained control over its police and judicial institutions, but is also responsible for exercising a wide range of powers relating both to the regulation of conduct and the provision of services. Among these may be mentioned the registration and inspection of nursing homes, the clearance of unhealthy areas, the treatment of venereal disease, the testing of gas and electric meters, the registration of employment agencies, analysis of fertilisers and feeding stuffs, the regulation of massage establishments, the licensing of petroleum storage, the enforcement of the Shops Acts, town planning, the protection of children, the verification of weights and measures, the provision of bridges, aerodromes, subways and street improvements.[1] Even where the City Corporation had managed to acquire by ancient privilege or special legislation rights and powers extending beyond their own territory, such as the right of holding markets within a radius of seven miles or the still wider jurisdiction as port sanitary authority, these were left untouched. The legislation of 1888, remarks Mr. Gardiner, made the mistake of "leaving the City like a sort of obsolete appendix at its centre."[2]

Whether the responsible Minister took the wiser course in evading the issue rather than meeting the full force of a struggle with the City Corporation, whether he should be blamed for being content to build round the ancient fortress instead of on its foundations, is a matter on which opinions may differ. Whatever may be said on that point, it is indisputable that in 1888 the metropolis received for the first time a municipal council based on a popular vote, possessing some degree of unity and a semblance of civic life.[3] Despite all the defects and limitations inherent in the reform of 1888, it was by far the most important event which happened to London during the 19th century.

The Metropolitan Board of Works, doomed to an ignominious death as a result of its own misdeeds, sought recklessly to make the short span of life which remained to it as full as possible of evil consequences to its successor. Its last days were occupied in making decisions committing the new Council to heavy expenditure in which the latter had no voice, in raising the salaries and pensions of its officers, in sanctioning an important encroachment in a main thoroughfare, in accepting a contract of vast dimensions for the construction of the Blackwall tunnel under the Thames. Throughout these pro-

[1] The City Corporation exercises all the extensive powers of a Metropolitan Borough Council in addition to its special powers. In certain other matters such as housing and parks, the City Corporation has concurrent powers with the London County Council which it seldom exercises directly.
[2] A. G. Gardiner: *Sir John Benn and the Progressive Movement*, p. 89.
[3] Sir Percy Harris: *London and its Government*, p. 40.

ceedings the Metropolitan Board of Works treated the nascent London County Council with neglect and contempt. The episode of the Blackwall tunnel was considered by the President of the Local Government Board to be so outrageous that he frustrated the intentions of the Metropolitan Board of Works by putting forward the "appointed day" when the London County Council took over, so that the contract could not be sealed. "That moribund and discredited body," remarked *The Times* of the Metropolitan Board of Works in a leading article,[1] "might have been allowed to expire quietly on the 'appointed day,' or, as Lord Rosebery put it, to 'wrap its robe round it and die with dignity,' if it had not resolved to flout its successor, to insult Parliament, to outrage public opinion and to defy the executive government. . . . Universal London will feel that it is well rid of a body so blind to its own dignity, so unmindful of the plainest precepts of duty, so indifferent, indeed, to ordinary restraints of decency, as the Metropolitan Board of Works has shown itself to be in the last few weeks. The gravest suspicions of corruption and malversation will attach to its memory. . . ."

The establishment of the London County Council was followed by a complete change of atmosphere in London government which was reflected by the personnel of the new Council. The first chairman was Lord Rosebery, a former Foreign Secretary and a man already marked out as a possible Prime Minister, whose name and personal qualities added greatly to the prestige of the assembly.[2] Among the members of the first Council were Sir John Lubbock, M.P., Sir John Sinclair (afterwards Lord Pentland), the Earl of Meath, Lord Hobhouse, G. W. E. Russell, Frederic Harrison, Sir Thomas Farrer, John Williams Benn and J. F. B. Firth—all men of outstanding ability and integrity. A few years later they were joined by Sidney Webb and a number of other Fabians or Progressives such as Sir William Collins, J. Ramsay MacDonald, Will Crooks, W. Stephen Saunders and Robert Donald. It is not too much to say that the London County Council attracted an entirely new type of man to municipal government in London.

---

[1] *The Times*, 20th March, 1889, p. 9.
[2] *Cf.* A. G. Gardiner: *op. cit.*, p. 92.

It was not merely on the Council itself that a new spirit was evoked. Among the teeming millions of the electors there began to emerge some signs of a civic awakening, despite the fact that the Londoner dwelling in the County of London was denied the name of citizen and relegated to the status of a mere "inhabitant." For the first time a great part of the metropolis, "deprived hitherto of its birthright by the little medieval fortress of privilege that absorbed its dignities and honour, found itself with a common life, and a common instrument of government popularly elected, aflame with the spirit of social reform."[1] The work of the Council was from the outset done in the full light of publicity; and for purity of administration its record was from the beginning unsurpassed. In all these features the London County Council presented a striking contrast to the old Metropolitan Board of Works, to the still existing vestries and to the City Corporation.

From 1889 to 1907 the Progressive Party was in power at Spring Gardens (the headquarters of the Council) except during the period 1895–8, when it shared power with other groups. The Progressive Party consisted of Liberals and Socialists in alliance. The Conservatives were organised in the Moderate Party, which in 1907 changed its name to the Municipal Reform Party and swept into power. It remained the dominant group until 1934, when the Labour Party gained a victory[2] which was repeated in 1937 and 1946.

During the early years of the London County Council's existence the party divisions were neither very deep nor clear-cut. The London County Council was a "progressive" body because municipal life in London had been starved and frustrated for decades and there was in consequence an immense amount of urgent work which obviously needed doing. In all the main departments of local government, education, public health, the conditions of life and labour, amenities of all kinds, the new forces made themselves felt.[3] The Progressives undoubtedly created a municipal spirit in London.[4] In this they

[1] A. G. Gardiner: *op. cit.*, p. iii.

[2] See my articles: "Thoughts on the London County Council Election," *Political Quarterly*, April–June, 1934; and "London and the London County Council Election," *Political Quarterly*, April–June, 1937.

[3] A. G. Gardiner: *op. cit.*, pp. 146–7.   [4] *Ibid.*, p. 94.

were greatly assisted by the League of Municipal Reform,[1] founded in 1875, which played an important part in the establishment of the London County Council; and also in a different way by the Fabian Society in its early years.[2]

During the 19th century all the public utility services in the metropolis were in private ownership; and since municipal gas, water, electricity and transport undertakings were a commonplace in the provincial towns, it was inevitable that a demand for municipal ownership should arise in London. It was only to be expected that the Progressives would before long turn their attention in that direction. When this actually occurred it led to great hostility on the part of Conservative interests and to an intensification of party conflict. It was, however, Conservative politicians who took the first decisive step of linking up municipal politics in London with the national party alignments. On 7th November, 1894, Lord Salisbury, then Prime Minister, made a violent attack on the London County Council at a meeting of the Metropolitan Union of Conservative and Unionist Associations. He denounced the London County Council as "the place where Collectivist and Socialistic experiments are tried . . . a place where the new revolutionary spirit finds its instruments and collects its arms"; and he advised his audience to throw their whole strength into the coming election. This was the first occasion on which the London County Council was definitely made an arena of national party conflict.[3] The election in 1895 resulted in a dead-heat for the opposing forces.

Four years later Mr. Sidney Webb wrote regretfully: "The record of the past three years is in many ways a disappointing one to those who desire social reform. This decline in municipal activity has been due, in great part, to introduction of considerations of Imperial party politics into County Council affairs. During the past few years the Council has not been allowed to take any important step for the benefit of London,

---

[1] *Cf.* John Lloyd: *London Municipal Government, the History of a Great Reform* (1910); and the works of J. F. B. Firth, its able and energetic chairman who literally gave his life to the service of London.

[2] Many of the early Fabian Tracts dealt with London; and Sidney Webb, one of the most prominent Fabians, was an influential member of the, London County Council for 15 years.   [3] A. G. Gardiner: *op. cit.*, p. 212.

without its first being considered what effect it might have on the battle between Home Rule and Unionism in the House of Commons. I have always protested against this degradation of the County Council. London's municipal interests are far too important to be made the shuttlecock of Imperial parties."[1]

In general, one may say that party politics have their uses no less than their abuses in local government; and on the whole the work of the London County Council has gained in vitality and coherence through the organisation of its members into clear-cut political parties. But while party politics may be a good thing in municipal life so long as the issues which divide the members are relevant to local government, party alignments and loyalties tend to become a public nuisance when they are linked up with matters which have no bearing on the work of the Council. This has happened from time to time in the history of the Council and is at all times undesirable.

The Local Government Act, 1888, had been introduced by a member of Lord Salisbury's second administration; yet paradoxically enough the London County Council, a by-product of that legislation, was confronted with persistent opposition by Parliament during the years between 1889 and 1902, except during the Gladstonian regime of 1892-5.[2] Thus Parliament refused to revise the "impossible rule" requiring every estimate in excess of £50 to come before the full Council after three days' clear notice; it refused to permit the London County Council the right to take over the water supply, although this was the common practice in provincial towns and was in line with proposals made by the Board of Works and approved by a Royal Commission; it rejected the Council's Tramways Bill enabling the tramcars from South London to cross Westminster Bridge and go along the embankment to Charing Cross.[3] Even though Parliament had agreed to the creation of a major local governing body for London it seemed incapable of rising above its own petty fears and jealousy of the new organ. It is not clear that it has risen above them today.

[1] Sidney Webb: *A Letter to the Electors of Deptford* (1899), p. 7.
[2] Lord Salisbury's administrations covered 1885-6, 1886-92 and 1895-1902. *Cf.* A. G. Gardiner: *op. cit.*, pp. 112, 212.
[3] A. G. Gardiner: *op. cit.*, pp. 148-9.

For the first year or two of its existence hopes were entertained, by leading members of the London County Council, that a solution could be obtained through friendly co-operation of the problem of bringing the City into organic relation with the County Council. Lord Rosebery no doubt had this object in mind when he offered himself as a candidate for the City at the first election of the new organ.[1] The City Corporation was itself sufficiently well disposed to the London County Council to permit it to use the Guildhall for its formal meetings during the first few months of its existence.[2] But these friendly gestures soon faded away as unification again became a dominant issue.

For the first ten years after it came into being—until about the end of the century—the leading group on the County Council continued to press relentlessly for the merging of the new county council with the ancient corporation. In 1893, the President of the Local Government Board announced that the Government had decided to appoint a Royal Commission "to consider the proper conditions under which the amalgamation of the City and the county of London can be effected, and to make specific and practical proposals for that purpose." The members included Sir Homewood Crawford, the Solicitor to the City Corporation.

The Royal Commission proposed the creation of a Corporation of the City of London to include both the old City and the County Council, and having as its title the Mayor and Commonalty and citizens of London. The Corporation was to take over all powers from existing bodies (except those relating to certain charities) and was then to devolve administrative functions to local bodies to the utmost possible extent, subject where necessary to the over-riding authority of by-laws made by the Corporation laying down broad principles of policy.[3] The local bodies were to be strengthened; and one would be established under the name of the Council of the Old City to deal with the square mile of the ancient Corporation. The City police were to be fused with the metropolitan police force;

---

[1] A. G. Gardiner: op. cit., p. 202.   [2] Ibid., p. 91.
[3] Report of the Royal Commission on the Amalgamation of the City and County of London, 1894, B.P.P. vol. xvii, pp. 16-21.

finance would be dealt with through one fund levied by the new Corporation. The Lord Mayor was to become the ceremonial head of the whole metropolis. He would be elected by the Council and was to be suitably provided with funds for his office. "We may look," said the Commission, "for the maintenance in the future of all the useful and many of the stately traditions of the past; and in particular the Lord Mayor may be trusted to represent before the world the great community of which he is head, with the splendour becoming his position."[1]

This plan, like so many that had preceded it, fell stillborn from the Commission's Report. Before the enquiry was complete the City withdrew their representative and refused further co-operation on the ground that due consideration was not being given to "the important interests" concerned. This was, of course, a sign of hostility to the proposals of the London County Council.[2]

The arguments in favour of requiring the City Corporation to adapt itself to the growth of London and to share its history, traditions and prestige with the rest of the metropolis have been and still are overwhelming, whether considered on the grounds of logic, constitutional law,[3] political right, social expediency or mere common sense; and this was as true in 1835 or 1894 as it is today. There has at no time been any greater justification for permitting the City Corporation to pursue its isolated existence within the narrow confines laid down many centuries ago than there was for permitting the corporations of the provincial boroughs to remain unreformed in 1835. Unfortunately, however, while in the latter case

---

[1] Report of the Royal Commission on the Amalgamation of the City and County of London, 1894, p. 21.

[2] *Ibid.*, p. 11; A. G. Gardiner: *op. cit.*, p. 203.

[3] Mr. Asquith laid it down as an incontrovertible proposition "both of constitutional law and common sense that the Corporation of the City holds its property and privileges in trust, not for that square mile of which the Guildhall is the centre, but for the 5,000,000 people dwelling in the 2,000 miles of streets who now constitute the real London, the London which is entitled to regard itself as heir to the property and to all those great traditions and associations of which the Corporation is the trustee." Hansard: Commons Debate, 23rd March, 1899, vol. 69 (4th series), col. 171.

modern needs were regarded as superior to ancient rights, in the case of the City Corporation political power was successfully invoked to protect the ancient rights, regardless of consequences to the rest of the metropolis.

Looking back at the events of the 1890's in the light of subsequent history, it appears, however, to have been a mistake for the London County Council to expend so much of its energies in the first decade of its existence on attempting to obtain amalgamation with the City. It was an error of judgment to attack the problem of unifying the metropolis at its strongest point. The London County Council would have done better to have concentrated its efforts on absorbing the area which lay outside its boundaries. This would have been an easier and more useful task to pursue and it would possibly have met with success at that time. But to recognise the importance of the out-county territory in, say, 1895, required a greater amount of foresight than was possessed by those in control at County Hall. Moreover, the leaders of the London County Council no doubt believed the City to be weaker than it actually was. Furthermore, if the Gladstone Ministry had not fallen in 1895 the City Corporation might have been brought into a general reorganisation of London government on the lines suggested by the Royal Commission of 1894.

Lord Salisbury, who succeeded Gladstone, not merely disliked the London County Council because it appeared to him to be a radical body, but also because he was a Little Londoner and unable to conceive the metropolis as a whole. At an Albert Hall meeting on 16th November, 1897, he explicitly stated that we should have obtained more efficient machinery if we had been content to look upon London not as one great municipality but as an aggregate of municipalities.[1] It was scarcely possible to argue on the subject with a statesman holding such an outlook on the civic affairs of the capital city.

The area within which the London County Council was given jurisdiction was the same as that in which the Metropolitan Board of Works had exercised its powers: a territory of some 75,442 acres (including the 671 acres of the City). The boundary of this area may possibly have formed a dividing

[1] A. G. Gardiner: *op. cit.*, p. 250.

line between town and country in 1855, or at any rate have served to indicate roughly the limits of the metropolis. But already in 1888, when the London County Council was established, it had long since ceased to have any significance from that or any other point of view. It certainly no longer corresponded with the social or economic realities of London life. The area of the London County Council was obsolete from the first moment of its birth.

Moreover, the probable future trends of population were clearly known. In 1885 there appeared in the *Journal of the Royal Statistical Society* an important article by H. Price-Williams on the population of London, 1801–81. The population of the registration district[1] (which coincided closely with the Metropolitan Board of Works area) contained at the census of 1881 3,816,000 persons. The results of the author's investigations led him to conclude that the point of maximum density had already been reached in many of the better-class suburban residential districts and that population would not in future continue to increase there to any appreciable extent. Hence, population as a whole within the registration district would be unlikely to continue increasing at the rapid rate hitherto observed.[2] Price-Williams then declared that although further increase within the present limits beyond about twice its average density was improbable, there was nothing to indicate that it would not continue to do so as rapidly as heretofore in the outlying suburban districts. "This overflow of population of London into the extra-metropolitan districts has already begun, and during the past two decades has been very appreciable"—and a number of places were mentioned where the increase had been conspicuous, such as Tottenham, Walthamstow, Leyton, Stratford, West Ham, Ealing, Wimbledon, Croydon, Barnes and Mortlake. "These large districts," he continued, "destined at no distant date to become part of the greater London of the future," have altogether an area of over 46,000 acres, and with an average density of 94 to the acre, could accommodate nearly 4,000,000 more inhabitants.[3]

[1] 75,362 acres.
[2] *Journal of the Statistical Society* (1885), vol. xlviii, p. 380.
[3] *Ibid.*, p. 382.

This forecast made in 1885 proved to be remarkably accurate. The population of the London registration area mentioned above rose to 4,211,743 in 1891 and reached its peak in 1901 at 4,536,063, while the out-county area has filled up in the manner prophesied and in 1931 contained 3,806,939 persons. Price-Williams estimated the population of Greater London would be 7,109,000 in 1917; it was actually 7,480,201 in 1921.[1]

The density of London was high; but the denseness of the legislature was even greater. Parliament ignored the statistical knowledge available to any well-informed person in Westminster or Whitehall, it ignored the facts of daily life which must have been evident to everyone living in London, it ignored the growth of London during the 19th century, it refused to contemplate the certainty of future growth. It insisted on legislating for London in terms of what might have been appropriate in 1855 but which was clearly unsuitable in 1888. The County Council which was called into being was scarcely even conscious of its own shortcomings. It turned its eyes towards the centre when it should have turned them to the periphery. It looked inwards instead of outwards.

As a result there was yet another embittered struggle with the City Corporation with nothing but defeat to show at the end of it; while on the outskirts of the metropolis there grew up, silently and unnoticed, a greater menace to the good government of London.

[1] *Journal of the Statistical Society* (1885), vol. xlviii, p. 431; *London Statistics*, 1934–6, vol. xxxix, p. 27.

CHAPTER X

THE METROPOLITAN BOROUGH COUNCILS

It has been pointed out[1] earlier that the Local Government Act, 1888, left all the pre-existing authorities intact except the Metropolitan Board of Works. The vestries and district boards in particular were affected only in a very minor degree. The failure of Parliament to deal with this aspect of the problem was, indeed, one of the principal grounds on which the Act of 1888 was criticised.

The defects of the vestries and district boards were notorious during the regime of the Metropolitan Board of Works,[2] and there was little improvement in their administration after 1888. In 1892 Mr. Sidney Webb complained that the 5,000 vestrymen in most cases practically elected each other; and he urged that since London was probably too vast and too heterogeneous for a unitary system of municipal administration, district councils should be set up to take over most of the work then done, or left undone, by the vestries.[3] Dissatisfaction of a more concrete kind was manifested in 1897, when Kensington and Westminster, the two wealthiest parishes in London, sought to acquire the status of boroughs under the Municipal Corporations Acts.[4]

In 1899 Mr. Balfour introduced a Bill to deal with the position of the minor authorities in London. In order to understand the genesis of the measure it is necessary to remember the avowed hostility of Lord Salisbury, the Prime Minister, to the London County Council, his conception of the capital city as a mere series of unrelated fragments situated in juxtaposition to one another, his decision to introduce national party politics into the London County Council elections, and the immense change in the spirit of municipal life in London which the London County Council had wrought.

The London Government Bill had as its main purpose the

---

[1] *Ante*, p. 81. [2] *Ante*, pp. 67–70.
[3] Sidney Webb: *The Reform of London* (published by the Eighty Club, 1892), pp. 7–8. [4] H. Finer: *English Local Government*, p. 472.

undermining of the interest and authority which the London County Council had aroused. It sought to strengthen and magnify the district councils to the greatest possible extent and to emphasise their independence of the larger body in subtle as well as in obvious ways. Every device which might tend to divide the allegiance and confuse the loyalty of Londoners was imported into the Bill; while at the same time nothing was done to ensure coherent administration or to give the London County Council power to over-ride parochial views in the interests of the metropolitan community as a whole.

The 28 districts into which the county was divided by this measure followed as closely as possible the lines of the old vestries; and little attempt was made to rationalise the areas in terms of size, population or rateable value.[1] In each district a metropolitan borough council was set up, consisting of a mayor, aldermen and councillors. The mayor and aldermen were to have robes of office, gilt chains, a mace and all the other insignia likely to encourage the feeling of their separate civic consciousness.[2] Westminster acquired the right to call itself a city; Kensington became a Royal borough. The new municipalities were to be "subordinate in point of area, but not subordinate in point of dignity."[3] There were to be 28 town halls and 28 town clerks. "There was to be not one London, but thirty Birminghams."[4] The Metropolitan Borough Councils were given the same powers of promoting and opposing Bills as Birmingham, and their Parliamentary powers were actually greater than those possessed by the London County Council. The approval of loans for the provisions of baths and wash-houses, public libraries, cemeteries, sanitary conveniences and various other works of public health was to be obtained from a central government department instead of from the London County Council.[5]

---

[1] The only principle was that no place was to become a municipality which had not a population between 100,000 and 400,000 or a rateable value in excess of £500,000. Mr. Balfour laid this down in his speech introducing the Bill. Hansard: vol. 67 (4th series), col. 364.
[2] Sir Percy Harris: *London and its Government*, p. 62.
[3] Balfour: Hansard, vol. 67 (4th series), cols. 366–7.
[4] A. G. Gardiner: *Sir John Benn and the Progressive Movement*, p. 256.
[5] *Post*, pp. 266 *et seq.*

Mr. Balfour's speech on the Bill in the House of Commons amply confirmed the fears of those who suspected the measure to be an attack on the growing power of the London County Council. The Minister first asked whether "these great municipalities—for great municipalities they will be"—should be linked with the London County Council, such as by making each mayor an *ex-officio* alderman of the latter body. Both organs would be dealing with the same population, and Mr. Balfour admitted that it would seem natural and plausible for there to be some official bond between them; "but the Government decided to reject the plan, in the interest of the new municipalities." The undesirability of reintroducing an element of secondary election could also be urged against such a proposal, but that was of minor importance. The real reason, explained Mr. Balfour, why he was unwilling to adopt any system of liaison was that, in the first place, "it would inevitably drag those councils into the political vortex in which the London County Council appears to flourish" (Interruption: "Whose fault is it?"). . . . "There is another reason and a still stronger reason which guides me in this matter. I look forward to these municipal boroughs having a great and most legitimate influence with the London County Council. I cannot think it would be otherwise. . . . For these reasons we have not thought it desirable to introduce in this Bill any formal machinery for officially linking together the new municipalities and the London County Council."[1]

The London Government Act was to a considerable extent a simplifying measure, for in establishing the 28 metropolitan boroughs it abolished a large number of pre-existing bodies, including not only all the administrative vestries and district boards but also 44 non-administrative vestries, 12 burial boards, 18 public library commissions, 10 commissions for baths and wash-houses, 2 market boards, 56 bodies of overseers and a score of bodies calling themselves trustees of the poor.[2] This was a highly desirable feature of the Act.

The City Corporation was left untouched by the legislation,

---

[1] Hansard: Parliamentary Debates, 23rd February, 1899, vol. 67 (4th series), col. 352 *et seq.*
[2] Albert Bassett Hopkins: *The Boroughs of the Metropolis*, p. 21.

and Mr. Asquith opposed the Bill with special vehemence on this ground, declaring it to be a scheme "to surround and buttress the unreformed City with a ring of sham municipalities; to impair and destroy in most material particulars the corporate and administrative unity of London as a whole." The Bill, he said, tended to delay and frustrate the legitimate aspirations of the greatest city in the world,[1] since the unity of London in any deep sense is not incomplete but impossible so long as the government of the City is left unreformed.[2] "When you call these new authorities municipalities you are giving them a false name . . . and the mischief of it is . . . that it proceeds on a false analogy and that it suggests a false ideal. What is a municipality as we here in England understand it? A municipality is a community of spontaneous growth, self-governed and self-contained; a whole in itself. There is only one community in the metropolis which answers that description, and that is London as a whole."[3]

This view of the London Government Act has been shared by most liberal critics of London government. Sir Percy Harris remarks that it set up authorities which in some ways regard themselves as rivals of the County Council.[4] Mr. A. G. Gardiner observes that the only conceivable purpose of the Balfour proposals was to prevent the voice of London as a community from being heard on any subject affecting its common interest, and to substitute a chorus of sectional and competing interests.[5] In order not to disturb the City, the measure of 1899 set up a system of new municipalities which left London a mosaic of unreal and arbitrary cities, and its essential unity unrecognised.[6] Mr. Herbert Morrison has recently pointed out that the motive of the Salisbury Government in 1899 was to apply the principle of divide and conquer to the London scene.[7]

If these were the hopes entertained by Mr. Balfour and his colleagues in promoting the London Government Act they

---

[1] Hansard: House of Commons, vol. 69 (4th series), col. 178.
[2] *Ibid.*, col. 171.   [3] *Ibid.*, cols. 174–5.
[4] Sir Percy Harris: *London and its Government*, p. 42.
[5] A. G. Gardiner: *Sir John Benn and the Progressive Movement*, p. 256.
[6] *Ibid.*, p. 263.
[7] Herbert Morrison: *How Greater London is Governed*, pp. 100–1.

were fulfilled to a most remarkable extent, as the next part of this work will show, although frequently in ways which can scarcely have been anticipated at the end of the 19th century.

The membership of the Metropolitan Borough Councils amounts to a total of 1,615 members, to which must be added 450 positions on committees, school management boards, etc., filled by persons other than members.[1] The councils inherited or subsequently acquired a large number of functions relating to public health, electricity supply, street maintenance and improvement, the scavenging and lighting of highways, the removal and destruction of refuse, the provision of parking places, baths and wash-houses, public libraries and museums, burial grounds, crematoria and mortuaries, parks and open spaces. They are the authorities for maternity and child welfare services, for providing tuberculosis clinics, for the registration of births, deaths and marriages, for supervising the sanitary condition of factories and shops, for inspecting canal boats, seamen's lodgings, cowhouses, dairies, food on sale, offensive trades, slaughterhouses, and a score of other matters.

It is probably desirable in a metropolitan centre of such gigantic proportions as London there should be a double-deck municipal structure of some kind, since it would be impracticable, or at any rate inadvisable, to administer every service through a single central organ. But to jump from this postulate to the other extreme of establishing 28 uncontrolled and unco-ordinated borough councils endowed with both the form and the substance of an entirely independent municipal status was an act which could scarcely fail to produce conflict and confusion. Some of the adverse consequences will be discussed on later pages. In the meantime, we may note that Mr. Herbert Morrison, writing from his long experience both on the London County Council and as Mayor and councillor of the metropolitan borough of Hackney, points out that there has been plenty of friction between County Hall and the town halls, in which not only members but also officers have played their part.[2] As long ago as 1908 the London Reform Union protested against the "absurd and suicidal waste of public money"

---

[1] *London Statistics*, 1934–6, vol. xxxix, p. 7.
[2] Herbert Morrison: *How Greater London is Governed*, p. 101.

involved in permitting the metropolitan boroughs to promote
Bills in Parliament or to oppose the Bills of the London County
Council or of one another.[1]

The attitude of the Salisbury Government is illustrated by
a further incident which took place when the London Water
Bill was passing through Parliament in 1902. The Water Board
which it was proposed to set up for the metropolis was to have
no less than 69 members. The London County Council
objected to its unwieldy size and suggested the omission of
representatives from the metropolitan borough councils and
the urban district councils outside the county of London. The
London County Council urged that the common interest would
be adequately safeguarded in the hands of the London County
Council and the other neighbouring county councils. The joint
committee of both Houses of Parliament which was considering
the Bill agreed with this view and struck out the local representation, thereby reducing the membership of the board to 35.
The Government was determined to maintain the status of the
recently established metropolitan boroughs and "intervened in
an unprecedented manner" to secure the restoration of the
constitution originally proposed for the Board, much to the
disappointment of the London County Council.[2] Incidentally,
a departmental committee appointed in 1920 to consider the
working of the statute in question recommended a reduction
in the membership of the Metropolitan Water Board.[3]

The friction and conflict of interest between district and
county authority in the metropolis has not been confined to
the sphere of local administration but has also been reflected
in Parliament itself on local government questions affecting
London. The metropolitan borough councils have often been
able to exert more influence with London Members of Parliament than the London County Council.[4]

Despite these set-backs and obstacles the London County
Council has grown both in scope and in power. Since the
beginning of the present century it has taken over the functions

[1] *London Today and Tomorrow* (published by the Union).
[2] Sir H. Haward: *The London County Council from Within*, p. 341.
[3] Departmental Committee to enquire into the effect of the Metropolis
Water Act, 1902 [Cmd. 845], 1920. B.P.P. vol. xxi, p. 7.
[4] Herbert Morrison: *op. cit.*, p. 101.

of the London School Board in every sphere of education. It has become the public assistance authority for the metropolis, thereby superseding twenty-five boards of guardians and the Metropolitan Asylums Board. It has become the town planning authority for the county of London.

CHAPTER XI

THE METROPOLITAN WATER BOARD

The manner in which the welfare of millions of Londoners was ignored, and the mismanagement of an essential service permitted to continue in the light of known evils and a certain remedy, is demonstrated by the history of the metropolitan water supply in the 19th century. The handling of this service would seem almost incredible were the facts not on record in a mass of official reports.

The legislative regulation of the London water supply goes back to 1543, when the Hampstead Water Act—the first local Act relating to the subject—was passed. In 1605-7 the City Corporation was authorised by Parliament to establish the New River undertaking, a task which the City Fathers handed over to a private citizen (Sir Hugh Myddelton) to carry out.[1] In 1619 the New River Company obtained a Royal charter embodying a partnership arrangement with the King. In the 17th and 18th centuries a number of other companies obtained powers to supply water in the London area.[2]

Early in the 19th century several more companies were formed and most of the smaller undertakings were ..bsorbed. By about 1830 there were 8 companies supplying London, 5 of them operating north of the Thames and 3 of them on the south side. Most of the companies were working under private Acts of Parliament; but in some cases their powers were derived from letters patent.

In 1827-8 a Royal Commission enquired into the quality and salubrity of the water supply as a result of a petition of complaint from residents in the south and west of London. The report showed the complaints to be well founded. The Commission said the supply was defective in purity and

---

[1] In 1581 Peter Morrys, a Dutchman, obtained consent from the City Corporation to erect a water-wheel in one of the arches of London Bridge to work a pump supplying houses in the City. This continued in operation for 200 years.

[2] For details, see the Return as to Water Undertakings in England and Wales, B.P.P., 1914, vol. lxxix, pp. ii–vii.

cleanliness, and required improvement. In 1828 a Select Committee of the Commons endorsed the views of the Royal Commission and recommended that Telford, the eminent engineer, should be invited to draw up a scheme to provide the whole metropolis with pure water. Telford submitted his report in 1834, and it was referred to another Select Committee of the Commons.[1] The ravages of cholera again drew the attention of the public to the subject and the House of Lords appointed a Select Committee to consider the London water supply in 1840.[2]

In 1845 the Royal Commission on the Health of Towns issued their celebrated report. In it the Commission called attention to the ruinous competition existing among the water companies supplying London. At that time statutory areas were not assigned to the various undertakings, and it was this which enabled competition to take place. In 1849 there was an outbreak of cholera in London which caused the death of 14,137 persons. The mortality from the disease was higher in London than elsewhere: i.e. 62 per 10,000 of the population as compared with 30 per 10,000 for England and Wales.[3] None of the special Acts empowering the companies laid down any conditions as to purity, and there was no supervision of the supply by a responsible public officer.

The following year (1850) the General Board of Health made a detailed investigation into the situation. They found that most of the water supplied for domestic purposes was drawn from the Thames, and more than half of it was delivered without any filtration at all.[4] Dr. Gavin, one of the witnesses, stated that the water supplied to the poor was so unpalatable and injurious to their health that they were unable to drink it and had in consequence made beer their common beverage.[5]

An important contributory factor in producing disease was the pollution caused by storing water in open house-cisterns, which were frequently in a filthy condition. Dr. Milroy, the

[1] Select Committee of the House of Commons on the supply of Pure Water to the Metropolis (1834), B.P.P., vol. xv, p. 3.
[2] B.P.P., 1840, vol. xii, p. 159.
[3] A. Bassett Hopkins: *The Boroughs of the Metropolis*, pp. 7–8.
[4] Report of the General Board of Health on Metropolis Water (1850), B.P.P. vol. xxii, pp. 11, 312.   *Ibid.*, p. 43.

well-known public health authority, gave the Board numerous instances where from this cause water had become unusable except with grave danger to health.[1] Storage was necessitated partly by the absence of piped supplies in the poorer quarters of the town,[2] and partly by the intermittent supply which the water companies provided at the tap or butt which often served a whole street of houses. "So disgusted are the inmates themselves of even the poorest dwellings with the water in their butts or cisterns," reported the Board, "that very frequently they will use it only for the purpose of washing, and, unless they can catch the water directly from the pipe when it is *on*, they are obliged either to beg it from some neighbour, or (as is frequently the case) get it from the public-house where they deal. This appears to be of very common occurrence indeed, even in some tolerably decent localities, and must be admitted to be a flagrant injustice, inasmuch as they are charged indirectly in their rent for what they have little or no benefit from."[3] In addition to the positive evil of dangerously impure water, there was in some districts, such as Bermondsey, the negative evil of a shortage of pure water during the cholera epidemics.[4]

The disadvantages from which the citizens suffered were due in some degree to the elementary stage which had been reached in the scientific analysis and treatment of water. For example, taste was still relied on in 1850 by the officers of the General Board of Health as a method of testing the purity of the supply; and the water was heated in order to bring out the full flavour of the impurities.[5] But the major cause of the dangers and discomfort to which the public was exposed was the simple fact that the water supply was in the hands of 8 companies whose principal interest lay in increasing the profits of the shareholders. This was quite clearly recognised by the General Board of Health. "From the whole of the evidence," they reported, "it is clear that the existing companies

[1] Report of the General Board of Health on Metropolis Water (1850), p. 26.
[2] 17,456 houses in the metropolis were unsupplied with water in 1850. This was 6 per cent of the total, but in some densely populated districts upwards of 18 per cent of the houses were lacking in piped supplies. *Ibid.*, p. 8.   [3] *Ibid.* pp. 25–26.   [4] *Ibid.*, p. 17.   [5] *Ibid.*, p. 35.

supply an insufficient administrative machinery for improvement of the supply of water, separately considered; for that supply can only be improved by measures in detail, from which *immediate* profit is not to be expected, while their eventual profit cannot easily be made clear to meetings of shareholders. This circumstance, which is prejudicial to the improvement of any water supply, is still more so to the required combination of water supply with drainage works, or to the improvement of the latter under any such connexion. The isolated position of many of the companies does indeed create an interest hostile to the proposed combination, as was shown by subscriptions entered into by some of them for the avowed purpose of raising up an opposition to a Public Health Act, which provided for the combination of local works under one and the same management."[1]

The Board of Health recommended that a number of improvements should be insisted upon, including a constant supply throughout the metropolis,[2] the abandonment of the Thames as a source of supply in favour of a supply from Hindhead, more stringent purification and the extension of a piped supply. They concluded that the only way of obtaining efficiency, economy and the necessary works within a reasonable time would be to consolidate all the company waterworks into a single public undertaking responsible both for water supply and drainage.[3] Under a consolidated management 5 or 6 of the 7 principal pumping establishments could have been closed down; and a saving of £80,000–£100,000 made in the cost of operation.[4] The companies were, of course, to receive reasonable compensation.[5]

It seems strange that after such lengthy and authoritative enquiries by disinterested bodies, and with clear-cut recommendations to go upon, Parliament should have deferred decisive action in regard to the London water supply for more than fifty years longer. Yet so it was. And even when the Metropolis Management Act was passed in 1855 water powers were not conferred on the Metropolitan Board of Works.

[1] Report of the General Board of Health on Metropolis Water (1850), pp. 165–6. [2] *Ibid.*, p. 314. [3] *Ibid.*, p. 319. [4] *Ibid.*, p. 322. [5] *Ibid.*, p. 272.

The sovereign legislature was scarcely unacquainted with the manœuvres and disputes of the water companies. Indeed, the General Board of Health ventured the remark that the money wasted in obtaining private Acts for the metropolitan water companies, and the unnecessary multiplication of capital works in the same field of supply, would have sufficed to provide London with works and aqueducts of even greater magnificence than those possessed by ancient Rome.[1]

During 1851 the House of Commons appointed a scientific commission of three chemists to enquire into the proposal to transfer the source of supply from the Thames to the Surrey heights and Watford. This commission reported adversely on the recommendation of the General Board of Health to abandon the Thames supply but advised that the water should be drawn off at a point beyond the tidal range of the river in order to avoid contamination.[2] They strongly advocated the acquisition of the Hertfordshire chalk springs under public ownership.[3] In the same year the Government introduced a Bill to amalgamate all the existing companies and to enable the Treasury to acquire the consolidated undertaking on giving six months' notice. The idea underlying this measure was to municipalise the water supply whenever an adequate municipal council should be established for the metropolis. The Home Secretary was to have power to designate the sources of supply. This Bill passed its second reading by a narrow majority and was then defeated.

It was followed by the Metropolis Water Act, 1852, the first public general statute dealing with water applying to London. This Act confirmed the companies in their functions. It did not seek to promote either unified management or public ownership. At the same time it introduced regulation both as to the quality and quantity of supply. No company was to draw water from the Thames below Teddington Lock. Reservoirs within 5 miles of St. Paul's were to be covered in. Water supplied for domestic use was to be satisfactorily filtered unless drawn direct from wells into a covered reservoir for distribu-

[1] Report of the General Board of Health on Metropolis Water (1850), p. 292.
[2] Royal Commission on the Water Supply, 1869, B.P.P. vol xxiii, p. xliv.
[3] *London Water Supply*, L.C.C. publication, No. 882, June 1905, p. 28.

tion. After a period of 5 years the companies were to provide a constant supply of wholesome water, but a proviso was added that no company should be bound to provide a constant supply in any district until four-fifths of the owners and occupiers of the houses in the district should require the company in writing to provide such a supply and until all the water fittings in the houses were constructed according to regulations prescribed by the company. This proviso was utterly absurd in view of the fact that most of London's poorer inhabitants were both illiterate and ignorant of the law. It was no doubt inserted at the request of the companies in order to render the obligation to provide a constant supply a nullity, and it had the desired effect. The Act was a failure.[1]

There were further outbreaks of cholera in 1854 in the neighbourhood of Golden Square; and these were traced to a well in close proximity to a sewer.[2] Other water-borne diseases such as diarrhoea and enteric fever continued to affect the health of Londoners for many years. In 1866 a very severe attack of cholera broke out in the east of London which led to a number of complaints against the East London Waterworks Company. It was alleged that the Company was drawing water from a point in the River Lea which received the sewage of five towns; that its reservoir at Old Ford, being below the level of the river, received the contaminated sewage; that the supplies were deficient in quantity and defective in purity; and that the water thus provided had been the principal or sole cause of the fearful mortality from cholera. The Board of Trade appointed Captain Tyler to enquire into the matter. His report is a remarkable document.

He found that the River Lea was admittedly contaminated and the first charge was therefore substantiated. Second, the water in the covered reservoir was proved to be dangerously

[1] The Select Committee on the East London Water Bill (1867), p. xxii, drew attention to the inefficiency of these provisions of the Act of 1852 and subsequent statutes and the inability of the poorer classes to enforce them. "Your Committee have come to the conclusion that the Act of 1852 has failed to secure for the inhabitants the advantage which they ought long since to have enjoyed of a well-regulated supply of water in their houses for domestic purposes" (p. xvi).

[2] Report on last two cholera epidemics (1856); Report of Royal Commission on the Water Supply of the Metropolis (1893), B.P.P. vol. xl, p. 64.

impure. Third, the Company was supplying unfiltered water from open reservoirs in flagrant breach of the Metropolis Water Act, 1852. Fourth, the mortality figures in the various districts gave strong support to the conclusion made by the Registrar-General that the East London Company's water was charged with choleraic poison and was the principal cause of disseminating the disease. Fifth, there was ample evidence to show that the supplies were often grossly deficient in quantity. There was no Sunday supply; and even on weekdays the water was often only turned on for a few minutes. For example, in a street called Butler's Buildings 250 people lived in 14 houses; and in Gibraltar Gardens 150 persons dwelt in 20 houses. Each of these two streets was provided with one pipe ¾ inches diameter and the regular time for turning on the water was from 7.10 to 7.35 a.m. in Butler's Buildings and from 4.35 to 4.55 p.m. in Gibraltar Gardens. During this time 200 gallons of water would be supplied for the use of 250 persons in the former case and slightly more in the latter; and none at all on Sundays. The Company received ten shillings a house per annum from the landlord—the full amount which would be payable if each house had had a pipe of the same size.[1] For this disgraceful state of affairs the Water Company was held chiefly responsible, but the parish officials and the landlords were also blamed.[2]

These revelations failed to produce drastic action on the part of the Government and yielded no more than the institution of yet another enquiry. A Royal Commission had been set up in December 1866 to ascertain whether a supply of wholesome water could be obtained by collecting and storing water in the high grounds of England and Wales for the use of the large towns; and if so, which places of this kind would be best suited for providing a supply to the metropolis. In April 1867 the terms of reference were widened so as to authorise the Commission to enquire into the existing water supply of the metropolis and to advise whether there were districts other

[1] Report Relating to East London Waterworks Company, 1867 (lviii) pp. 2, 5, 6, 7, 8, 15 and 23.
[2] The Metropolis Local Management Act, 1862, gave power to the vestries to instal water fittings at the landlord's expense in certain houses. See Select Committee on East London Water Bill, 1867, B.P.P. vol. ix, p. xx.

than the high grounds from which London might obtain pure water.[1]

The Royal Commission reported in 1869 to the effect that the whole principle of private ownership and operation was wrong, and that the admitted evils of the existing situation could be remedied only by public management. The expediency and advantage of consolidating the water supply under public control, they declared, are manifest on many grounds. It is the only means of ensuring a constant supply, and especially of providing proper facilities of the poor. It would lead to an improvement in the quality of water by ensuring filtration and purification of the mains. It would facilitate the use of water by the fire brigades. It would promote economy.[2] The Commission remarked that the duty of supplying the inhabitants of a city with water had been regarded as a municipal function from very early times, and that the supersession of the municipalities by joint-stock companies was a comparatively modern innovation. "Of late years," they continued, "many towns in England have come to the conclusion that the new practice was a fundamental error" and had gone back to the older system of municipal ownership.[3] With this tendency the Commission signified their full agreement. "We believe the public management to be far more correct on general principles than the supply by joint-stock organisation, which is obviously only applicable to those cases in which a fairly remunerative return may be anticipated for the capital expended. But a sufficiency of water supply is too important a matter to all classes of the community to be made dependent on the profit of an association."[4]

One would have thought that with the publication of this report an overwhelming case had been made out for public

[1] The primitive state of the water supply in the third quarter of the 19th century is illustrated by the fact that there were no waterworks serving Hammersmith in 1870. The inhabitants relied mostly on the rainfall or shallow surface wells or on the Thames. A contemporary writer tells us that one of his earliest recollections was seeing the family's daily supply of drinking water delivered by men and women carrying cans suspended from a yoke across the shoulders. G. A. Sekon: *Locomotion in Victorian London* (1938), p. 4.
[2] Royal Commission on Water Supply (1869), p. cxxii, Section 248.
[3] *Ibid.*, p. cxx, Section 246.　　　　[4] *Ibid.* p. cxxii, Section 249.

ownership and unified management. The Royal Commission was, indeed, merely repeating the advice given by the General Board of Health in 1850, but the wisdom of that advice or, rather, the dangers of ignoring it, had been demonstrated only too vividly by the death, sickness and discomfort which tens of thousands of Londoners had suffered in the intervening years from the evils it sought to remedy. But Parliament was apparently not to be convinced.

A Government Bill was introduced in 1871 to give effect to the recommendations of the Royal Commission. It originally contained a clause giving a power of compulsory purchase of the company undertakings, but this was withdrawn in the face of strenuous opposition. The Bill professed to have as its objects the provision of a constant supply of wholesome water, and the audit of the companies' accounts. But it gave dissatisfaction both to the local authority chiefly concerned and to the water companies. The Metropolitan Board of Works objected because it did not go far enough in recognising municipal responsibilities, while the companies opposed on various grounds, including an objection to the strictly limited powers to be conferred on the Metropolitan Board of Works.[1]

In the result, the Metropolis Water Act, 1871, introduced a further measure of regulation, but omitted to transfer the duty of supervision from the Board of Trade to the newly established Local Government Board.[2] The Select Committee of the House of Commons to which the Bill was referred reported in discouraging terms on its proposals. "Upon full consideration under their notice," they said, "your Committee have arrived at the conclusion that the supply of water in large cities should be either actually provided by municipal authorities or regulated under their control, if provided by private enterprise; and that the supply of water to this great and growing city cannot be placed upon a proper footing until Parliament shall have determined upon the proper municipal administration of the metropolis. In the meantime your committee recommend

[1] Special Report of the Select Committee on the Metropolis Water (No. 2) Bill, 1871, p. v.
[2] Under the Act a water examiner was appointed to make monthly tests of the water supplied by the companies. He was appointed by the Board of Trade and paid by the companies.

this Bill as the best measure available, under present circumstances, for securing to the inhabitants of the metropolis a constant supply of water.[1]

The failure of Parliament to take the obvious course in dealing with so serious a menace to the welfare of the capital appears even more strange when we recall that by this time most of the important provincial cities had already been granted power to provide or acquire municipal waterworks. Liverpool and Manchester inaugurated their own water supply in 1847, Leeds in 1852, Bradford in 1855, Huddersfield in 1869. Birmingham obtained power in 1876, and Sheffield was practically the last in 1888. Yet public ownership in London was to be delayed for more than 30 years longer despite the many protracted and strenuous attempts which were made in the last quarter of the 19th century to acquire the water companies. The water companies certainly did not continue their baneful existence by virtue of any merit which they possessed. Indeed, the glaring defects of their service had been repeatedly exposed at official enquiries. What, then, could have been the reason for their extraordinary longevity? The answer is probably to be found in their powerful and persistent lobbying activities; and perhaps in other more clandestine and less reputable methods of persuasion, which they may have used in common with the other public utility interests. There can be little doubt that the public utility companies were greatly concerned to delay and defeat the establishment of a municipal council for London, knowing that their own interests would be jeopardised by its creation and they must have worked actively against such an eventuality.[2]

---

[1] Special Report of the Select Committee on the Metropolis Water (No. 2) Bill, 1871, p. xviii.

[2] When the London Council was actually established the water interests began to bore from within. The following comments by Sir John Benn, for many years a member of the London County Council, on the fate of a series of dead water Bills, in March 1897, are significant in this connection.

"It had become a farce to see the County Council laboriously drafting these water Bills at Spring Gardens while one of their number was acting as chief executioner outside. Would it not be better if the Parliamentary Committee went to Mr. Boulnois? The Progressive Party had long since learned that Lord Onslow was only the ornamental head of his party. In any case he found that the County Council Bills were dealt with at West-

Whatever the political causes may have been, the actual situation continued almost unchanged for some years.[1] In 1872 there was a serious failure in the supply of water to Bermondsey, which lay within the area of the Southwark and Vauxhall companies. The complaint was investigated by Lt.-Colonel Bolton, the water examiner appointed under the Act of 1871, who reported that the grievances were well founded and largely due to the bad state and small size of the pipes and fittings supplying the poorer quarters of the district.[2] In 1877 a Select Committee drew attention to the disadvantages and dangers attendant upon divided management of the related services of water supply and fire brigades. The unsystematic arrangements existing in London, said the Committee, whereby the fire brigade was under the Metropolitan Board of Works, the police forces under two separate bodies of commissioners, and the water supply drawn from 8 companies, "does not furnish adequate protection to life and property, and contrasts un-favourably with provincial systems where the fire brigade, water supply and police are under one authority."[3] The water supply could not be made to afford the full measure of pro-tection that the metropolis should possess against fire until it was consolidated in the hands of a single authority con-ducting the business not for immediate profit but for public convenience.[4]

In 1880, Lt.-Colonel Bolton once more emphasised in his annual report the numerous advantages which would accrue if the water service were in the hands of one authority instead of many. "There is nothing in the character of such an organisa-tion," he remarked, "that a public authority, invested with

minster quite irrespective of Lord Onslow's wishes. All he asked was that it should be clearly understood that the water policy of the London County Council was being dominated by the chairman of one of the water com-panies." Quoted by A. G. Gardiner: *Sir John Benn and the Progressive Movement*, p. 275. Mr. Boulnois was a member of Pa.liament and openly represented the water interests there.

[1] Report on the application of the Constant Service System of Water Supply, 1872, B.P.P. vol. xlix, p. 7.

[2] Report to the Board of Trade on the failure in the supply of water to Bermondsey and other parts of the metropolis, 1872, pp. 4 *et seq.*

[3] J. F. B. Firth: *Reform of London Government*, pp. 54–5.

[4] Metropolitan Fire Brigade Select Committee, 1877, B.P.P. vol. xiv, p. xxii.

stringent powers, could not administer more efficiently and more economically than it is possible for private associations to do." He then went on to show that works under consideration by the companies were unnecessary, and demonstrated once again the wastage and variety of standard resulting from the existing system.[1] He estimated a saving of £500,000 would be gained by amalgamation. A Government Bill was introduced by Mr. Cross, the Home Secretary, in 1880. This Bill provided for the creation of a water trust to acquire and manage the companies' undertakings. Before it could be passed the Government fell. The new Ministry appointed a Select Committee under Sir William Harcourt to consider the question, and once again public ownership was strongly recommended. Yet still no action was taken.

The Metropolitan Board of Works promoted Bills in 1878, 1884, 1885 and 1886 to enable them either to provide a subsidiary supply, to purchase the company undertakings, or to acquire power to introduce Bills dealing with the water supply of the metropolis.[2] Bills were also introduced by the City Corporation and the vestries. All these failed to pass through Parliament. In 1899 the Water Committee of the London County Council complained of the gross partiality shown by Parliament towards the water companies, pointing out that during the preceding ten years, while the proposals of the elected representatives of the people were being summarily rejected, no less than 16 private Acts of Parliament had been obtained by the water companies.[3]

The London County Council, during its early years under the leadership of the Progressives, had confidently believed that they could obtain municipal ownership of gas, water, electricity and transport services,[4] and the majority party had vigorously advocated these objectives. Eight Bills were introduced by the Council in 1895 to acquire the several water companies' undertakings. They passed the second reading and

[1] Annual Report of Lt.-Colonel Bolton, the Water Examiner appointed under the Metropolis Water Act, 1871, dated 31st January, 1880, p. 11.
[2] Return as to Water Undertakings in England and Wales, 1914, B.P.P. vol. lxxix, p. vii; *London Water Supply*, L.C.C. publication No. 882, 1905, p. 9.
[3] A. G. Gardiner: *Sir John Benn and the Progressive Movement*, p. 279.
[4] *Ibid.*, p. 112.

were referred to committee; but Parliament was dissolved before the committee stage had been completed.

In 1892 another Royal Commission was appointed to enquire into the water supply of the metropolis. This body pointed out, among other things, that there was extensive pollution of both the Thames and Lea rivers and that the River Pollution Act, 1878, was virtually a dead letter. It also reminded the public that the areas over which the 8 water companies had parliamentary powers bore no relation to either the administrative county of London, Greater London, or the Metropolitan Police District—or, indeed, to any other area. They recommended that no area smaller than the City and Metropolitan Police Districts should be considered for dealing with the water needs of London.[1] The deliberations of this body, like those of so many of its predecessors, were entirely wasted and led to no action of any kind.

In 1895 a serious crisis appeared to be imminent in the water situation. The companies seem to have arrived at the end of their resources and the vast population of London was living on "the brink of a water famine."[2] Two years later, indeed, a quarter of the entire population of London was subjected to water famine conditions through the default of the East London Waterworks Company.[3] The situation called for immediate action, and in 1896 the Government introduced a Bill to establish a Water Board. The London County Council was to be given representation on this board, but on a basis below that to which it was entitled according to its population and rateable value. This board was to have power of controlling the water service and was also to be authorised to promote Bills to purchase the company undertakings.[4] There was very great opposition to the measure from the counties bordering

---

[1] Report of the Royal Commission to enquire into the Water Supply of the Metropolis (1893), pp. 5, 11.

[2] A. G. Gardiner: op. cit., p. 267. A member of the London County Council gave an illustration of the supply during the famine. A court in Poplar where 86 persons lived in 10 rooms was supplied by a single tap, which worked for only 4 hours a day—and then the water was dirty. Ibid., p. 275.

[3] London Water Supply, L.C.C. publication No. 882, 1905, p. 18.

[4] Neither of the two Bills before the Committee appeared to them to offer satisfactory solutions. The date of the Report was 14th July, 1891.

London and a clause was inserted in the Lords enabling any metropolitan county to declare itself outside the jurisdiction of the new water authority. So great was the pressure that the Government was obliged to drop the Bill.

In 1897 the London County Council promoted a Bill to purchase the undertaking of the Chelsea Waterworks Company, to be followed by similar Bills dealing with the other undertakings. The President of the Local Government Board (Mr. Chaplin) opposed the second reading on the ground that a further Royal Commission should first be appointed to consider various matters before a scheme for purchase could be sanctioned.[1] The Bill was defeated.

The opposition of the counties was a new factor which made the situation far more difficult to deal with. In 1891 a Select Committee on the London Water Commission Bill had recommended in a special report that, assuming the desirability of establishing a single public representative water authority for the metropolis, the London County Council should be empowered to promote legislation to make themselves the responsible body for London; that they should be permitted to purchase the company undertakings; but that outlying local authorities with municipal supplies, such as Croydon and Richmond, situated within the metropolitan water area, should be assured of their continued independence.[1] Ever since 1888 it had been assumed that, if and when municipal ownership of water came about, the London County Council would become the responsible body; and the London County Council had been considering purchase from the beginning of its existence. But the statute which established the London County Council also created the county councils in the neighbouring shires and their resistance to any extension of the London County Council's powers over their territory was the decisive factor which prevented the London County Council from becoming the water authority.

In 1897 the Government appointed yet another Royal Commission on the metropolitan water supply, within the limits of the companies' areas. The representatives of the

---

[1] Royal Commission on the Metropolitan Water Supply (1899), Minutes of Evidence, B.P.P. 1909, vol. xxxviii, p. 245.

county councils expressed views before this Commission which refused to admit the possibility of administration over a wider territory. The witness for the Middlesex County Council, for example, stated that his county objected to the purchase of the water undertakings by anybody; but if a purchase was to be made, no scheme would be acceptable unless it provided for "absolute non-interference with Middlesex either as to rating or otherwise." The representative of the Surrey County Council explained that his council "had always been opposed to any scheme which would put Surrey under the control of the London County Council."[1] The County Councils of Kent and Essex expressed similar views. The London County Council did not attempt to resist these claims. The Commission considered, therefore, that inasmuch as all the metropolitan counties except Hertfordshire were bent on demanding what the London County Council was pledged to concede—i.e. complete county autonomy—the purchase of the water undertakings by the London County Council would necessarily be followed by a severance of the supply and distribution works into 5 distinct portions. This appeared to be so objectionable as to be practically inadmissible; and on that ground, among others, the Royal Commission concluded that the London County Council should not be the purchaser.[2]

Ultimately, after a further three years of delay, which witnessed another water famine and the promotion of several more Bills by the London County Council, all of which failed to pass through Parliament,[3] the Metropolitan Water Act, 1902, was passed, setting up the Metropolitan Water Board. In introducing the Bill Mr. Walter Long, the President of the Board of Trade, said that any public body to administer the water supply would have to be representative of the whole area concerned. The London County Council had been suggested, but in his view while the London County Council represents inner London, it could not be considered to represent that

[1] Royal Commission on the Metropolitan Water Supply (1899), Minutes of Evidence, B.P.P. 1900, vol. xxxviii, Pt. I, pp. 57–8.

[2] *Ibid.*, p. 58, Section 140.

[3] For details, see *London Water Supply*, published by the London County Council, No. 882, 1905, pp. 19–21. One of these Bills enabled the L.C.C. to provide a supplementary supply from Welsh sources.

"outer larger London" which is vitally concerned in the question. A new body covering the whole area had therefore to be found.[1] The new body established by the Bill was authorised to acquire by purchase the undertakings of the 8 metropolitan water companies and those of 2 local authorities.[2] The constitution of the Metropolitan Water Board was that of an indirectly elected authority and its membership composed of representatives of the following bodies:[3] the London County Council (14); the county councils of Middlesex, Essex, Surrey, Kent and Hertford (1 each); the Common Council of the City of London (2); the Westminster City Council (2); the 27 other metropolitan borough councils (1 each); the county boroughs of West Ham (2) and of East Ham (1); the municipal borough councils of Leyton, Tottenham, Willesden and Walthamstow (1 each); joint committees of certain borough and urban district councils (7 in all); the Thames and Lee conservancies (1 each). The total membership is thus 66 representatives, who need not be members of the councils which appoint them. They hold office for 3 years.

The London County Council was naturally disappointed at the failure of its persistent efforts to obtain control of the metropolitan water supply on the lines of ordinary municipal ownership. But even apart from that aspect of the matter it took strong objection to the Government Bill on the ground that the constitution of the proposed water board was such as to preclude any effective public control; that "a mere mockery of representation" was given to each outside area, whilst the vastly preponderating interest of London was denied the benefits of proper control; that an irresponsible body would be authorised to levy rates and incur debt; that the Board was far too large for its function; and that the inclusion of representatives of the metropolitan boroughs side by side with those of the London County Council ran counter to the recognised principle that matters of common interest to the whole of

[1] Hansard: Parliamentary Debates (4th series), vol. 101, 1902, col. 1384.
[2] The municipal undertakings were those of the Tottenham and Enfield Urban District Councils. The Staines Reservoirs Joint Committee was also absorbed.
[3] The composition of the Board has been varied since it was created, but its total membership is unchanged.

London should remain in the hands of the central representative responsible authority.[1] A joint committee of the Lords and Commons was set up to consider the Bill after its second reading, and this committee decided at first to reduce the size of the Water Board to about 35 members by deleting the representation of the metropolitan boroughs and of the urban districts and boroughs in the counties of Essex, Kent, Middlesex and Surrey. The Government refused, however, to accept this amendment; and the committee thereupon decided not to adhere to its own decision by means of an extraordinary interpretation of Parliamentary procedure on the part of the chairman (Lord Balfour of Burleigh). The Bill therefore passed in its original form so far as the constitution of the water board was concerned.[2]

The statutory area of supply given to the Metropolitan Water Board is 573 square miles,[3] containing at the present time about 7½ million consumers. This territory extends from Sunbury in the west to Gravesend in the east—a distance of 34 miles; and from Ware in the north to Westerham in the south. It includes the whole of the City and the County of London, together with large parts of the 5 home counties. The Board was constituted in 1903 and took over the company undertakings in June 1904.

The story of this belated transfer to public ownership would not be complete without an account of the compensation which was paid in respect of it.[4]

The rate of charge for a domestic supply varied among the different companies, but it was in all cases based on the annual rateable value of the premises.[5] As a result, with every increase in the value of the house property in London the companies were able automatically to increase their revenue without necessarily increasing the supply. It appeared from a return

[1] *London Water Supply*, L.C.C. publication, No. 882, 1905, p. 25.
[2] *Ibid.*, pp. 26–8.
[3] The area actually supplied is about 437 square miles.
[4] For a full analysis of the arbitration award, see the Report by the Comptroller of the London County Council on the financial aspects of the acquisition in *London Water Supply*, published by the L.C.C., No. 882, 1905, pp. 29–56.
[5] Royal Commission on Metropolitan Water Supply (1899), B.P.P. 1900, vol. xxxviii, Pt. I, pp. 16–17.

made to Parliament in 1886 that the average amount of water supplied daily to each house by 6 of the companies was actually less than it had been in 1872, yet the average water rental had increased from £1 18s. 1d. for each house in 1872 to £2 5s. 3d. in 1883.[1] Neither the Metropolitan Board of Works nor the City Corporation made any attempt to ensure that the companies should not increase their charges unless they also increased the supply. During these years the 8 companies were all paying dividends at rates varying between £5 10s. per cent and £11 18s. 8d. per cent.[2]

The share capital of the companies rose from £8,769,514 in 1867 to £12,330,830 in 1871.[3] By the end of 1883 it had increased to £14,719,565, and its market value stood at nearly £25 millions. In 1880 Lord Cross proposed to buy out the companies at a sum of about £29 millions.[4] This compensation was then regarded as exorbitant and the proposal was dropped.[5] In 1897 the share and loan capital totalled £16,432,284 nominal value, with a market value of £41,705,443.[6] The ordinary stock was subject to a maximum rate of dividend, and where this limit was not paid in any year the shareholders were entitled to back dividends in subsequent years. The claims of the companies for back dividends amounted to nearly £20 millions when they were first formulated in 1880. In 1897 the companies rendered an account to the Royal Commission showing £5,732,856 still outstanding in respect of unpaid back dividends. In the case of 6 of the companies the back dividends in arrears amounted to 81 per cent of the ordinary share capital.[7]

The London County Council had been considering the possible purchase of the companies' undertakings from 1889

[1] J. F. B. Firth: *The Reform of London Government*, p. 82.
[2] Return Relating to the Metropolitan Water Companies, 1884.
[3] J. F. B. Firth: *op. cit.*, p. 84; Royal Commission on the Water Supply (1867), vol. xlix.
[4] A. G. Gardiner: *op. cit.*, pp. 267–8.
[5] The Cross proposals, in the opinion of Sir H. Haward, had a lasting effect on the position and reputation of the companies. Prior to 1880 their stocks were a comparatively unknown security, but subsequently their market values were enormously enhanced, despite Government warnings against speculation in them. See Report of the Comptroller in *London Water Supply*, published by the L.C.C., No. 882, 1905, p. 31.
[6] Royal Commission on the Metropolitan Water Supply (1899), p. 4.
[7] *Ibid.*, p. 5.

onwards, but both its Water Committee and the Parliamentary Committee objected to the terms on which Parliament was accustomed to pay compensation in the case of municipal acquisitions of utility undertakings.

As the law then stood the Stock Exchange valuation of the stock would have formed the basis of an arbitration award. But the London County Council pointed out to the Royal Commission of 1899 that there were objections to this method of calculation. In the first place, much of the plant was obsolete and no longer capable of producing revenue. Second, a limit of time—a period of 6 years was suggested—ought to be put on the liability to pay back dividends. Third, an over-valuation of the companies' assets had followed the automatic increase in their revenues consequent on the quinquennial increase in the rateable value of premises supplied. Fourth, charges would certainly be revised by Parliament or the central government in the near future, and the legality of the companies' scales was in some respects dubious. Fifthly, the Stock Exchange valuation was augmented by the unduly high dividends which had been paid in some cases, made possible only by a failure to make proper provision for renewals and depreciation.[1] In short, the London County Council politely indicated that the water supply in London was a commercial racket, and deserved to be dealt with as such.

When the transfer actually took place the companies claimed nearly £51 millions as compensation, irrespective of their liabilities in regard to debenture stocks and mortgage loans amounting to £11 millions, and excluding costs. The aggregate claim was thus about £62 millions in respect of undertakings upon which a total sum of less than £23 millions had been expended on capital account.[2] They were awarded by the arbitration tribunal a sum of £30,662,323, in addition to which the Metropolitan Water Board took over responsibility for debenture stocks amounting to about £12 millions.[3] There was a further sum of £219,287 awarded as compensation to the

---

[1] Royal Commission on the Metropolitan Water Supply (1899), pp. 11–21.
[2] *London Water Supply*, published by the L.C.C., No. 882, 1905, p. 38.
[3] The exact items were: Redeemable Debenture Stocks, £7,217,838; Irredeemable Debenture Stocks, £4,365,110. For the latter, £6,060,165 of Metropolitan Water (A) Stock was issued in substitution.

directors of the water companies for loss of office. The enormous costs of the arbitration made another item of £254,829![1] The total compensation payable was about £43 millions in cash, the stock equivalent of the Board being £47 millions.[2] This sum was more than £2 millions in excess of the average Stock Exchange valuation of the companies' undertakings prevailing between 1900 and 1902. Since the capital monies expended by the companies amounted in all to £23 millions, the public had to pay no less a sum than £20 millions in order to buy back the statutory rights and privileges which Parliament had conferred upon their proprietors.[3] "With this millstone around its neck the Metropolitan Water Board started its life," writes the clerk and parliamentary officer to the Board in a recent article entitled (one hopes satirically) "The Romance of London's Water Supply."[4]

The debt charges resulting from this huge obligation, together with those caused by subsequent capital expenditure of £12 millions, consume 9s. in every £ of revenue received by the Board, or 45 per cent of its total expenditure. A departmental committee appointed in 1920 to report on the working of the Metropolis Water Act, 1902, pointed out that "the transfer took place at a time when the business of the companies was in a specially favourable position, and as this position has not been maintained the Board has been placed at a disadvantage."[5] It is difficult to believe that this was due to mere chance. In any case it is certain that the long and unnecessary procrastination of Parliament was highly advantageous from a financial point of view to the companies, and equally unfavourable to the public interest.

[1] The Board's costs were £142,645; the companies' costs £91,168; and payments to the Court of Arbitration £21,016.

[2] A. V. Huson: *Londoners and their Water Rates*: special supplement to the *Morning Post*, 9th December, 1935. The earnings of the companies on their capital of £20 millions at the time of transfer were equivalent to the income to be derived from £47 millions of stock bearing interest at 3 per cent—a good indication of the profits then being made from the water supply.

[3] *London Water Supply*, published by the L.C.C., No. 882, 1905, pp. 51, 54.

[4] G. F. Stringer in a special supplement to the *Morning Post*, 9th December, 1935.

[5] Departmental Committee on the Metropolis Water Act, 1920, p.

I have recounted at some length the story of the metropolitan water supply during the 19th century for two reasons. First, because the facts are not as well known as they deserve to be, and must be followed in detail to be appreciated. Second, the almost unbelievable neglect and disregard of the public welfare manifested by Parliament in regard to the water supply is highly characteristic of its attitude towards London government as a whole. The apathy towards the considered advice of a multitude of Royal Commissions and official committees, the refusal to apply proved remedies to known evils, the indifference to the discomforts of the poor no less than to the dangers of the rich, the utter disregard for economy or efficiency in administration, the feeble compromise at the last moment, the frittering away of golden opportunities for reform, the readiness to give way to vested financial interests, the absence of any enthusiasm for the public welfare, above all the incapacity to conceive the problem of London government as a whole: all these tendencies are perfectly displayed by the discreditable history of London's water supply in the 19th century. The City Corporation during this period abandoned its control of the Thames, made no attempt to reacquire the water supply, and behaved generally "not as a local government authority, but as a party of private gentlemen looking after their own interests."[1]

[1] Sir G. Laurence Gomme: *London in the Reign of Victoria*, p. 38.

CHAPTER XII

THE DRAINAGE AND SEWERAGE SERVICES

The efficient sewerage and sewage disposal of the metropolis was from the beginning of the 19th century closely connected with the purity of the water supply. But this simple fact was consistently ignored by Parliament and successive Governments.

For several centuries sewers had been constructed in London by commissioners of sewers for particular districts. There were several of these in the early 17th century. Their work consisted of making sewers to convey rain and surface water from the streets, and water from the houses, as rapidly as possible to the Thames.

The normal method of sanitation was by means of cesspools. The water-closet was invented about 1810 and began to be installed to an appreciable extent after 1830. But the cesspool continued to be predominant for many years longer; and in 1841 it was estimated that the 270,859 houses in the metropolis were fitted with 300,000 cesspools.[1]

Prior to 1815 it was a penal offence to connect a cesspool to the sewers, but in that year it was made permissive, and in 1848 compulsory to do so. This change in the law, combined with the introduction of the water-closet, transformed the whole main drainage problem in London. It led to the necessity of covering over the sewers, which were usually open before then; it also gave urgent and immediate importance to the question of the pollution of the Thames.[2] "The main natural drainage artery of London, the Thames, had now become the main sewer; and one, owing to tidal action, of a particular obnoxious type."[3]

The dangers inherent in this situation manifested themselves

[1] Mr. Phillips, surveyor to the Metropolitan Sewers Commission, stated that in 1848 "there is scarcely a house without a cesspool under it" and a large number had more than one. Third Report of the Metropolitan Sanitary Commission (1848), p. 26.

[2] Royal Commission on Metropolitan Sewage Discharge (1884), First Report, B.P.P. vol. xli, pp. xi–xii.

[3] Report on Greater London Drainage (1935), p. 15.

in frequent outbreaks of Asiatic cholera. London had an epidemic in 1831–2 when there were 10,182 cases of cholera and 4,885 deaths from the disease;[1] and there were further outbreaks in 1848–9 and 1853–4. "The strong suspicion prevailed," observed the Royal Commission on Metropolitan Sewage Discharge in 1884, "that defective drainage had contributed to the alarming mortality."[2]

In 1847 a Royal Commission was appointed to enquire into the sanitary condition of London. There were at this time in existence eight separate commissions of sewers. Each of them consisted of a board with almost unlimited powers in its own district and none at all outside of it; each had its own peculiar mode of conducting business, its own staff of engineers, surveyors, clerks and other officers; its own regulations as to the size of drains, their rates of inclination, mode of execution and cost. Under such divided management it was utterly impossible to expect an efficient result. It is scarcely surprising that "the evils of the system, as manifested in the defective drainage of the metropolis, were beginning to be seriously felt."[3]

The Metropolitan Sanitary Commission, as the Royal Commission was called, undertook as its first task a consideration of the causes which might lead to a further outbreak of cholera in the near future. They found that the disease was particularly liable to spread in those districts where the water supply was insufficient, where sewers and cesspools were inadequately cleansed, and at points where the contents of sewers were emptied into the river.[4] They recommended that complete measures for the prevention of cholera depended on works that required further investigation; but immediate steps were called for in respect of cleansing the entire system of sewers, the scavenging of cesspools, and the extended application of the water carriage principle of sewage removal through the provision of more abundant water supplies.

The Metropolitan Sanitary Commission further declared that

[1] Metropolitan Sanitary Commission, First Report, B.P.P. 1847–8, vol. xxxii.
[2] Royal Commission on Metropolitan Sewage Discharge (1884), First Report, p. xv. [3] Ibid., pp. xi–xiii.
[4] Report of Metropolitan Sanitary Commission (1847), vol. xxxii, p. 21, and evidence passim.

the control of sewers should be placed in the hands of a single board.[1] The division of natural drainage areas among several district authorities working independently of each other was extravagant and inefficient; and it was impossible for improved works of drainage to be put in hand in such circumstances. The work of the existing sewer authorities was often ill-advised and wasteful.[2] Moreover, the business clearly needed the regular attention of paid and competent officials instead of having to wait for court days on which the local sewer commission met, at infrequent intervals, when the work was rushed through, without proper consideration, by "the honorary and irresponsible members who casually attend."[3] Much of the business coming before these district courts, meeting weekly, fortnightly, monthly or quarterly, arose from the very defects of their own imperfect plans and works; and it was thought that under a better system of administration it would diminish. The Commission condemned the widespread practice of including among the members of these sewer boards eminent public men occupying high positions such as the Duke of Wellington, the Lord Chancellor, the First Lord of the Treasury and so forth, who it was known could never attend the meetings. The execution of public works requiring technical knowledge could not be dealt with as a side-line to another occupation, but called for the undivided practical attention of well-qualified paid officers.[4]

This Metropolitan Sanitary Commission also emphasised the desirability of combining responsibility for the water supply with control over the drainage, sewage and refuse disposal services. They further drew attention to the importance of considering the whole question of areas, in order that the sewage authority should have jurisdiction over a territory which was also suitable for the administration of these related services.[5]

The Report was acted upon promptly, but in the narrowest possible way. An Act was passed in 1848 to consolidate and unify the metropolitan commissions of sewers except those operating in the City districts. Twelve Commissioners were appointed, together with 5 *ex officio* representatives of the City

[1] Report of Metropolitan Sanitary Commission (1847), vol. xxxii, p. 49.
[2] *Ibid.*, p. 49.   [3] *Ibid.*, p. 51.   [4] *Ibid.*, pp. 50–1.   [5] *Ibid.*, p. 50.

Corporation[1] (who would vote where the City was affected). This body was empowered to keep the sewers in order, to make new ones acquired "for effectually draining the area within the limits of the Commission," and to divide the area into separate sewage districts on each of which a rate could be levied.[2] The effect of this limitation was that the new body had no power to provide for the removal of sewage from the neighbourhood of London to a more remote place. Three of the largest London sewers at that time had their outlets in the Thames between Battersea and Vauxhall Bridges.[3]

The Sewers Commission was superseded by a similar body in 1849. This second Commission considered that sewage should be kept out of the Thames altogether. They advertised an announcement asking for competitive plans for draining London, but were unable to decide what form of drainage the town ought to have. Thus matters ambled slowly on, one Commission succeeding another—the sixth was appointed 1854 —without any fundamental improvement being achieved. In the meantime the state of the river was becoming "very alarming."[4] In 1848 a severe outbreak of fever attacked the boys of Westminster School as a result of defective sewers and cesspools in the Abbey Precinct.[5] The consolidated Commission of Sewers may have been somewhat better than its predecessors,

---

[1] 11 & 12 Vic., c. 112. The Act provided there should be one Commission for Westminster, Southwark, and such other places in Middlesex, Surrey, Essex and Kent not more than 12 miles from St. Paul's (but not being within the City of London) as may be named in such Commission. A local and personal Act for the City was passed in the same year, 11 & 12 Vic., c. clxiii. "To provide for the sanitary improvement of the City of London . . . and for the better cleansing, sewering, paving and lighting thereof." This was really the first modern Public Health Act for the City: it enabled a medical officer to be appointed. It provided for the Commissioners of Sewers in the City to be appointed by the Common Council and for the Lord Mayor to be their chairman. The Metropolitan Commission of Sewers could require the City Commission to do any works in the City, or act in default.

[2] Cf. Royal Commission on Metropolitan Sewage Discharge (1884), First Report, p. xiii.

[3] Ibid., Second Report, p. 4.   [4] Ibid., First Report, p. xv.

[5] Metropolitan Sanitary Commission (1848), Third Report, p. 14; Dr. Sutherland's Report to the General Board of Health on Westminster, showing the Results of inefficient Drainage in the Cloisters and Precincts of Westminster Abbey, B.P.P., 1854, vol. lxi, p. 11.

but it was nevertheless a highly inefficient body. The Victoria Street sewer which it constructed fell to pieces owing to defective workmanship.[1]

The London highways in the middle of the century were almost indescribably filthy. The streets of Bermondsey were said by the medical officer to be "a disgrace to the civilised world." Deptford was described as "perhaps the worst-regulated town in the Empire from a sanitary point of view." In the streets of Whitechapel, where disease had wrought great havoc among the people, there was neither sewerage, drainage, cleansing, paving nor a good supply of water. In Wandsworth there was no drainage and the cesspools and privies were constantly overflowing. In Hampstead the sewerage and drainage were lamentably defective, while in Hackney the characteristics of the streets were stated to be the presence of overflowing cesspools and privies, cowyards and piggeries "with a loathsome ditch which has been causing disease for 20 years."[2]

These conditions were not confined to the poorer districts. In the wealthiest parts of the City the increase of horse-drawn traffic had produced a serious sanitary nuisance which it was apparently beyond the wit of the feeble local authorities to deal with. In 1850 the General Board of Health remarked that "strangers coming from the country frequently describe the streets as smelling of dung like a stable yard."[3] It was estimated that 20,000 tons of manure were dropped every year. The effect of this on the surface water which drained into the Thames can well be imagined.

In 1855 the Metropolitan Board of Works was established.[4] It will be remembered that this body superseded the Commissioners of Sewers and was expressly authorised to construct a system of sewers to prevent all or any part of the London sewage from flowing or passing into the Thames near the metropolis.[5] The new Board immediately proceeded to have a lengthy dispute with referees appointed by the Government

[1] Royal Commission on Metropolitan Sewage Discharge (1884), First Report, p. xv.
[2] Metropolitan Sanitary Commission (1847), Second Report, p. 17.
[3] Report of General Board of Health on Metropolitan Water (1850), pp. 235–6.
[4] *Ante*, p. 59.   [5] Metropolis Management Act, 1855, Section 135.

on the question where the sewage outfall should be located. The Metropolitan Board of Works favoured Barking, while the referees considered that the sewage should be taken down to the sea. The Metropolitan Board of Works was unwilling to carry the sewage so far without a grant-in-aid from the Exchequer.

Meantime, the stench from the Thames grew to such an extent that at last it reached the nostrils of the legislators themselves. Mr. Brady, M.P., reminded the House of Commons in 1858 that "it was a notorious fact that honourable gentlemen sitting in the committee rooms and the library were utterly unable to remain there in consequence of the stench which arose from the river."[1] There was, indeed, almost a panic among the governing class in the summer of 1858, caused partly by the unusually dry season which occurred but chiefly by the deterioration in the general sanitary condition of the Thames. Mr. Owen Stanley stated in the House of Commons that the nuisance had become intolerable both in the committee rooms and in the chamber itself. He disclosed that Mr. Gurney, the official in charge of ventilation, had written to the Speaker of the House of Commons saying that he could no longer be responsible for the health of Members of Parliament owing to the impossibility of dealing with the menacing odours which invaded the building from every side.[2] The Law Courts were in an equally serious plight and were the subject of constant complaint by litigants, witnesses and judges. The Lord Chief Baron in the Court of Exchequer remarked during a trial that "the stench from the river is most offensive, and I think it right to take public notice of this, that in trying this case we are really sitting in the midst of a stinking nuisance."[3]

So critical was the situation felt to be that there was talk of Parliament either abruptly terminating its session or removing to a more healthy place. Neither of these courses was adopted,

[1] Royal Commission on Metropolitan Sewage Discharge (1884), First Report, p. xxvii.

[2] Hansard, 3rd series, vol. 151, cols. 421–3, 25th June, 1858. Stanley continued: "The stench has made most rapid advance within two days, that up to Tuesday he (Gurney) got fresh air draughts from the Star Chamber Court; but that when night came the poisonous enemy took possession of the Court, and so beat him outright." [3] *Ibid.*

but the session was wound up more speedily than would normally have been the case.[1] A Bill was hurried through Parliament by the recently formed Derby Ministry to extend the powers of the Metropolitan Board of Works for the purification of the Thames and the main drainage of the metropolis The Queen's Speech proroguing Parliament included a statement that "the sanitary condition of the metropolis must always be a subject of deep interest to Her Majesty, and Her Majesty has readily sanctioned the Act which you have passed for the purification of that noble river, the present state of which is little creditable to a great country, and seriously prejudicial to the health and comfort of the inhabitants of the metropolis." The new statute relieved the Metropolitan Board of Works of the obligation to submit their plans to the Government, in consequence of which the Barking scheme was put in hand without further controversy. The work was completed by about 1864.[2]

When the Act of 1858 was being passed it was stated that the Metropolitan Board of Works would treat the sewage with a process of deodorisation before discharging it into the river, for at least some months of the year. Indeed, Ministerial pledges were given in Parliament to this effect on behalf of the Metropolitan Board of Works.[3] But this undertaking was never carried out. A number of experiments were conducted in regard to the utilisation of sewage, and one of these proved an effective method of deodorisation, yet it was abandoned as unprofitable. "There was no proof that the process could be adopted with any hope of profit to the ratepayers, and as this hope of profit appeared to be their only inducement in defecating sewage, they dismissed the Company and made them at once remove their works and plant from the ground."[4] The Board's sense of civic responsibility for the health of the people could scarcely have sunk to a lower point.

About this time the recently established Thames Conservancy Board became concerned about the state of the river from the standpoint of navigation, and a further voice was thus added

[1] *Annual Register* (1858), p. 215.   [2] *Ante*, p. 62.
[3] Royal Commission on Metropolitan Sewage Discharge (1884), First Report, p. xxix.   [4] *Ibid.*, p. xxxii.

to the mounting chorus of protest directed against the prevailing methods of sewage discharge. The origin and constitution of the Conservancy will be described in the next section. For the moment it will suffice to say that the Conservators' primary duties relate to the prevention of pollution, the maintenance of navigation and land drainage. Their work was therefore seriously affected by the discharge of large quantities of crude sewage into the Thames. They complained again and again to the Metropolitan Board of Works of the silting up of the river with sewage mud and the continuous pollution of the water, but without effect.

In 1869 the condition of Barking gave rise to serious complaints in the House of Commons and produced a strongly worded memorial from the Vicar and other inhabitants describing the dire consequences resulting from the discharge of sewage in the locality. "The nuisance and danger are far in excess of anything implied in the statement made in the House of Commons before referred to," declared the memorialists. "The filth and refuse of the largest city in the world is concentrated in all its horrors and abominations in the immediate vicinity of the dwellings of your Memorialists, and they feel that while every town and village, and private house above London, is forbidden to pollute the river in the interests of the great Metropolis, yet that they and the inhabitants of all the populous towns below London are completely sacrificed to the comfort of the latter. There are banks within a few hundred yards of the houses of some of your Memorialists composed of solid sewage, six, eight, and ten feet deep; and the backwater of nearly undiluted sewage sweeping up Barking Creek is so great that it must infallibly, beyond the possibility of doubt, breed a pestilence sooner or later."[1]

An enquiry was set on foot as a result of this complaint, but no action was taken as the allegations were considered to be exaggerated. In 1877 the Thames Conservancy submitted a report to the Metropolitan Board of Works requiring them to deodorise sewage before discharging it into the river. This

[1] Memorial from the Vicar and other inhabitants of Barking . . . in consequence of the Discharge of Sewage through the Main Outfall Sewers of the Metropolitan Board of Works (1869).

matter was referred to arbitrators, who upheld the contention of the Metropolitan Board.[1]

Nevertheless, as the years went by the volume of complaint against the Metropolitan Board of Works in regard to the noisome state of the Thames increased. Between 1880 and 1882 complaints were lodged by the Thames Conservancy Board, by the Captain of the *Warspite*, by the London Steam Boat Company, by the Erith Local Board, by the Local Boards of Woolwich and Plumstead, by the Town Clerk of the City of London, by a number of city business men, and by a memorial signed by 13,000 persons begging the Government to take action.[2] In 1884 a Royal Commission was appointed to enquire into the matter.

The Royal Commission on Metropolitan Sewage Discharge decided that the nuisances of which so much complaint was made, had a "real existence."[3] In the summer of 1884, five members of the Commission embarked on a Conservancy boat at Woolwich in order to conduct a personal examination into the state of the Thames. The river was black with sewage which the tide had brought up as far as London Bridge. "We found a condition of things," stated the Commission in their Report, "which we must denounce as a disgrace to the metropolis and to civilisation."[4] They advised that the existing system called for immediate remedy and that the discharge of crude sewage into any part of the Thames was unjustified.[5]

Out of this report grew the modern methods of dealing with the problem. At Barking and Crossness, where the main outfalls are situated, the sewage is treated by elaborate plant and machinery; the effluent is discharged into the river and the sludge carried out to sea by a fleet of municipal steam vessels working night and day.[6]

As I shall show later, the drainage and sewage problems of London have by no means been completely solved. There are,

---

[1] Royal Commission on Metropolitan Sewage Discharge (1884), First Report, p. xxxiii.
[2] *Ibid.*, pp. li–lxi.
[3] *Ibid.*, p. lxii.
[4] *Ibid.*, Second Report, p. ix.
[5] *Ibid.*, p. xlvi.
[6] Herbert Morrison: *How Greater London is Governed*, pp. 85, 91; Percy Harris: *London and its Government*, pp. 114–15; *London Statistics* (1934–6), vol. 39, pp. 126–7.

9

indeed, some very urgent matters pressing to be dealt with. Here I have dealt only with the historical development during the past century. In regard to this uninviting but most essential service we can observe the same tendencies as those which have been at work in almost every department of London government: a complete lack of forethought or imagination, a failure to grapple with the problem before it reached the emergency stage, a pitiful absence of leadership among the municipal bodies, a lack of courage on the part of the central government. Everything was left undone or badly done until the resulting evils became intolerable. At no point was there any sign of a civic sense at all comparable to that existing in the vigorous provincial towns.

CHAPTER XIII

THE PORT OF LONDON AND RIVER AUTHORITIES

The handling of the various administrative functions connected with the river to which London owes so much of her commercial supremacy provides a further illustration of the piece-meal manner in which the government of the metropolis has been fashioned.

Conservancy functions over the Thames between Staines in Middlesex and the sea were said to have been exercised from time immemorial by the City Corporation. The upper river was placed in charge of a body of commissioners in the 18th century. The Thames Commission was large, unincorporate and inefficient.[1] In 1857, after a prolonged dispute between the Crown and the City Corporation over the latter's claim to ownership of the bed and soil of the river, the Thames Conservancy Act[2] was passed constituting a body of conservators consisting of the Lord Mayor, 2 City aldermen and 4 common councilmen, together with 5 representatives of the Admiralty, Board of Trade and Trinity House. The estate of both the Crown and the City Corporation in the bed and soil of the river was vested in the new body, which was charged with maintaining the Thames in a navigable condition, preventing pollution and ensuring land drainage.

In 1866 the Conservancy was reconstituted on an enlarged basis to take over from the Thames Commissioners the upper reaches of the river.[3] In the following year another Act was passed to extend from Staines to the metropolis the existing provisions against pollution, and gave to the Conservators an exclusive right to dredge the river while prohibiting others from doing so. Large sums were expended on this work. This body controlled the whole navigable river to Yantlet Creek.

The composition of the Thames Conservancy has been

[1] Fred S. Thacker: *The Thames Highway* (1914, privately printed), p. 123.
[2] 20 & 21 Vic., c. cxlvii.
[3] Thames Navigation Act, 1866; Fred S. Thacker: *op. cit.*, p. 239.

changed from time to time. It now consists of 34 members appointed by interested Government departments (such as the Ministry of Agriculture and Fisheries, the Board of Trade and the Ministry of Transport), the Port of London Authority, the Metropolitan Water Board, and specified groups of county borough councils, non-county boroughs and urban district councils.[1] The members hold office for a term of three years. They elect their own chairman each year.[2]

The original jurisdiction of the Thames Conservancy was from Staines to the estuary at Southend. This was extended in 1866 to Cricklade in Wiltshire and for a distance of three miles up all tributary streams. This distance was subsequently extended. The Thames Conservancy Act, 1894, widened the powers of the Board to cover the entire course of the river and all its tributaries; but in 1908, when the Port of London Authority was established, the tidal part of the river from Teddington to the Nore was given into the keeping of the new authority.[3] The Conservancy is now responsible for a stretch of river 136 miles long.[4]

The functions of the Conservancy were formerly wider in scope than they now are, and included in 1902 the regulation of all vessels within the port, the improvement of navigation by dredging and other means, the appointment of harbour masters, the removal of wrecks and obstructions, the licensing of docks, piers and embankments, the maintenance of naviga-tion beacons, the enforcement of the legislation regulating

[1] For detailed information, see *London Statistics* (1934-6), vol. 39, p. 6.
[2] Another body of a similar kind is the Lee Conservancy Board. There are 15 members of this body appointed by the City Corporation (1), by the London County Council (2), by the county councils of Bedfordshire, Essex, Hertfordshire and Middlesex (1 each), by the West Ham County Borough Council (1), by the Metropolitan Water Board (2), by certain local authorities within the counties of London, Essex, Middlesex and Hertfordshire (4 in all) and by barge owners (1). The Board was constituted by the Lea Conservancy Act, 1868, and has a jurisdiction from the source of the river to the boundary marks at Bow Creek and over all tributary streams. With certain additional members the Board functions as the Lee Conservancy Catchment Board and is in that capacity responsible for land drainage.
[3] Thacker: *op. cit.*, Chapter X.
[4] Inter-Departmental Committee on the Thames and Lee Conservancies (1923), p. 5.

PLATE III. The Thames at Blackfriars Bridge. The fine embankment west of the bridge was carried out by the Metropolitan Board of Works (1855–88). The lower reaches to the east of the bridge have not been embanked by the City Corporation and the mud flats can be seen in front of the wharves and warehouses

(*Aerofilms*)

explosives and petroleum in respect of the river, and so forth.[1] But at no time has its jurisdiction been fully comprehensive; and it was, of course, relegated to a minor position when, after the formation of the Port Authority, it was left with only the pleasure and residential reaches of the river to look after. In 1930, however, the Land Drainage Act vested in the Conservators the duties of catchment board for the Thames Catchment Area, a region covering about 3,800 square miles[2] and comprising parts of fifteen counties.

Responsibility for the hygiene of the port was placed on the shoulders of the City Corporation, which in 1872 was made the Port of London Sanitary Authority. In that capacity it was charged with preventing the introduction from overseas of cholera and other diseases and also with carrying out, in its area, the provisions of the Nuisances Removal Acts and the other statutes relating to public health.[3] The area assigned to the Port Sanitary Authority extends from Teddington Lock to the North Foreland, a distance of some 88 miles.

The policing of the Thames was entrusted to the Metropolitan Police Commissioner, who maintains a special river force for that purpose. The Thames Conservancy has no police powers and employs no police officers to enforce its bye-laws. The protection of the docks might be supposed to rest with the Metropolitan police force; but the Commissioner regards the docks as private property and takes the view that his men cannot be stationed within the dock area unless their expenses

[1] Report of the Royal Commission on the Port of London (1902), p. 33.

[2] See Thames Conservancy Act, 1932, for the principal provisions relating to the constitution, powers and duties of the Conservators, which are briefly as follows: The construction and maintenance of locks, weirs and all other works necessary for the carrying on of the navigation; the establishment and maintenance of ferries; the appointment of water bailiffs for the protection of the fisheries; the regulation of the water levels; the removal of sunken vessels; the removal of obstructions from the river and towpaths; dredging for the purpose of maintaining and improving the navigation; the maintenance of the flow and the prevention of pollution of the river and all tributaries and streams connected with it; the granting of licences for works in the river; the registration of steam launches, houseboats and pleasure boats, and the regulation of these vessels; the levying of tolls on vessels; the making of bye-laws for a number of purposes; the regulation of the navigation; the prevention of pollution from vessels.

[3] H. Jephson: *The Sanitary Evolution of London*, pp. 239–40; Report of the Royal Commission on the Port of London (1902), p. 59.

are paid by the proprietors, or unless rioting is imminent.[1] In consequence, the Port of London Authority has to maintain its own private police force, the second largest body in the country.

Another authority with important duties is Trinity House, a body descended from an ancient guild of pilots and seamen.[2] Trinity House is in charge of pilotage,[3] lighting and buoying; and it also examines the qualifications of persons seeking to become dockmasters.

A fifth authority is the Watermen and Lightermen's Company, which also originated in a craft guild dating from the 14th century. The Company was regulated by legislation in the 19th century; and the result is that persons who are not freemen of the Company, apprentices, or "contract service men"[4] are prohibited from navigating any craft for hire on the important parts of the river. The tugs, barges and lighters that travel up and down the Thames are almost entirely manned by this privileged fraternity, either as employees or owners.

The distribution of power between the Thames Conservancy, Trinity House and the Watermen's Company was declared by a Royal Commission in 1902 to be contrary to the interests of the port as a whole.[5] Among other disadvantages they noted the danger of errors and omissions in buoying owing to the fact that the dredging authority (the Conservancy) had no connection with the buoying authority (Trinity House).[6] In Liverpool, the Mersey Docks and Harbour Board controls not only the docks, but is also in charge of conservancy, pilotage, lighting and buoying. But London has never recognised the need for even this degree of simplicity and unification.

Of greater moment than the lack of co-ordination among the various authorities was the feeble manner in which the Thames Conservancy carried out its primary duty of dredging and

---

[1] Report of the Royal Commission on the Port of London (1902), p. 60.
[2] The governing body of Trinity House are the Elder Brethren, who elect their members from the Younger Brethren. The latter are chosen by the former from persons outside the fraternity.
[3] Pilotage is compulsory in the London pilotage district. For details, see *London Statistics* (1934–6), vol. 39, p. 309.
[4] Report of the Royal Commission on the Port of London, p. 55.
[5] *Ibid.*, p. 111.   [6] *Ibid.*, p. 47.

safeguarding the navigability of the river. By the opening of the 20th century, the neglect and the incompetence of the conservators were a serious menace to the international status of London as a port.[1]

The dominant fact in the situation was, however, the deplorable condition into which the docks had been permitted to fall. Unlike Liverpool, where the docks had been constructed at the beginning of the 18th century by the municipal corporation, and managed from their inception either by the City Council or (from 1857) by the more broadly based Mersey Docks and Harbour Board,[2] the London Docks had been farmed out by Parliament from 1789 onwards to a number of private companies trading for profit. These companies were at first granted monopolies with statutory restrictions as to maximum charges and dividends. From 1825 the principle of monopoly was abandoned in order that the public might derive the full advantages of competition.[3] The remainder of the century was occupied with an economic struggle between the various dock companies that so exhausted their resources and their credit that they sank into a state of quiescence; and from 1885 no further capital development whatever took place despite a vast increase in the shipping entering the port. The management of the companies was both incompetent and of doubtful integrity; but the fundamental trouble was that the whole principle of their existence was wrong.

In 1900 a Royal Commission was set up to enquire into the causes and cure of this deterioration in the port facilities. The Commission recommended the establishment of a non-profit making public trust, on the lines of the Mersey Docks and Harbour Board, in order that unified control of river conservation and dock administration might be achieved. The new authority was also to take over the licensing of watermen and lightermen and the lighting of the harbour.[4] The London County Council and the City Corporation were to provide

[1] Lincoln Gordon: *The Port of London Authority* in *Public Enterprise* (edited by W. A. Robson), pp. 19–20.   [2] *Ibid.*, pp. 13–15.
[3] *Ibid.*, pp. 17–18; Report of the Royal Commission on the Port of London, pp. 63, 64.
[4] Report of the Royal Commission on the Port of London (1902), pp. 112–14; Lincoln Gordon: *op. cit.*, p. 20.

£2,500,000 of the £7 millions of new capital expenditure required to modernise the port, and they were to receive substantial representation on the governing body.[1] In 1903 the Government introduced a Bill to carry out these proposals of the Royal Commission, but with the Progressives in power the London County Council looked too lively a body in the suspicious eyes of the City; and the opposition which the Bill aroused there led to its withdrawal. The London County Council then attempted in 1905 to establish a municipally appointed port authority, but this project was unacceptable to the Government. The following year saw the formation of a powerful Liberal Government, and Mr. Lloyd George as President of the Board of Trade was responsible for the Bill which eventually passed into law as the Port of London Act, 1908.

The Port of London Authority created by the Act consists of 28 members, of whom 17 are elected by the immediate users of the services provided by the Authority. The composition is as follows: there are 8 members elected by shipowners, 8 by merchants, and 1 by river craft owners; 1 is elected by wharfingers. Of the 10 remaining members 1 is appointed by the Admiralty, 2 by the Minister of Transport, 4 by the London County Council (2 from among their own number and 2 from outside persons), 2 by the City Corporation (1 from among their own number and 1 from outside persons) and 1 by Trinity House. The Chairman and Vice-Chairman are appointed by the Port Authority from their own members or from other persons. The term of office for the entire body is 3 years.[2]

The Port Authority exercises jurisdiction over the tidal portion of the river and its estuary: that is, from Teddington Lock to a distance seawards of 57 miles from London Bridge (69 miles in all). It took over the whole line of docks and was authorised to construct new ones and to improve those already

---

[1] The London County Council was to have 8 places and the City 2.

[2] For an excellent account of the election and working of the authority see Lincoln Gordon: *The Port of London Authority* in *Public Enterprise* (edited by W. A. Robson). I am greatly indebted to Mr. Gordon's able description. A longer account is contained in *The Public Corporation in Great Britain* by the same author.

existing. It is responsible not only for the administration of the entire dock system, but also for a vast series of warehouses, refrigerators and other ancillary establishments. It superseded the Thames Conservancy in regard to conservancy functions in the tidal portion of the river, and has thus acquired extensive powers in regard to dredging and the deepening of the Channel, the removal of obstructions, and the regulation of navigation. It also carries out the registration and licensing of rivercraft, and the licensing and control of lightermen and watermen.[1] On the other hand, Parliament did not permit the Port of London Authority to take over any of the privately-owned wharfingers who compete with the docks for general business. Moreover, as though deliberately to make the task of the Port of London Authority more difficult than it otherwise need be, these rival wharfingers were given representation on the Port Authority. They were also permitted to send lighters or similar craft to load or discharge goods oversides from vessels using the docks, without payment of dock charges—a continuance of an old privilege which was originally justified as a condition for granting exclusive monopolies to profit-making dock companies[2] but which in the changed circumstances of today has become an indefensible abuse. Even when Parliament decided at long last to organise the London docks on a public service basis it was unwilling to make a complete job of the matter.

The compensation given to the companies in exchange for their derelict property, when it was transferred in 1908, was on the basis of 26 years purchase of existing revenue: a method of calculation which was in any case excessive, but especially so in view of the state of disrepair into which the docks had been permitted to fall and the inability of the companies to maintain their revenue without further large expenditure.[3]

Opinions differ as to the merits of the principle on which the Port Authority is founded. Dr. Lincoln Gordon, an able American investigator, considers the almost complete absence of municipal representation on the governing body to be a good

---

[1] The complicated relations between the Port of London Authority and the Watermen's Company are governed by Section 11 of the Port of London Act, 1908.

[2] Lincoln Gordon: *op. cit.*, p. 18.   [3] *Ibid.*, p. 43.

feature on the ground that localism in connection with ports may be detrimental to national transport as a whole. The interests of the country may require a national authority in supreme control of all the ports. At the same time, Dr. Gordon recognises that the work of the Port of London Authority must affect municipal services such as housing and public assistance; and hence by implication he admits the difficulty of separation.[1] Mr. Herbert Morrison, on the other hand, regards the Port of London Authority as a "capitalist soviet," and deplores the domination of a public authority by private interests.[2]

Whichever of these two views may be the correct one, it is indisputable that the London County Council was deliberately sidetracked when the Port of London Authority was set up in 1908. The legislation is significant as an example of the slight attention paid even by a Liberal Government to the claims of the principal municipal authority in London. The insufficiency of its area was not an insuperable difficulty as was shown by the creation of the City Corporation as Port Sanitary Authority in 1872.

Nevertheless, it is agreed on all hands that the Port of London Authority has a fine achievement to its credit and has brought the dock and harbour facilities in London to an unsurpassed level of efficiency.

[1] Lincoln Gordon: *op. cit.*, p. 28.
[2] Herbert Morrison: *How Greater London is Governed*, p. 138.

CHAPTER XIV

HIGHWAYS, TRAFFIC AND TRANSPORT

In this section I shall deal with a group of topics which are essentially related—highways, bridges, traffic and transport. Their proper management is one of the most vital elements in the good government of a large city. The way in which these services are administered reacts inevitably on the conditions of work and play, on housing and town planning, on health and general welfare.

Prior to 1835 the situation regarding highways in London had developed on promising lines. The squares which we still recognise as the finest example of planned development and architectural harmony in the metropolis had been constructed in the 18th century; and Nash's great scheme embodying Regent Street and its extensions into Regent's Park and the Mall had been achieved in all its splendour and unity.[1]

In 1787 all London roads south of the Thames were placed under one authority, while in 1826 the Metropolitan Road Board was set up to deal with those north of the river. This body consisted of the members of Parliament for the City, the county of Middlesex and the City of Westminster, together with 40 other individuals. They appointed Macadam as their Surveyor General[2] and his work greatly improved the surface of the roads.

The placing of all the turnpike roads in Greater London under one authority was almost brought about in 1833.[3] But the introduction of the railways, by diminishing the importance of the highways, changed the outlook and destroyed the prospect of consolidating the network of metropolitan roads into a unified system.[4] Nevertheless between 1831 and 1851

---

[1] The best account of London's physical development is contained in S. E. Rasmussen: *London: The Unique City.*

[2] London Traffic Branch of the Board of Trade, Third Annual Report (1910), Appendix H., p. 179 *et seq.*

[3] For an account of the turnpike roads, see Sir G. Laurence Gomme: *London in the Reign of Victoria,* p. 36.

[4] *Ibid.,* pp. 2–3; G. A. Sekon: *Locomotion in Victorian London,* p. 26 and Chapter VIII.

the necessity for street improvements in London was fully recognised by Parliament, and about a dozen Select Committees were appointed to consider plans for the improvement of the metropolis and to advise as to the best means for carrying them out.[1]

The first suburban railways in London were two short lines sanctioned by Parliament in 1833 and opened three years later. One ran from the Minories to Blackwall Pier and the India Docks[2] via Stepney and Limehouse; the other connected Greenwich with the City. The latter enabled Londoners to live in Deptford and neighbouring suburban retreats. In 1838 the first main line railway reached the Metropolis, terminating at the outskirts of the town at Euston Station. In the following year Croydon was linked to London by railway. By the middle of the century the principal terminal stations had been constructed at Waterloo, King's Cross, Paddington and Fenchurch Street; and the main goods yards were also laid down.[3] From then onward the problem of planning London became infinitely more urgent and important, but no steps were taken to bring it about.

A great opportunity was lost in these early years of the railway age. Between 1831 and 1851, the Government or Parliament could have formulated a plan to provide a unified network of railways entering the capital that would have effected an enormous saving in time, space and money over the haphazard arrangements which were permitted to develop. The question whether a great central terminus should be built was actually referred to a Royal Commission appointed in 1846; but the Commission was opposed to the project on the ground that the railways were principally useful for carrying goods. The vast increase in passenger traffic, and especially suburban traffic, was not foreseen. The Commission declared that the convenience of passengers did not require the prolongation of the railways into the heart of London—

[1] Sir G. Laurence Gomme: *London in the Reign of Victoria*, p. 154.
[2] J. F. P. Thornhill: *Greater London: A Social Geography* (1935), pp. 58–9.
[3] The best account of railway development in the metropolis is to be found in G. A. Sekon: *Locomotion in Victorian London*. Chapter VIII.

which, indeed, should be forbidden—and still less the establishment of one great central terminus into which the railways from all parts of the country would run. Such an arrangement, said the Commission, would merely be for the convenience of the small number of passengers who passed through London and the advantage would therefore be small in comparison with the many difficulties arising from divided management.[1] The lack of imagination shown by this Royal Commission in failing to anticipate the immense expansion in railway passenger traffic is indeed remarkable when the revolutionary change in speed introduced by the trains is recorded. In 1821 the celebrated horse coach *Vivid* took 5¼ hours to travel from London to Brighton, a distance of 51 miles. In 1851 the Brighton express was doing the journey in 1 hour 15 minutes.[2]

The Royal Commission on Railway Termini made the important recommendation that street improvements and terminal developments should be carried out under the exclusive authority of a central Government department in order that the ends of utility and beauty might be served by a single well-considered scheme, whereas if these matters were left to originate with separate railway companies without any supervision other than the cursory examination provided by Parliamentary committees that object would not be attained.[3] The Commissioners of Woods and Forests might well have been employed for this purpose had any notice been taken of so intelligent a suggestion. The Commissioners of Woods and Forests had under the Metropolitan Improvement Act, 1846, been empowered to make Battersea Park from former waste lands; they had built a fine bridge and embankment, and made a new road from Sloane Street by which to approach it.[4] Their functions could easily have been extended to comprise railway terminals and highways in London and their reputation

[1] Report of the Royal Commission on Metropolitan Railway Termini (1846), B.P.P., vol. xvii, p. 6.

[2] This was stated by the Chairman of the railway company at the half-yearly meeting on 24th January, 1851. According to J. F. P. Thornhill: *Greater London* (1935), p. 59, the time taken was only 1 hour 10 minutes.

[3] Report (1846), p. 8.

[4] Fifth Report of the Commissioners for the Improvement of the Metropolis (1846) p. 4. London Traffic Branch of the Board of Trade, Third Report (1911), Appendix H, p. 194.

justified further responsibilities. But the proposal was never adopted.

By 1855 traffic congestion in the metropolis was becoming a serious problem. About 200,000 persons entered the City each day by various means. The passengers arriving by rail at London Bridge had increased from 5,558,000 in 1850 to 10,845,000 in 1854. In the same period of 4 years the numbers travelling to London on the South-Western line had risen from 1,228,000 to 3,308,000. In 1854 the passengers arriving at Shoreditch Station numbered 2,143,000, at Euston 970,000, at Paddington 1,400,000, at King's Cross 711,000, at Fenchurch Street 8,144,000.[1] About 15,000 persons travelled to work by the river steamers. The omnibuses were performing an aggregate of 7,400 journeys through the City each day, and there was also a great concourse of cabs, carts, carriages, wagons and other street traffic.[2]

The turnpike system was still in operation at the middle of the century and the turnpike tolls of the metropolis were still let by auction to the highest bidder, although some of the more progressive parishes had taken over the roads in their areas and abolished the tolls.[3] In 1855 a Select Committee of the House of Commons was appointed to consider metropolitan communications. This committee reported that little could be done by Parliament to improve matters until an authority were established on a sufficiently comprehensive basis to plan and carry out improvements adequate to meet the existing and prospective traffic needs of the capital.[4] The same year saw the passing of the Metropolis Management Act setting up the Metropolitan Board of Works, which was given extensive powers to make, widen or improve highways to facilitate communication between different parts of London.[5] At the

[1] Report of the Select Committee on Metropolitan Communications (1855), B.P.P., 1854–5, vol. x, p. iii.

[2] The horse bus started in 1829 and occupied all the main roads within a range of 4 miles from the centre of London. F. Pick: "The Organisation of Transport in London," *Journal of the Royal Society of Arts*, vol. lxxxiv, p. 212.

[3] Twentieth Report of the Commissioners of the Metropolis Turnpike Roads North of the Thames (1846), B.P.P., vol. xxiv.

[4] Select Committee of the House of Commons on Metropolitan Communications (1855), Report, p. iv.   [5] *Ante*, p. 59.

same time the local streets were left in the hands of the vestries and district boards. Sir Benjamin Hall, when introducing the Bill in the Commons, laid great stress on the bad state of the thoroughfares under the multitudinous *ad hoc* bodies then existing. At that time the short stretch of road between the beginning of the Strand and Temple Bar (1,336 yards in length) was divided among 7 separate paving boards.

The Metropolitan Board of Works constructed a number of important new main highways[1]; it also acquired under the Metropolis Toll Bridges Act, 1877, and a later statute the interests of private companies and of H.M. Office of Works in 10 cross-river bridges, which it proceeded to free from tolls. Prior to 1877 all the cross-river communications, with the exception of London, Westminster and Chelsea bridges, had been constructed and maintained by private companies.[2]

The other event of importance in the latter part of the 19th century was the appearance of the tramway. The first horse-drawn tramway was opened in 1870.[3] The Tramways Act, 1870, authorised the Board of Trade to confer powers to construct a tramway either on a local authority or on a company or person who had obtained the consent of the local authority. The Act made the Metropolitan Board of Works and the City Corporation the tramway authorities for their respective areas; and the Board then consented to concessions being granted to a number of private companies for 21 years. The London County Council began to acquire the tramways when these franchises ran out but continued to lease them to the companies for short terms. In 1896 the Council obtained powers to work tramways themselves and four years later to operate them by electricity. Hence in 1899, when part of the tramways south of the river were acquired from the London Tramways Company, the London County Council decided to operate them as a municipal undertaking. The remaining tramways on both sides of the Thames were taken over for direct operation during the next seven years.

There was never a coherent system of tramways in London even at the time of their greatest popularity. Under the Tram-

---

[1] For details, see *ante*, pp. 62–3.
[2] Royal Commission on Cross River Traffic (1926), Report, pp. 3–6.
[3] Royal Commission on London Transport (1905), Report, p. 29. An experimental line was permitted for a few months in 1861. G. A. Sekon: *Locomotion in Victorian London*, p. 92.

ways Act the consent of the minor London authorities was required before a line could be run through their areas[1]; and one-third of the frontagers in a street could veto any proposal to establish a tramway in that street. The Royal Commission on London Transport in 1905 reported that tramway development had undoubtedly been seriously checked by the operation of these vetoes which were sometimes exercised without regard to the importance of establishing adequate transport facilities.[2] One result was that within the administrative county there were 3 distinct tramway systems, all owned and operated by the London County Council, but separated from each other by long distances.[3] Outside the boundaries of the administrative county the trams were in the hands of private companies. This would not have mattered if the company systems had been worked in harmony with the municipal lines, so that through services were provided. But this was not the case. The in-county and out-county systems were not physically connected even where they met at the frontier, and in 1905 the Royal Commission on London Transport reported that in no case were through running facilities provided. In consequence every passenger was forced to change cars.[4] When the relatively small size of the administrative county is recalled, the inconvenience of this will be appreciated. Even by 1923 the difficulty had not been overcome in all districts.[5]

The underground railways, which have become of such importance in our own day, evolved from a series of shallow railways which were laid down from about 1863 onwards. They were originally scarcely differentiated from suburban lines attached to the main railway companies which linked up outlying suburbs such as Clapham, Brixton, Streatham and Tooting with the central core of London.[6] It was not until electric traction had become a practical method of operation that the

---

[1] Tramways Act, 1870, Section 4. Where the proposed tramway passes through the area of more than one local authority the Minister can make an order if the authorities along two-thirds of the proposed tramway consent.
[2] Royal Commission on London Transport (1905), Report, p. 54.
[3] Ibid., p. 42.   [4] Ibid., pp. 41–2.
[5] Cf. Royal Commission on London Government (1923), passim.
[6] F. Pick: "The Organisation of Transport in Greater London," Journal of the Royal Society of Arts, vol. lxxxiv, p. 212.

"tubes" could be bored deep down below the surface. This development began in 1890 but did not reach fruition until the opening years of the 20th century.[1]

The underground railways were designed and constructed by independent companies operating for profit, who promoted special Acts of Parliament in accordance with their own plans. There was no comprehensive examination of transport needs in the public interest at any stage during the process, and there was no continuous control by any supervising authority.[2] The London County Council had no power to intervene. It could do no more than merely oppose Bills brought forward by the promoters.[3] A Joint Select Committee of both Houses on London Underground Railways in 1901 agreed with the view expressed by the London County Council and the City Corporation that there should be a more direct control and supervision over all projects for underground railways in London; and they recommended that a Government department such as the Board of Trade, or a body of Commissioners, or a Joint Select Committee, would be the appropriate body.[4] They also strongly advised that the London County Council, the City Corporation, and the other county councils concerned, should be empowered to construct or assist in the construction of London underground railways.[5] Nothing was done in regard to either of these useful recommendations. The London underground railways were permitted to grow up higgledy-piggledy without any sort of co-ordination or general plan being devised. It has cost many millions of pounds of State-guaranteed expenditure, much unnecessary effort and the establishment of a statutory monopoly to remedy the resulting defects and deficiencies.[6]

In 1905 a Royal Commission was appointed to consider the best methods of improving London transport both above ground and below; and, in the second place, to enquire whether it was advisable to set up a central authority to control all

[1] G. A. Sekon: *Locomotion in Victorian London*, p. 159, and Chapter VIII, *passim*.
[2] Royal Commission on London Transport (1905), Report, p. 28.
[3] *Ibid.*, p. 29.   [4] *Ibid.*, p. 96.   [5] *Ibid.*, pp. 27–8.
[6] To take one example out of many, the recent merging of Holborn and British Museum Stations is said to have cost £500,000.

schemes of tramway and railway construction of a local character.

The Royal Commission on London Transport made the most thorough investigation into London's transport problem which had ever been attempted. Their report explains that the existing streets are in many instances survivals of village roads and lanes which were placed in charge of small and independent local authorities whose views did not extend beyond the requirements of the limited areas over which they had control. The lack of street planning, the irregular street widths and other causes of traffic congestion was ascribed chiefly to the fact that "there did not exist in the past any municipal or other authority having jurisdiction over the whole area, and possessed of sufficient power and resources to enable it to deal satisfactorily with the problem of locomotion, and other questions allied thereto."[1] The Commission obviously deplored the manner in which the interests of London had been neglected in this matter.[2] At the present time—1905—they noted, the width of suburban roads was being defined merely with reference to local convenience and the wishes of the owners of building sites. In consequence, the same want of forethought and the same lack of provision for arterial necessities which had produced the chaos from which the metropolis was then suffering were being repeated in the outlying areas slightly further away from the centre of the town.[3]

The Royal Commission declared that a permanent body was required to deal with questions of locomotion in Greater London.[4] It was not possible, however, in their opinion to confer on the London County Council the necessary powers to deal with Greater London traffic in view of the friction which would arise with other local authorities in the area. They did not suggest an enlargement of the territory of the London County Council or the setting up of a Greater London Council, but proposed the creation of a permanent traffic board consisting of 5 members, including possibly one representative from the London County Council and one from the City Corporation. The functions of this board would include the examination

[1] Royal Commission on London Transport (1905), Report, p. 17.
[2] Ibid., pp. 95–6.   [3] Ibid., p. 40.   [4] Ibid., pp. 1–2.

and revision of all public or private Bills relating to transport or communications in London; the improvement of main roads out of London; the supervision of bye-laws regulating building in undeveloped districts in Greater London; the consolidation and amendment of traffic laws and regulations; the promotion of amalgamations and joint-working between passenger transport undertaking; and the better ordering of the arrangements relating to the breaking up of streets.[1]

The proposal to create an *ad hoc* body to deal almost entirely with traffic and locomotion is particularly surprising in view of the fact that the Royal Commission fully appreciated the intimate connection between transport and housing. They drew attention to the fact that the railway companies were neither expected nor required to provide cheap trains in order to open up new working-class residential districts, and emphasised that in order to relieve overcrowding in the central area it would be necessary to provide improved means of transportation into and out of London, both as regards speed and cheapness.[2] Yet the relation between town planning, housing, the provision of open spaces and the other social services for which the London County Council or the local authorities were responsible on the one hand, and the functions appertaining to highways and transport on the other, seem never to have been understood by the Commission.

The other great mistake which the Royal Commission made was in failing to envisage the potential importance of the petrol motor. They considered the question of retarding tramway development for the time being in order to test the possibilities of the new method of traction; but in the end they jumped to the conclusion that "tramways will continue to be the most efficient and the cheapest means of street conveyance, and we cannot recommend the postponement of tramway extension in London on the ground of any visible prospect of the supersession of tramways by motor omnibuses."[3]

The motor omnibus appeared on the London streets in 1899; and there were nearly 250 of them in service when the Royal

[1] Royal Commission on London Transport (1905), Report, pp. 97–102.
[2] *Ibid.*, p. 16.   [3] *Ibid.*, p. 42.

Commission issued its report. In 1907 the number rose to about 800; and in 1908 to 1,000.[1]

The administrative proposals of the Royal Commission were ignored, but in 1907 the London Traffic Branch of the Board of Trade was instituted. Its functions were limited to the consideration of new schemes of locomotion seeking statutory authority, so far as they might come within the scope of the Board of Trade; the collection of information and the presentation of an Annual Report to Parliament on the subject of London traffic; and the carrying out of any additional duties with which it might be entrusted.[2] This was a small step which did little to achieve the comprehensive purposes at which the Commission had aimed. Even the costly and obstructive nuisance of unnecessarily frequent road upheavals was not dealt with. The London Traffic Branch pointed out in 1914 that the right to break up the London streets was not confined to highway authorities but was also possessed by the Postmaster-General, the Metropolitan Water Board, certain water companies, 7 gas companies, 15 electric light companies, and 17 tramway companies, operating under 14 separate Acts of Parliament containing divers procedures.[3]

Thus matters were allowed to drift on into the motor-car era without any effective steps being taken to deal either with the communications of the metropolis or the organisation of its public transport on bold or imaginative lines. The early railway age undoubtedly missed a great opportunity for planning a unified railway system for London with interconnected main terminals and goods yards.[4] The early petrol age no less certainly missed another great chance for planning a fine system of highways and encircling thoroughfares to carry the new vehicular traffic with speed and convenience.

It cannot be said that the failure to act was due to ignorance or unawareness of the need. Already in 1911 the London Traffic Branch of the Board of Trade was complaining in its Annual Report that the defects of the road system of the

---

[1] London Traffic Branch of the Board of Trade, First Annual Report (1908), p. 22.   [2] Ibid., p. vi.
[3] Ibid., Sixth Annual Report (1914), B.P.P., vol. xli, p. 11.
[4] J. F. P. Thornhill: Greater London, p. 105.

metropolitan area was due to want of foresight "and from neglect to take timely steps, such as have been taken in other great capitals, to provide for the needs of the future by laying down a plan to which expansion could be made to conform." The process of repairing avoidable mistakes, continued the Branch, "which is slow and enormously costly, furnishes a warning which ought not to be disregarded in view of the fact that the mischief is still growing, but no plan has yet been made, and that the same difficulties which have arisen in the centre of London are being reproduced to-day in the out-skirts."[1] One would have thought that influential Englishmen travelling abroad could scarcely have failed to notice the superiority in this respect of foreign cities such as Paris or even Barcelona, and been spurred to demand drastic action on their return to London. But no emulation of this kind took place; and the London Traffic Branch of the Board of Trade, like so many Royal Commissions and other advisory bodies in the past, was allowed to remain a voice crying in the wilderness.

In 1923 the Royal Commission on London Government[2] recommended that an advisory committee should be set up to advise the appropriate Government department and to co-ordinate schemes of transport, town planning, housing and drainage. In 1924, after the first Labour Government took office, an omnibus strike occurred, in the course of which the great privately owned traffic combine undertook to concede the workers' demands provided that the companies were protected from further competition on the part of the so-called pirate motor-buses owned by small proprietors, who were boring in on the most profitable routes in the centre of London. The London Traffic Act was then passed in partial fulfilment of the Royal Commission's recommendations.[3] It set up the London and Home Counties Traffic Advisory Committee to report to and advise the Minister of Transport on questions relating to

---

[1] London Traffic Branch of the Board of Trade, Third Annual Report 1910), B.P.P., vol. xxxiv, p. 2.

[2] Report of the Royal Commission on London Government (1923), pp. 75–9. See below, p. 308–9.

[3] For an account of the original constitution of the Committee, see Percy Harris: *London and its Government*, pp. 158–65.

traffic and transport within a large area covering more than 1,800 square miles.

The most significant feature of the Act was the power it conferred on the Minister of Transport to schedule certain routes as restricted streets for the purpose of limiting the number of omnibuses which could ply for hire thereon.[1] This was the beginning of the monopolistic restriction of surface transport in London, a step which was claimed to be necessary in order to relieve traffic congestion and to maintain prosperity among transport operators so as to enable them to provide better conditions of employment.

The whole of London's passenger transport system was constructed and managed by profit-making companies with the exception of the tramways, most of which were owned by the London County Council and other urban local authorities in the outlying areas. The London County Council had endeavoured to run omnibuses to connect with their tramways, but had been restrained by the Courts on the ground that they were acting in excess of power;[2] and the Council never obtained Parliamentary authority to enter this field.

From the time when the London and Home Counties Traffic Advisory Committee was established the continuance of the limited amount of competition which then existed was doomed to extinction. During the next few years the great amalgamation of companies known as the traffic combine became the protagonists of the idea of unified management,[3] a principle capable of many different interpretations. In 1931 Mr. Herbert Morrison, Minister of Transport in the second Labour Government, introduced the Bill which ultimately became the London Passenger Transport Act, 1933. It was piloted through its subsequent stages after the formation of the National Government first by a Liberal Minister and later by a Conservative Minister of Transport; but with the exception of one or two important changes the measure became law in substantially the same form as that in which it was introduced.

[1] London Traffic Act, 1924, Section 7.
[2] *London County Council* v. *Attorney-General*, [1902] A.C., p. 165.
[3] Percy Harris: *op. cit.*, p. 161; Ernest Davies: "The London Passenger Transport Board" in *Public Enterprise* (edited by W. A. Robson), pp. 157-9.

The Act set up a board to take over and administer the whole mass of London passenger transport undertakings, including the underground railways and tubes, the Metropolitan railway, the motor-buses owned by the London General Omnibus Company, Thomas Tilling and the smaller concerns, the Green Line and other motor-coach companies, the municipal tramways owned by the London County Council and thirteen other local authorities, and also those belonging to the combine. More than £100 millions of capital equipment was thus transferred. The only forms of public transport excluded from the new organisation were the main-line railways, with whom a scheme for pooling receipts from local traffic was required to be arranged, the taxicabs and private hire motor-cars, and the motor-coach lines running to provincial towns. Apart from these exceptions the London Passenger Transport Board was given a monopoly in respect of all passenger transport within its area.

The Board consists of a chairman and 6 other members appointed by a body of Appointing Trustees. These trustees are the Chairman of the London County Council, a representative of the London and Home Counties Traffic Advisory Committee, the Chairman of the Committee of London Clearing Bankers, the President of the Law Society and the President of the Institute of Chartered Accountants in England and Wales. The members of the Board are required to have had experience and shown capacity in transport, industrial, commercial, or financial matters or in the conduct of public affairs. Two of their number must have had not less than 6 years' experience in local government within the area of the Board.[1]

There is much that can be said and has been said concerning this Board and the constitutional development of which it is a part.[2] And—as in the case of the metropolitan water and dock undertakings—the compensation paid to the shareholders of

[1] Considerable changes were introduced in the status and method of appointment of members by the Transport Act 1947.

[2] See Ernest Davies: "The London Passenger Transport Board" in *Public Enterprise* (edited by W. A. Robson), pp. 154–208; and the concluding chapter to that volume; Terence O'Brien: *British Experiments in Public Ownership and Control*; Herbert Morrison: *Socialization of Transport*.

the companies was clearly exorbitant. This, however, is not the place to discuss these matters at length. It is necessary to introduce the London Passenger Transport Board into this study because it touches a vital public service in London, but at this stage I shall do no more than draw attention to two points deserving attention. The first is that the Board was given a very large area of operation extending over nearly 2,000 square miles. The second is that the connection between the elected local governing bodies in this area and the appointed Board is so slender as to be almost negligible. All that is required is the almost nominal qualification of 6 years' experience in local government *at any time* on the part of two members of the Board, together with the inclusion of the Chairman of the London County Council among the appointing trustees.

It is interesting to recall that this experiment was the work of Mr. Herbert Morrison, most of whose public life has been spent in local government in London and whose devotion to democracy and to local government is unquestionable. "Had there been a general municipality for Greater London," writes Mr. Morrison, "it would have been possible for the provision and management of public transport to be vested in it," although he adds that it is not certain that such a huge commercial undertaking would be appropriate as part of the ordinary machinery of local government.[1] However that may be, it is clear that the existing organisation of local authorities in the metropolitan area in 1931 (or at any other date for that matter) made it impossible for a moment to consider transferring the privately-owned transport undertakings to municipal ownership and control.

In the process of establishing the London Passenger Transport Board Mr. Morrison took the opportunity of enlarging and making permanent[2] the London and Home Counties Traffic Advisory Committee, and of giving it a predominantly municipal character. Its 40 members hold office for a term of 3 years and are appointed by the following authorities: By the

[1] Herbert Morrison: *How Greater London is Governed*, p. 130.
[2] The Committee had been established in 1924 for a limited number of years.

Home Secretary (1), by the Minister of Transport (1), by the London County Council (6), by the City Corporation (1), by the City of Westminster (1), by the other Metropolitan Borough Councils (6), by the County Councils of Middlesex (2), Essex, Kent and Surrey (1 each), and Bucks and Herts (1 together), by the county borough councils of Croydon, East Ham and West Ham (1 each), by the London Passenger Transport Board (2), by the amalgamated railway companies (2), by the Minister of Labour to represent the interests of labour engaged in the transport industry within the London Traffic Area (5), by the Minister of Transport to represent the interests of (a) persons providing or using mechanically-propelled road vehicles, (b) persons providing or using horse-drawn road vehicles and (c) the taxicab industry within the area (3). There are also 3 members appointed to represent the various police forces concerned with traffic regulation. Two of these are chosen by the Home Secretary to represent respectively the Metropolitan Police Force and the extra-metropolitan police forces within the area; while the third is appointed by the City Corporation to represent its own police.

The statutory functions of the Advisory Committee were also extended so that they can now advise and report to the Minister of Transport on any matters relating to traffic, transport and highways within the London Traffic Area and make representations to the London Passenger Transport Board on any matter connected with the Board's services and facilities within the Traffic Area which in their opinion should be brought to the Board's notice.[1]

Under the earlier Act of 1924 the Advisory Committee were charged with considering the co-ordination of the various forms of transport services, the causes hindering the free circulation of traffic, the comparative desirability of different means of transport, and other specific matters which the Minister might refer to them.[2]

The London and Home Counties Traffic Advisory Committee group their reports under four heads, dealing respectively with road and bridge improvement works, traffic control,

[1] London Passenger Transport Act, 1933, Section 59; see also London Traffic Act, 1924.   [2] London Traffic Act, 1924, Section 2.

road safety and passenger transport. Their recommendations cover a wide variety of topics, ranging from the position of public telephone kiosks to the provision of pedestrian guard rails. They include such matters as the cross river facilities in East London, the closing of streets for repair works, the stopping places for omnibus and trolley cars, traffic light controls, the delivery of coke, parking places, the working of one-way streets and roundabouts, the setting aside of play streets for children, the fares charged on a particular omnibus route, and minimum fares for school children.[1]

In its earlier years the Traffic Committee dealt with some of the larger and more important questions relating to public transport within the metropolis, but in recent times it has tended to confine its attention to relatively minor points of detail. In its Report for 1932-3 the only matters dealt with in this sphere were the provision of special travelling facilities for blind persons on tramcars and for old age pensioners in West Ham, and the renaming of Enfield West railway station.

In regard to highways the Committee has put on record in its Report for 1934-5 a notable opinion concerning its own ineffectiveness. After expressing the view that the road facilities in Greater London have in the past decade not been commensurate with the remarkable expansion of population in the metropolis,[2] the Committee declared: "We cannot feel that the results of our labours are wholly satisfactory. We are charged with the duty of reporting to you and advising you on any matters relating to traffic within the London Traffic Area which in our opinion ought to be brought to your notice, and in this connection we feel it our duty to inform you that we are of opinion that a complete plan or programme of improved means of road communication is a matter of urgent necessity."[3]

Shortly after this confession of failure, the Minister of Transport announced that he had appointed Sir Charles Bressey, the chief engineer of the Roads Department at the Ministry, to devote himself solely to the problems of highway

---

[1] See, for example, the Tenth Annual Report of the Committee, 1934-5 (H.M.S.O., 1936), *passim*; Ernest Davies: *op. cit.*, pp. 178-9.
[2] Tenth Annual Report of the London and Home Counties Traffic Advisory Committee, 1934-5, p. 5.   [3] *Ibid.*, p. 19.

development in the metropolitan region and to prepare a highway plan covering the next twenty or thirty years. This was the origin of the Bressey Report.

Sir Charles Bressey conferred with the Advisory Committee during his deliberations, and the Minister stated that the Bressey Report would be referred to the Committee when it was prepared. But no amount of "consultation" of this kind can hide the fact that the Traffic Advisory Committee has not proved successful as a device for solving the essential problems of highways and traffic in Greater London.

The proposal advocated by the Royal Commission on Transport for a comprehensive road authority responsible for the whole network of highways in Greater London has thus never been achieved. But the creation of a statutory committee to advise the Minister and the appointment of Sir Charles Bressey indicate a strong tendency for centralised devices to replace the machinery of local government in this branch of administration owing to the failure to adapt the structure of London government to the facts of social change.

Discussion of the Bressey Report will be reserved to a later page.[1] In the meantime, we may note there is yet another authority in London concerned with traffic regulation: namely, the Metropolitan Traffic Commissioner appointed under the Road Traffic Act, 1930. His principal functions consist of granting public vehicle licences for tramcars, trolley-buses, motor-omnibuses and coaches, and in issuing road service licences to operate vehicles over approved routes. He is endowed with wide discretionary powers to attach conditions to such licences. The licensing of taxicabs and cab-drivers, as well as of omnibus drivers and conductors, is in the hands of the Commissioner of Metropolitan Police.

[1] Post, p. 420 *et seq.*

CHAPTER XV

A WORD ABOUT OMISSIONS

Considerations of space make it impossible for me to deal with all the other services which have suffered in one way or another from the mismanagement and neglect of London by Parliament during the 19th and 20th centuries. It is necessary to select; and I have therefore concentrated on some of the principal functions—in particular, those in connection with which a public authority has sooner or later been established.

Much might be said, however, of some of the other services, such as gas supply, where municipal activity was effectively excluded.

The behaviour of Parliament towards the metropolis during the 19th century is difficult to explain except in terms of the lobbying activities of the public utility companies which plundered the consumer in the wealthiest city in Europe. It was obvious to the astute business men who directed the fortunes of these companies that if London were to obtain an organ of local government suited to its needs or in any way comparable with the municipal council of a great provincial city the privately-owned utility undertakings would almost certainly be municipalised. Hence they not unnaturally did their utmost to prevent such a body from coming into existence.

This aspect of the situation was perfectly well understood by politicians and students of public affairs conversant with the position during the latter part of the 19th century. *The Times*, for example, in a principal leading article on the reform of London government published on 7th October, 1874, wrote that the gas supply in London was admittedly bad. Various remedies had been proposed, among them the establishment by the Metropolitan Board of Works of a rival public supply. This, said the newspaper, was not a desirable solution save as a menace to be held over the existing gas monopolists. "A better and safer remedy," continued *The Times*, "would be that the metropolis should . . . buy up the gas companies and itself conduct their business. The supply of good gas and good water to the metropolis ought to be the concern of a municipality, and neither work could properly be carried out except by the City of London and the metropolis combining to carry it out. Though we care very much for bright gas and cheap gas, it is, indeed, chiefly from this point of view that we value the present movement against gas as it has been. Our present sufferings, however unpleasant at the time, will not have been without their use if they give an impetus to the pending project for a reform of the municipal government of London." Mr. Gladstone, speaking as Prime Minister in support of Harcourt's Bill for the reform of London government in 1884, reminded the House of Commons that the supply of water and gas, "two of the most elementary among the purposes of municipal government, have been handed over to private corporations for the purpose of private profit because you have not chosen to create a complete municipality for the metropolis."[1]

I mention these matters only to guard myself against criticisms that I have omitted to deal with this or that service. I cannot claim to be exhaustive. I have merely endeavoured to trace the main trends in the development of London government; and to show how the various authorities emerged.

Without some knowledge of the historical evolution of these institutions during the past hundred years it is impossible to understand the situation as it exists today in London. To a consideration of that situation we can now turn.

[1] Hansard: *Commons Reports* (1884), vol. 290, col. 550.

Part II

THE PRESENT

CHAPTER I

CONFLICT AND CONFUSION

London is by far the largest and most important capital city in the world.[1] Her primacy is based on a singular combination of circumstances not to be found elsewhere. She is at once the political capital of Britain and the British Empire, the judicial centre, the cultural centre, the commercial centre, the greatest manufacturing city in England, the world's largest port, and—despite several vicissitudes since the War of 1914–18—the world's principal financial centre.

Hence one finds in London an extraordinary concentration of wealth and fashion, industry and commerce, art and science, political power and administrative authority. There are Parliament, the Cabinet and the great executive departments of state. There is the Court at Buckingham Palace and the entourage clustering round St. James's. There are the Bank of England, Lloyd's, the Royal Exchange, the Baltic and the other celebrated markets in which business is conducted on a world-wide scale. There are the British Museum, the Royal Society, the Record Office and the other precincts of learning. There are the great teaching and research hospitals, the Royal Colleges of Physicians and of Surgeons, the concentration of leading physicians and surgeons in Harley Street. There are the colleges of the University, scattered through the city but now converging on the fine central site which American generosity has provided for the University headquarters. There are the arenas of sport at Wimbledon, Wembley, Henley, Ranelagh, Hurlingham, Lords, and the Oval. These are but some of the worlds which revolve and have their being within the solar system of the English metropolis. In their several ways they make London the centre of gravity in English life for commerce, finance, government, art, literature, law, medicine, entertainment, sport and social life generally.

Having said so much, can one go further? Can one say that

[1] New York is excluded since it is not a national capital. As a metropolitan centre, however, New York is London's only serious rival.

London is a city of grace and delight and convenience, a city of noble streets and splendid buildings, a place to inspire enthusiasm and admiration in all who behold it? Can one say that London is a well-governed city, a model of civic pride and virtue?

The answer must be a negative one. London is indubitably the largest and most important capital in the world. She is equally clearly one of the worst planned, or rather most unplanned of cities; and her local government administration is unsatisfactory as regards its organisation and basic principles.

In the first part of this work I described the creation during the past century of the various administrative organs, and gave some account of the historical setting which formed the background. In this second part I shall deal with the contemporary activities of these institutions, and analyse the government of London as it exists today.

CHAPTER II

A JUMBLE OF AUTHORITIES

At the last census in 1931 the population of Greater London[1] was 8,203,942 and that of the administrative county 4,397,003. By 1936 the total was estimated to have risen to 8,575,700. The number of persons living in the administrative county was estimated to have fallen to 4,141,100, which was less than the number in outer London (4,434,600) by nearly 300,000. This compares with a population in Greater Paris of 4,933,855 (1931), of which 2,891,000 is contained within the city walls. The population of Berlin in 1933 was 4,190,847;[2] that of New York was 6,900,000 in 1930,[3] and estimated to be 7,434,346 in 1937. Greater London contains more persons than any one of 15 European states, including Holland, Belgium, Greece, Finland, Bulgaria, Norway, Portugal, Sweden, Switzerland and Denmark. It comprises approximately one-fifth of the population of Great Britain. In 1801 there were fewer persons living in the whole of England and Wales than are today inhabiting Greater London.

In 1837 the area covered by London, including the outlying arms stretching northwards, was about 22 square miles.[4] Today the metropolis presents a continuous built-up territory exceeding 400 square miles, extending beyond Enfield in the north to Croydon in the south, and from Hounslow in the west to Ilford in the east; while beyond this there are numerous expanding suburban settlements and dormitory towns situated

[1] Greater London comprises the Metropolitan and City Police districts.
[2] *London Statistics* (1934–6), vol. 39, pp. 35–6; Sarah Greer: *Outline of Governmental Organisation within the Cities of London, Paris and Berlin*, prepared for the New York City Charter Revision Commission by the Institute of Public Administration, N.Y. (1936).
[3] The comparative rate of growth between these four cities is as follows:

	1800	1880	1910
Paris	647,000	2,200,000	3,000,000
London	800,000	3,800,000	7,200,000
Berlin	182,000	1,840,000	3,400,000
New York	60,000	2,800,000	4,500,000

Sir Walter Besant: *London in the Nineteenth Century*, p. 4.

along the railway lines.[1] The outer London districts began to leap ahead about 1881, a few years before the London County Council was established. They have grown rapidly ever since; with the result that there are now a larger number of Londoners living outside the County of London than within it.[2]

The area of the administrative county consists of less than 117 square miles, and is almost the same as that which was given to the Metropolitan Board of Works in 1855 and its predecessor the Commissioners of Sewers for the Metropolis in 1848. The county measures $16\frac{1}{4}$ miles in extreme length from east to west; and $11\frac{3}{4}$ miles in extreme breadth from north to south.[3] In more than 90 years there has been no extension of boundaries, and the London Government Act, 1899, actually caused a reduction of the area by 621 acres.[4] The boundaries of the present county did not correctly represent the limits of the metropolis even when they were first laid down in 1848. They were certainly already obsolete in 1888, when the London County Council was established. The London County Council's area is about one-sixth of the Metropolitan Police District—i.e. Greater London.

We may consider now the areas of some of the other local governing bodies operating in the region. There is the Metropolitan Police District, an area of 692 square miles, situated within a radius of between 12 and 15 miles from Charing Cross, and containing 8,192,943 persons. It extends from Cheshunt in the north to Epsom in the south, and from Staines in the west to Romford in the east. There is the area of the Metropolitan Water Board, 573 square miles with a population of 7,332,400, extending from Ware to Sevenoaks. Although smaller than the Police District it includes certain regions lying beyond it. There is the area of the London Passenger Transport Board, 1,986 square miles in extent and containing 9,140,000 persons, of which 1,551 square miles (about 80 per

[1] J. F. P. Thornhill: *Greater London*, p. 72; H. P. Clunn: *The Face of London*, p. 13.
[2] *Twenty-five Years of London Government*, L.C.C. Publication, No. 3117/1935, p. 11.
[3] *Facts for Londoners*, Fabian Tract, No. 8.
[4] Report of the Royal Commission on London Government (1923), p. 2.

cent) form the "special area" within which the Board has exclusive transport rights.[1] It stretches 32 miles to north and south of the centre and 25 miles to east and west. There is the area of the London and Home Counties Traffic Advisory Committee, 1,820 square miles, containing 9,144,000 persons, and that of the Metropolitan Traffic Commissioner, 2,417 square miles, containing 9,462,000 persons. The catchment area under the Thames Conservancy Board comprises 3,843 square miles with an unstated population, while the Lee Conservancy Board exercises its functions over a catchment area of 548 square miles. "Drainage London"—the area served by the main drainage system operated by the London County Council —covers 178 square miles and includes 5,529,500 persons. It extends from Wood Green to Penge, and from Acton to Woolwich. It is about half as large again as the administrative county. "Electricity London," the area of the London and Home Counties electricity district, comprises 1,841 square miles and reaches out to Reigate and Hertford. It contains 9,088,600 persons.[2]

The area known as Greater London[3] contains the following medley of local government bodies: the London County Council; the City Corporation; 28 metropolitan borough councils; the Metropolitan Water Board; the Port of London Authority; the Thames and Lee Conservancy Boards; 5 county councils (Essex, Herts, Kent, Middlesex and Surrey); 3 county borough councils (Croydon, East Ham and West Ham); 35 municipal borough councils; 30 urban district councils; 4 rural district councils and 6 parish councils.[4] The total is 117 separate

[1] "The population of the London Passenger Transport Board area is now estimated at 9,500,000, and is thought to be the largest metropolitan aggregation in the world, unless one includes in greater New York certain parts of New Jersey. The 1,551 square miles of the London Passenger Transport Board territory in which it has exclusive powers is known as its 'Special Area.' " F. Pick: "The Organisation of Transport," *Journal of the Royal Society of Arts*, vol. lxxxiv, p. 208.

[2] For full information, see *London Statistics* (1934–6), vol. 39, pp. 12–17. A. Emil Davies: *The Story of the London County Council*, p. 13.

[3] i.e. the Metropolitan Police area described on p. 164.

[4] For details, see *London Statistics* (1935–7), vol. 40, pp. 1–6, 28–31. Of the local authorities mentioned above, 2 boroughs, 8 urban districts and 3 rural districts are partly within and partly without the area of Greater London.

organs, of which all but 4 are directly elected.[1] I have taken no account of numerous derivative bodies constituted for particular purposes, such as the assessment committees, county valuation committees and local pension committees, which would make a formidable addition.

Of the total, there are 30 elected bodies within the London County Council's area and 83 outside it. The county councils of Hertfordshire, Surrey, Essex and Kent have areas extending far beyond Greater London.

There is, however, no particular reason why one should confine the list to Greater London; for Greater London, as officially defined, like the City of London and the County of London, has in its turn become obsolete.[2] As I have shown, much larger areas have been carved out for such functions as passenger transport, traffic control and electricity supply. Suburban development has now reached the coast. More than 10,000 passengers travel to the metropolis every week-day from Brighton, and 20,000 from Southend.[3] A writer with a speculative turn of mind asks whether the aeroplane may lead Londoners to dwell as far away as the coast of France, which would then become dotted with London suburbs.[4] It used to be said that before 1855, outside the narrow limits of the City Corporation, London was nothing more than a geographical expression.[5] The same may be said today of London outside the narrow limits of the administrative county.

I do not propose to inflict upon the reader a catalogue of the powers and duties of the various classes of local authority operating in the metropolitan region; nor with a description of the great diversity in size, resources and character between the various classes of councils and also among authorities belonging to the same constitutional category. There are reliable text-books and manuals for those who seek information on these matters. My concern is with the fundamental problems

[1] Of these, 4 county councils, 3 municipal boroughs, 7 urban district councils and 3 rural district councils exercise jurisdiction outside Greater London as well as within it.

[2] Herbert Morrison: *How Greater London is Governed*, p. 122.

[3] F. Pick: *op. cit.*, p. 215.

[4] J. F. P. Thornhill: *Greater London*, p. 73.

[5] Sidney Webb: *The Reform of London* (pamphlet), p. 4.

relating to the metropolis, and I wish to avoid burdening the discussion with other matters.[1]

The day population of the County of London, and especially the more central portions of it, is very different from the resident night population. On every week-day an enormous army of workers invades the industrial and commercial heart of the metropolis, hurling itself through the traffic arteries by train, underground, omnibus, tramcar, motor-car and bicycle. The size of this army was found to be in excess of three-quarters of a million in 1921; today it probably numbers at least a million.[2] During its daily sojourn this vast influx has to be provided with all kinds of costly municipal services supplied by the London County Council or the metropolitan borough councils—fire brigades, street cleansing, ambulances, highways, bridges, sanitation, to mention only a few. At the moment when occupation involves liability to contribute to municipal taxation, these million or so workers pass over the border to the out-county areas where they reside and thereby avoid payment of rates in respect of their daily sojourn. The glaring injustice which results from casting most of the financial burden on the inhabitants of the administrative county is unaffected by the argument that these migrating workers help to produce the wealth on which the County of London depends.

Anomalies of this kind must always occur where the work-places are municipally separate from the areas of residence. Hence it is a principle of good government that administrative areas should become larger as the means of communication improve, in order that the areas of political organisation should comprehend the areas of diurnal movement made by the people.

[1] For a full discussion of the general problems arising out of the present organisation of local government, see *The Development of Local Government*, by W. A. Robson, Part I.

[2] Census, England and Wales, 1921. General Report, 1927, pp. 193–7. See also workplaces volume of the 1921 Census. Similar information was not given in the 1931 Census. The actual figure in 1921 was in excess of 800,000. These persons came into the area of the City Corporation and the Inner Ring of metropolitan boroughs to work. Their places of residence lay either outside the County or in the Outer Ring of metropolitan boroughs. By now, however, transport facilities have so greatly improved that a much larger number comes in daily from outside the County. *Cf.* my *Development of Local Government*, pp. 97–102.

There has been no attempt during the past three-quarters of a century to apply this principle to the general authorities responsible for local government in the metropolis. A series of legislative convulsions has, however, thrown up a number of *ad hoc* bodies of much greater dimensions than the general local authorities, to deal with water, police, the docks, transport and the like. There have, in consequence, been two opposing tendencies operating to produce the situation which now confronts us. On the one hand, a merging tendency which led to the London County Council absorbing the work of the Poor Law guardians, the Metropolitan Asylums Board, and the London School Board. On the other hand, a tendency towards separation of function reflected in the establishment of the Metropolitan Water Board, the Port of London Authority, and the London Passenger Transport Board.

Owing to the fact that Parliament has failed to deal with the general organisation of London government in a coherent way, a number of extraordinary expedients have been devised to overcome some of the serious disadvantages of the present situation. Thus, the London County Council is empowered to provide open spaces outside its own area and has purchased a number of parks beyond the county boundary. Among them are Finsbury Park (115 acres), Golders Hill (36 acres), Hainault Forest (1,108 acres), Hampstead Heath extension (80 acres) and Marble Hill (66 acres).[1] The City Corporation has acted in a similar manner and acquired such important open spaces as Epping Forest (6,000 acres), Burnham Beeches (480 acres), a stretch of downs and commons at Coulsdon and Purley (391 acres), and several smaller parks and recreation grounds.[2]

In regard to main drainage, the London County Council has been authorised to offer facilities to out-county districts, and at the present time no less than 24 local authorities outside the county have arranged for sewage from their areas to be admitted to the London County Council's main drainage system, in exchange for payments aggregating £115,000 a year.[3]

Most important of all is the way in which the London County Council has been driven to build great housing estates outside

[1] *London Statistics* (1934–6), vol. 39, pp. 170–1.   [2] *Ibid.*, p. 172.
[3] *Ibid.*, p. 127.

its own area, owing to there being no vacant land within its territory suitable for the purpose. The principal example of this is the estate at Becontree in Essex, 2,770 acres in extent, on which the London County Council has provided housing for 115,000 persons. Here is a new town as large as Bournemouth equipped with playing fields, open spaces, public, houses, cinemas, shops, churches, schools and hospitals. A number of large factories have been built in the neighbourhood.[1]

The London County Council has also developed large housing estates at other places outside the county, such as St. Helier (843 acres), Downham (522 acres), Mottingham (202 acres), Bellingham (252 acres), Watling (390 acres), Roehampton (147 acres), Wormholt (68 acres) and Castelnau (51 acres). In making these municipal settlements the London County Council has become in effect a colonising power, like ancient Rome; pouring out the treasure and labour of her citizens in order to make homes for them in foreign lands. It is highly unsatisfactory for the inhabitants of the administrative county to be asked to spend millions of pounds in this way to build up rateable value in the areas of outlying local authorities. Those outlying local authorities are in their turn faced with immense administrative problems, since they are required to provide schools, hospitals,[2] roads, and many other municipal services for masses of working-class Londoners who are suddenly dumped upon them by the London County Council and who are frequently regarded as unwelcome both for financial and social reasons. Conflict and friction easily arise in such circumstances between the London County Council and the other local authority concerned. Alderman Emil Davies tells us that even when the L.C.C. develops a housing estate within its own area, it may have to fight on hundreds of details with the borough council which has control over that particular district.[3] There is much greater disharmony when the estate lies outside the county boundaries, since the London County Council is regarded as an interloper and resented accordingly.

---

[1] *Twenty-five Years of London Government*, L.C.C. Publication, No. 3117/1935, p. 15. For a discussion of Becontree and Dagenham see *post* p. 431 *et seq.*

[2] Occasionally the London County Council may assist with a financial contribution, e.g. towards the cost of hospitals.

[3] A. Emil Davies: *The Story of the London County Council*, p. 15.

There are various other matters in regard to which the London County Council has been asked to provide services, or to make a financial contribution, for the benefit of out-county areas. For example, pupils from out-county districts are admitted to the technical institutes maintained by the London County Council; and the London County Council has contributed to the cost of main roads outside its borders.

The London County Council is not the only authority to administer services outside its own area. The City Corporation has the right to own markets within a radius of seven miles, and in pursuance of this privilege it maintains the Spitalfields fruit and vegetable market in Stepney and the metropolitan cattle market and abattoirs in Islington. The City Corporation is the sanitary authority for the whole port of London, extending from Teddington to the sea.

The existing areas into which the London Region is divided are not merely far too small in many instances to provide optimum efficiency; they are also irregular, illogical and inconvenient. The Edgware Road, for example, lies both in Middlesex and the county of London, with the result that cinemas on one side of the street may be open on Sunday while those on the other side may be closed.[1] The metropolitan boroughs range from small districts such as Finsbury, which contains a population of 69,000 living in an area of 587 acres, or Holborn, with 38,860 population in 406 acres, or Stoke Newington, with 51,208 in 864 acres, to much larger areas like Hackney, which has 215,000 persons and 3,287 acres of territory, Lambeth with 296,000 persons and 4,083 acres, and Wandsworth, whose population reaches 353,000 in an area of 9,107 acres.[2] The financial rescources of these districts vary enormously.

Even the great new areas that have been recently laid down for particular purposes can scarcely be regarded as conceived on scientific lines. The London Passenger Transport area, observes Mr. Frank Pick, the Vice-Chairman of the Transport Board, "is an accident arising out of the way in which the various undertakings serving it in the past were affiliated and

---

[1] Royal Commission on London Government (1923), Minutes of Evidence, p. 372. Sir Herbert Nield. [2] *Post*, p. 359.

grouped. It is not a logically defined or conceived area."[1] The boundary of the Metropolitan Police District appears to be drawn in the most arbitrary manner. I understand on high authority that the area assigned to the Metropolitan Water Board is considered unsatisfactory from a technical point of view. It comprises 573 square miles but should, it is said, be extended to cover the Thames and Lee Catchment areas, which cover respectively 3,843 square miles and 548 square miles.

The distribution of powers among all the multitudinous authorities which litter the metropolitan region proceeds on no orderly plan. I shall deal with this at length later on. Here I will do no more than mention one example. Is it not anomalous that four of the cross-river bridges should be maintained by the City Corporation, and the remainder by the London County Council, while the roads leading from them are under the control of yet other local authorities?[2] A similar confusion exists in regard to the demarcation of areas. The municipal housing estate at Becontree is situated in the areas of three contiguous councils—Barking, Ilford and Dagenham. Hence there is no single authority responsible for water supply, main drainage, parks and recreation grounds, education and amenities. Moreover, different requirements are laid down by the by-laws of these three authorities.[3]

In the circumstances, it is scarcely surprising that interest in civic affairs should run low in the metropolis. The percentage of electors voting in the London County Council elections in recent years has been as follows: 36.8 in 1922, 30.6 in 1925, 35.6 in 1928, 27.8 in 1931, 33.5 in 1934, 43.4 in 1937, and 26.4 in 1946.[4] In the metropolitan borough council elections the percentage was a trifle higher: 36.4 in 1922, 42.5 in 1925, 32.3 in 1928, 31.3 in 1931, 34.3 in 1934, 35.4 in 1937, and 35.1 in 1945. In the ancient City a contested election is a rare event. The election of common councilmen takes place each year.

[1] F. Pick: "The Organisation of Transport," *Journal of the Royal Society of Arts*, vol. lxxxiv, p. 208.
[2] A. Emil Davies: *op. cit.*, p. 15.
[3] Royal Commission on London Government (1923), Minutes of Evidence. Sidney Webb, p. 819.
[4] *London Statistics*, vol. 39, p. 21; Return of General Election of London County Councillors, 1937, No. 3270. For 1946, see a similar Return No. 3570. For Metropolitan Borough Councils see Return No. 3526 of 1946.

There was no contest in 1935, and in 1934 a contest took place in only one ward, where there were ten candidates for eight seats. Prior to that there had been no contest since 1927.[1] In extra-London the figures are equally uninspiring. The percentage voting at the election for the three county boroughs in 1935 were Croydon 35·5, East Ham 36·3, West Ham 27·2, an average of 33·6. The average percentage voting at the 1935 elections in municipal boroughs was 32·9, and in urban districts 27·8.[2] The corresponding figures for some of the provincial cities are given by way of comparison.

	Percentage of Electorate Voting[3]					
	1932	1933	1934	1935	1936	1937
Birmingham ..	36·2	36·0	33·0	36·26	31·6	32·6
Liverpool ..	42·5	40·6	36·8	43·9	42·9	51·8
Manchester ..	42·4	40·1	38·6	41·2	39·8	39·3

It is a well-known fact that public spirit in local government and interest in civic welfare is often found to a lesser degree in the great capitals than in provincial towns. There are many causes for this, one of the most important being the distracting effect of national affairs which tends to absorb the political attention of the people. It is, therefore, important that every effort should be made to concentrate and strengthen the limited municipal interests of the electorate in a great metropolitan centre. In London exactly the contrary course has been adopted. The chaos of areas and authorities confuses the voter, divides his allegiance, disperses his interest. He can neither comprehend the patchwork quilt of municipal councils and *ad hoc* bodies, nor recognise the part he as a citizen is expected to play in the general scheme of things. When a Londoner enquires or

---

[1] *London Statistics*, vol. 38, p. 23.　　　　[2] *Ibid.*, p. 26.
[3] The electors in wards in which there are no contested elections are not included in the electorate for the purposes of this table. The variation in the size of the electorate as between different years is therefore considerable.

complains or makes a suggestion about the work of a local authority he is frequently told that the matter concerns some other organisation. There is neither the concentration of responsibility nor of interest which characterises municipal government in Manchester or Birmingham, Leeds or Sheffield. There is virtually no recognition of the unity of the metropolis as a whole. Most people have come to think mainly of their own district, while even the London County Council refers in its official publications to "the urban agglomeration known as 'London.' "[1] Imagine Pericles inviting his fellow citizens to treasure the virtues of "the urban agglomeration known as Athens"!

The present chaotic condition of London government has certainly not come about by itself. There is nothing Topsy-like in its growth. It has been created step by step by the deliberate action of Parliament. Whether the final result as it now exists was ever intended or envisaged I will not undertake to say. It is certain, however, that Parliament has for more than a hundred years persistently rejected every opportunity to provide the capital with rational and comprehensive municipal institutions. "For some curious reason," writes Mr. Herbert Morrison, "Parliament has seemed to have a fear of order, dignity and cohesion in the local government of the metropolis. Even today one hears reports—which I hesitate to take seriously—to the effect that Parliamentarians are apprehensive at the growing powers of the London County Council. . . . Possibly this explains why successive Governments and Parliaments have consistently refrained from giving Greater London a comprehensive and simple system of municipal administration."[2] And in a later passage he remarks: "The muddle of local government in Greater London either means that Governments or Parliament wish it to be a muddle, on the divide and conquer principle, or that they have possessed neither the initiative nor the courage to grasp the problem boldly and settle it. If the local authorities of Greater London had been united, if the

---

[1] *Twenty-five Years of London Government*, L.C.C. Publication, No. 3117/1935, p. 11.
[2] Rt. Hon. Herbert Morrison, M.P.: *How Greater London is Governed*, p. 98.

whole area had been animated with a sense of civic dignity and healthy municipal independence, Parliament would have been bound to give Greater London as much democratic self government as is possessed by large cities like Manchester and Birmingham. But, in the past at any rate, the London local authorities have never been happier than when they were having a scrap with each other . . ."[1]

[1] Rt. Hon. Herbert Morrison, M.P.: *How Greater London is Governed*, pp. 122–3.

CHAPTER III

UNPLANNED DEVELOPMENT

By far the most serious consequence of the disintegration of power among 120 or more separate local authorities is the entire absence of any plan to guide the development of the region. The metropolis has been expanding continuously for the last hundred years until it has reached its present elephantine size; yet at no point has any governmental body, central or local, stepped in to regulate the manner and direction of its growth.

The need for some restriction on the encroachment of open spaces by building development or the construction of railways was publicly enunciated again and again during the past seventy years. In 1866 Sir William Fraser, M.P., declared that there were no means available by which a single acre of suburban land could be preserved from the devouring influence of the railways. "The tired mechanic, the Sunday wanderer, even the more hardly worked lawyer, or man of the desk," he complained, "will soon find not a spot of green grass near London."[1] J. F. B. Firth, another Member of Parliament and a devoted friend of London, drew attention in 1888 to the need for acquiring as open spaces and recreation grounds the 10,000 or more acres of commons and green fields which were then available outside the area of the Metropolitan Board of Works within a radius of fifteen miles from Charing Cross.[2]

In 1908 a further warning was uttered by the London Reform Union: "London is being closed in. All round the county area, districts that used to be suburbs are becoming nothing less than great towns, and are cutting off London from the country."[3] Walthamstow, Willesden, East Ham and Ilford were mentioned as examples. Apprehension of the result was accompanied by a clear understanding of the cause, which was correctly ascribed to the absence of any common plan or system in the growth of these outposts of the metropolis. The London

---

[1] Sir W. Fraser: *London Self-Governed* (1866), p. 24.
[2] J. F. B. Firth: *The Reform of London Government*, p. 59.
[3] *London To-day and To-morrow* (1908), pamphlet, p. 16.

A CENTURY OF BUILT-UP LONDON

These maps and those on the two following pages illustrate the growth of the built-up area.

1845

○Enfield   ○Loughton
○Edgware   Tottenham   Woodford○
○Finchley
Romford○
○Harron
on the Hill   ○Ilford
Hampstead
Wembley○   Kilburn○   ⌁Hachney   ○Stratford
Notting
Hill
Brentford○   Woolwich
Kew   Greenwich
Peckham   ○Eltham
Twickenham   Richmond
Streatham
Wimbledon○   Bromley○
Hampton
Court   Kingston
on Thames
○Orpington
○Croydon
Sutton ○

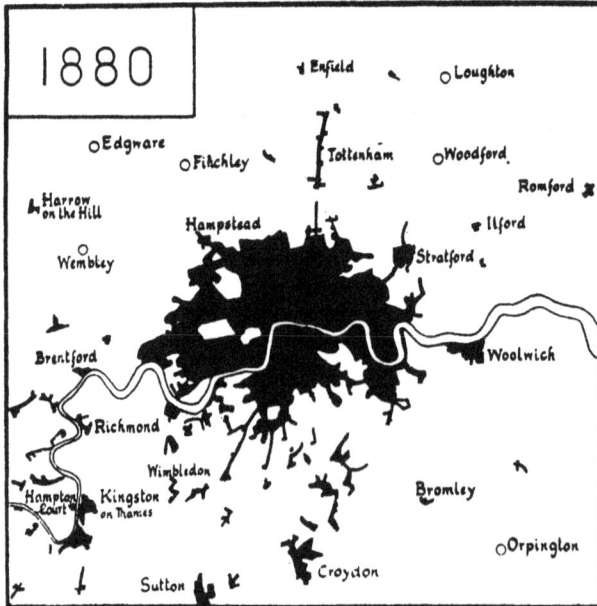

1880

⌁Enfield   ○Loughton
○Edgware   Tottenham   ○Woodford
○Finchley   Romford ⌁
Harrow
on the Hill   ⌁Ilford
○
Wembley   Hampstead   Stratford
Brentford   Woolwich
Richmond
Wimbledon   Bromley
Hampton
Court   Kingston
on Thames
○Orpington
Sutton   Croydon

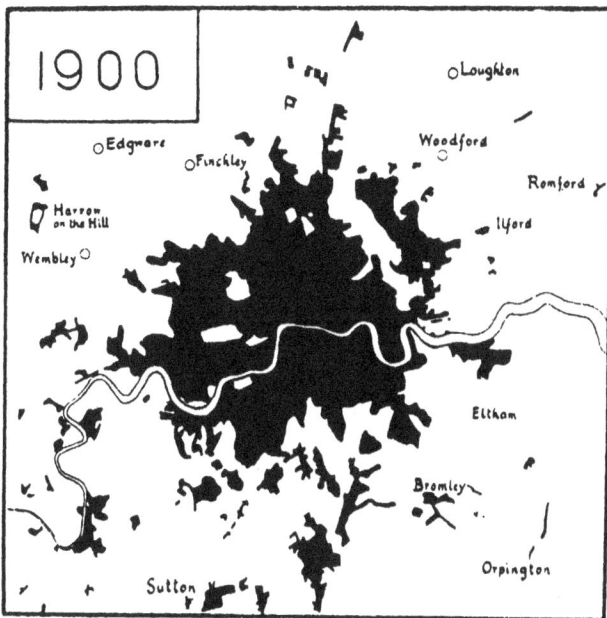

1900

Loughton

Edgware

Woodford

Finchley

Romford

Harrow
on the Hill

Ilford

Wembley

Eltham

Bromley

Orpington

Sutton

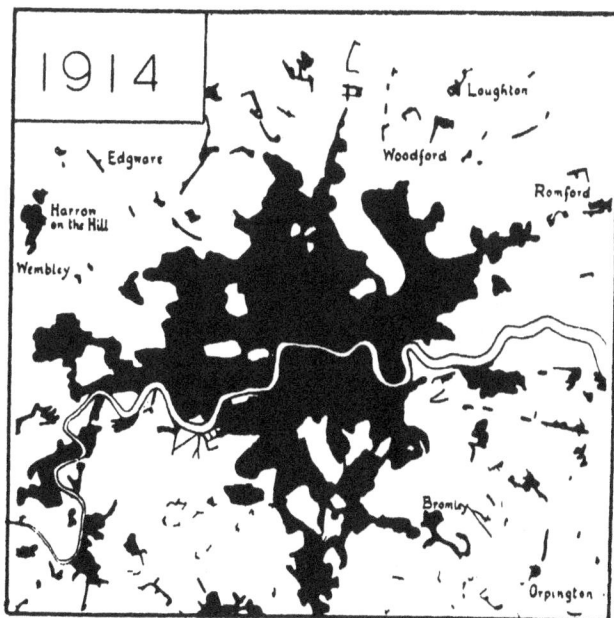

1914

Loughton

Edgware

Woodford

Harrow
on the Hill

Romford

Wembley

Bromley

Orpington

12

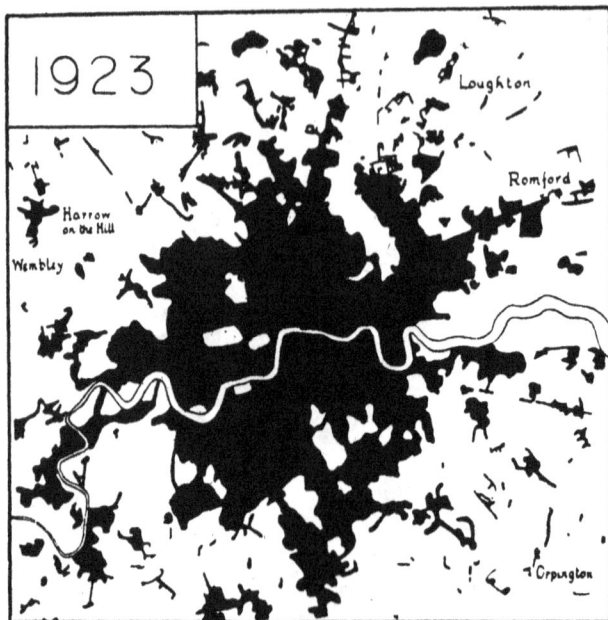

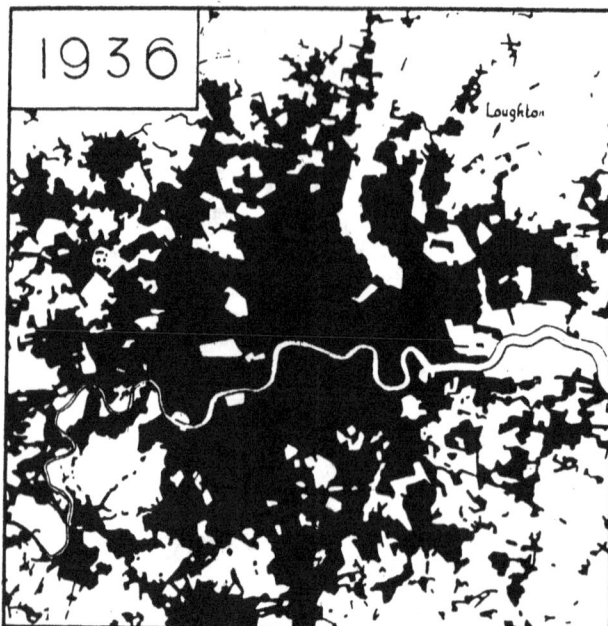

Reform Union lamented the absence of any organ representative of Greater London as a whole and deplored the fact that no power existed either to supervise the direction of roads leading in and out of London—the Petrol Age was just commencing—or the great new systems of electric trains running underground and on the surface from the centre of the metropolis to the outlying districts.[1]

This intelligent diagnosis was ignored, despite the fact that the Housing Act passed in 1909 contained the first town planning provisions to be embodied in English legislation. Two years later, the recently established London Traffic Branch of the Board of Trade was clamouring for better transport facilities in order that the many tracts of vacant land lying within a radius of fifteen miles from Charing Cross could be made available for the erection of dwellings.[2] The traffic men at the Board of Trade were absorbed exclusively with the notion of expansion based on transport, without giving any thought to the emerging idea of territorial planning.

The forces which were already at work in the first decade of the 20th century gathered momentum at an immense rate in the post-war period. The economic prosperity which during the past fifteen years has made London an exceptionally favoured place in comparison with most of the rest of the country, gave an unprecedented stimulus to the building of dwellings in the neighbouring countryside. Tens of thousands of houses of all descriptions, from bungaloid structures of the "Mon Repos" type to Tudoresque villas complete with garage and lounge hall, had descended like a blight on the delightful environs which the metropolis was fortunate enough to possess. No man's hand attempted to stay the devastating onrush of the speculative builder, aided and abetted in one notorious instance by a railway company turned landowner. No municipal authority even pretended to regulate or guide for the common good. Improved transport facilities, underground railways prolonged as surface lines, electrified main railways, motor-coaches

---

[1] *Loc. cit.*
[2] London Traffic Branch of the Board of Trade, Third Annual Report (1911), p. v. For an analysis of the forces making for decentralisation in London, see Fourth Annual Report, p. 8.

and cheap motor-cars all helped to make possible the helter-skelter rush of the tired business men and overwrought clerks or artisans seeking relief from the racket, the smell and din of the town amid the quiet and peace of rural surroundings. The fact that collectively the migrating households in many instances destroyed the very conditions which each of them was seeking individually, was a consequence which only an intelligent legislator or a far-sighted administrative authority contemplating the whole metropolitan region could have foreseen. And neither the one nor the other was to be found in the body politic of the nation.

The South is accustomed to pride itself on being the most civilised portion of Great Britain; but nothing could be more barbarous than the post-war misdevelopment of the metropolitan region.[1] Not only residential building but also industrial structures have wrought the utmost havoc. The Edgware Road, to take one notorious example, less than twenty years ago ran through pleasant fields and a charming pleasure lake. Today, the traveller along that road is confronted with an industrial shambles that disgusts the eye and is clearly obstructive from a traffic standpoint.

The evils arising from the absence of planning in most of the housing activity of recent years in the London Region are not merely aesthetic in character. Other disadvantages of a more practical character have occurred, of which the following examples may be mentioned.

At Slough, industrial development on a large specially equipped site has been for long hampered and delayed by the lack of housing for the workpeople, although the conditions are favourable for developing a convenient self-contained town. The local authorities on the spot have made no effort to use the opportunity and the area is presumably too far away for the London County Council to acquire for colonising purposes.

---

[1] Those who speak disrespectfully of the North of England as being imbued with a materialistic spirit indifferent to the higher values would do well to note that nothing in the south can touch Wythenshawe garden suburb of Manchester as a piece of imaginative municipal planning and construction full of material and spiritual possibilities. For an account of its construction, see Sir Ernest and Lady Simon's pamphlet, *Wythenshawe*, reprinted from *The Rebuilding of Manchester*.

PLATE IV. The Edgware Road at Staple's Corner. The traffic congestion resulting from the failure to plan this still partly undeveloped area is clearly shown. The frontages along both main highways are a jumble of factories, shacks and dwellings

(*Aerofilms*)

The London County Council estate at Becontree is open to grave objection from a planning point of view. In the first place, the site chosen is on land suitable for market gardens, of which there is a scarcity in the London Region, which means that fresh fruit and vegetables must be brought from longer distances and sold at higher prices. If the London County Council had selected a site a little further to the north or south, this would have been avoided. Secondly, the estate was designed primarily as a dormitory suburb for white collar workers, who must travel long distances to and from their employment on highly congested traffic routes. The position of Becontree would have made it an excellent residential settlement for Dockland workers. Failing that, the London County Council might have pursued a vigorous policy to attract metropolitan industries to the neighbourhood; but the Council appears to have done little if anything in this direction. Despite this, a number of large factories (including the important Ford plant) have recently been established there, yet the dwellings are not now available in many cases for the workers employed in them.[1]

The Great West Road, one of the principal arteries leading out of London, has been hopelessly damaged from the point of view of safety, speed and amenity by the lack of any intelligent planning or restriction of the frontage development. Factories, housing estates and gravel pits jostle one another in a disorderly array; and once again some of the best market-garden land in the region has been misused. The Kingston By-pass road presents similar defects; and so, too, does the Barnet By-pass road near Hatfield now in process of mis-development. On one side of the road is an aircraft factory, which virtually stops all traffic along the highway when work starts and leaves off. The frontage on both sides then proceeds in the form of houses, flats, cinemas, shops and garages. "The amenities of the area are being destroyed," states the National Housing Committee, "and the local authority is seriously

---

[1] Housing and Planning Policy: Interim Report of the National Housing Committee (February 1936). This is an unofficial body whose offices are at 5, Duke Street, Adelphi, London, W.C.2. I am indebted to the report for the following examples also.

embarrassed by the sudden and unforeseeable demand for services. The neighbourhood would have been quite suitable for a self-contained unit, had the development been adequately planned in advance with due regard for the arterial road."[1]

The influx of residents and of industrial enterprises in the territory which lies between the frontier of the administrative county and the outer boundary of Greater London has been the most striking feature of London's growth in recent years. In the last ten years an extra million persons have arrived in this territory (which we may call Outer London) either by birth, by moving outwards from Inner London, or by migration from other parts of the country.[2] About 200,000 of them appeared in the three years between 1931 and 1934.[3] The great majority have gone to dwell in the belt lying on the outer edge of the part that was already built up ten years ago. Between 1921 and 1931, 180 factories moved out of central London to the outer zone. No less than 53 factories, employing about 11,000 workers, have recently been established along two miles of the Great West Road.[4] In 1932–4 out of 470 net additions to factories in Greater London, 103 were transfers from Inner London.[5] It is the uncontrolled building of houses, industrial premises, shops, garages and so forth consequent upon this inrush and outrush of population and industry which has been most injurious to the interests of the metropolis as a whole. The process of industrial decentralisation is now proceeding more rapidly than ever. Less than half of the new factories are being constructed on trading estates, the remainder being distributed without regard to the amenities, convenience or health of the community.

The smothering of the entire area by buildings is a certainty

[1] Housing and Planning Policy: Interim Report of the National Housing Committee (February 1936).
[2] About 500,000 have come from other parts of the country.
[3] Third Report of the Commissioner for the Special Areas (England and Wales), [Cmd. 5303] 1936, pars. 20–4 and Appendix I. F. J. Osborn: *London's Dilemma*, published by the Garden Cities and Town Planning Association.
[4] A lecture by Dr. D. H. Smith given at a Conference held at Welwyn Garden City by the Garden Cities and Town Planning Association, June 1937.
[5] Board of Trade Surveys of Industrial Development, 1932–4.

PLATE V. The North Circular Road near Neasden. Another example of the misuse of costly main high-way facilities through a complete absence of planning or restriction

(Aerofilms)

if economic forces are left to work themselves out. The super-imposition of one railway system on another, writes Mr. Pick, the Vice-Chairman of the London Passenger Transport Board, must have the effect of leading to the complete occupation of land not definitely secured against exploitation within a twelve mile zone.[1] "So the solid core of London if enlarged may be more than is healthy or advantageous, and tends to stretch to a bulk twenty-five miles across, corresponding closely to the Metropolitan Police District."[2] There is no reason why the process should end there.

The disappearance of land suitable for open spaces has already proceeded at an alarming rate. In 1927 there were 32,000 acres available for the purpose within eleven miles of Charing Cross. By 1933 the amount had dwindled to 8,000 acres. According to the *New Survey of London Life and Labour*[3] there were (in 1930) within the Survey area 13,000 acres of public open spaces or private playing fields, while of the land still undeveloped only 5,000 acres was physically suitable for games, and a large proportion of this would certainly be built on. Outside the Survey area and accessible to its inhabitants there were said to be 28,000 acres of suitable ground within a radius of ten miles, out of a total acreage of 64,000 still undeveloped. But the *New Survey* found that the built-up area had increased by 2,500 acres in the preceding three years and therefore prophesied that a great part of this vacant land would be covered with buildings within the next decade. These and many similar facts make it abundantly clear, observed Sir H. Llewellyn Smith, the Director of the Survey, in his opening chapter, that the demand for playing fields "has outgrown the possibilities of supply within the limits of the open spaces now available."[4] The position is certainly much worse now than when he wrote.

---

[1] F. Pick: "The Organisation of Traffic," *Journal of the Royal Society of Arts*, vol. lxxxiv, p. 212.   [2] *Ibid.*

[3] Vol. i, p. 285. The whole subject was admirably dealt with by Sir Raymond Unwin in the First Report of the Greater London Regional Planning Committee, published in 1929. A clear warning was uttered in this report that "the lands which are urgently needed for playing fields and other open spaces are being rapidly over-run or disfigured by sporadic buildings"; p. 5.   [4] *Ibid.*, p. 47.

But whether the land is available any longer or not, the existence of open spaces on an ample scale for purposes of rest and recreation is essential to the health and welfare of a highly urbanised metropolitan community. As the people become deprived of easy access to the countryside so does the need increase. The *New Survey of London Life and Labour* points out that the "disamenities of traffic danger, noise and grime accentuate the importance of open spaces, parks and gardens where persons of all ages may find a safe and quiet refuge.[1] Every consideration, the Survey concludes, points to the need for far-sighted and comprehensive planning to meet London's growing needs in respect of healthy outdoor life.[2]

In a sense it is too late to act effectively now even if we had the will and the intelligence. We shall not recapture the lost open spaces nor recover for health and recreation what the builder has claimed for his own. Any belated action which is now taken will certainly be far less excellent, by the very nature of things, than the action which could so easily have been taken in 1921 or 1925 or even in 1930. If the Government or the local authorities had acquired for planning purposes the whole of the land in the Outer London Region, at any time since the turn of the present century, it would have proved a magnificent investment both from a social point of view and a financial standpoint. The City Council of Stockholm took action of this kind from 1904 onwards; why could not we in London also have looked forward a little? Now at the thirteenth hour there is much ado about improving the physical fitness of the nation. But one of the essential conditions of a healthy life—opportunity for fresh air and exercise—has been allowed to disappear before our very eyes as regards the fifth of the nation which lives in the metropolis.

The one striking contribution to the preservation of amenities was the decision of the London County Council in 1935 to establish, so far as it could, a green belt around London. For this purpose the London County Council announced its readiness to make grants to county and county borough councils outside the administrative county towards the approved cost

[1] *New Survey of London Life and Labour*, vol. i, p. 46.
[2] *Ibid.*, pp. 47, 285.

of acquisition or sterilisation of such lands. The London County Council's total commitment was limited to £2 millions and the offer was open for a period of three years. The latest figures show that by July 1938 some 68,000 acres (about 100 square miles) have been acquired or preserved under these arrangements.

The plan for a green girdle encircling the town is a splendid conception for which the greatest credit is due to Mr. Herbert Morrison and his colleagues at County Hall.But whilst admiring the idea we should at the same time recognise the difficulties which stand in the way of its attainment. In the first place, there is no such thing as a green belt still in existence to be acquired. Second, the negligence of past decades will have to be paid for by having the girdle at a more remote distance from the centre of London than is desirable or would have been necessary if more prompt action had been taken. Third, the absurdity of the present organisation is demonstrated by the fact that although the London County Council provides this very large sum of money, the green belt itself lies far outside its territory. Fourth, although the original initiative has come from County Hall, the project must be carried out by the neighbouring county councils. The London County Council is entirely dependent on their goodwill, ability and interest in attaining the object in view and is unable itself to propose any particular item in the programme. Moreover, since the grants from the London County Council do not in any case exceed 30 per cent of the cost and are usually considerably less, the success of the scheme depends primarily on the financial resources of the county councils and their willingness to expend large sums of money on this purpose. Fifth, the county councils concerned are not town-planning authorities, and there has been little attempt by the councils which are responsible for planning to prevent misdevelopment and speculative building in the vicinity of the green patches which constitute the belt.[1]

The need for territorial planning of the metropolitan region is apparent not only in regard to housing, highways, open spaces and amenities, but also in connection with transport. Mr. Frank Pick voices the concern of the London Passenger

[1] *Post* pp. 335, 409, 415.

Transport Board when he states that each means of transport creates its own pattern of housing development, "one overlying another until all is confusion, creating ever fresh problems with regard to all public utility services and even with regard to transport itself. All these are patterns calling for wise and careful treatment as parts of this whole called London. Yet incidentally there is no organisation able to control or direct their development to any predetermined end or result. All is accident. Large populations are dumped down in places where transport facilities are inadequate or unsuitable, as for example, at Becontree . . . or, for example, at North Ilford, where a patchwork of housing by enterprising builders has covered down over two square miles or more of agricultural land, and brought into being an additional population of about 40,000, requiring public services of all sorts. Such developments as this last are almost analogous to cancerous growths. The cells multiply without any organic framework or pattern, without differentiation of function, without the discipline of any control. As a consequence, vast sums of money, more than would ever be wanted under a planned expansion, must now be sunk in a scarcely remunerative enlargement of the transport facilities, pointing to the conclusion that with transport unified and co-ordinated there will still be no acceptable solution of this metropolis, unless the unification of the other essential functions keeps equal pace."[1] Mr. Pick later remarks that traffic considerations are being partly subordinated to other considerations which still operate on a basis of competition, since local authorities vie with one another in seeking expansion.[2] And he emphasises the obvious truth that there cannot be a competitive basis to the intelligent planning of a city.

It must not be thought that there are no town-planning authorities within the London Region merely because the area is an outstanding example of haphazard growth. On the contrary, there are too many of them. Under the Town and Country Planning Act, 1932, the City Corporation is the planning authority for the ancient city and the London County Council for the rest of the administrative county. Outside the

---

[1] F. Pick: "The Organisation of Transport," *Journal of the Royal Society of Arts*, vol. lxxxiv, pp. 213–14.   [2] *Ibid.*, p. 216.

county, power is conferred on councils of county boroughs and county districts (i.e. municipal borough councils, urban district and rural district councils). This means that in Greater London alone there are 77 separate town-planning authorities, and if the actual limits of the metropolitan region were taken, or the area of the London Passenger Transport Board, the number would be very much larger.

What can such a multitudinous assortment of authorities, many of them small, incompetent and devoid of imagination, hope to accomplish in so great a task as that of designing the basic scheme for London? The answer is obviously nothing. A day's drive through almost any part of Outer London will provide abundant evidence of the complete failure of the local planning authorities to cope with their problems. They cannot properly be expected to do so. The example I gave on an earlier page of the misdevelopment on the Barnet By-pass road near Hatfield is occurring in an area covered by a regional planning committee and by local planning powers.

In 1927 a body was set up with the hopeful name of the Greater London Regional Planning Committee. It was a composite body consisting of 45 members representing the London County Council, the City Corporation, the Councils of the six Home Counties, the Standing Joint Committee of the Metropolitan Borough Councils, the three county borough councils, and 126 other borough, urban and rural district councils.[1] Most of the latter were assembled in town planning groups. These groups were small collections of local authorities which had joined together for local planning purposes. No less than 152 separate local authorities were represented on the Committee.

The Greater London Regional Committee was constituted under the auspices of Mr. Neville Chamberlain, who was then Minister of Health. Its functions were purely advisory. Its area was that of the London and Home Counties Traffic Committee and covered more than 1,800 miles within a radius of 25 miles from Charing Cross. It had as its secretary Mr. Montagu Harris, at that time an official of the Ministry of Health and a past president of the Town Planning Institute; and as

[1] See the First Report of the Committee, Appendix I, for details.

technical adviser Sir Raymond Unwin, the distinguished town planner.

The Regional Planning Advisory Committee issued its first report in 1929, containing a series of brilliantly presented and vividly illustrated memoranda by Sir Raymond Unwin on the subjects of open spaces, ribbon development, and sporadic building, with many constructive proposals and a statement of the additional powers needed for effective regulation. The Committee stated that they had themselves been compelled to realise that the purely advisory powers with which they had been endowed made it impossible for them or their constituent bodies to give effect to the measures they recognised as both necessary and urgent. Joint or regional schemes had to be either purely advisory, dependent for their enforcement on the willingness and ability of each local authority to carry out the scheme in its area; or they had to displace the local schemes and deal with all the minor details for each district as well as the larger questions. Neither alternative was appropriate for a large and thickly populated region. The Committee therefore urged upon the Government the need for establishing a Joint Regional Planning Authority which should have executive powers as regards the larger regional matters while at the same time leaving to the existing authorities the making of local plans dealing with matters of local concern. The county councils were to take the leading part in the Regional Planning Authority.[1]

Nothing whatever was done to put into practice these urgent representations, although a deputation was received in November 1929 by Mr. Arthur Greenwood, Minister of Health in the second Labour Government. In 1931, following the financial emergency, among the "economies" imposed at the behest of the National Government, was the stopping of the grant for the professional staff employed by the Regional Committee. The sum involved was about £3,000 per annum, provided by the municipal authorities in London.[2] The result was

---

[1] See the First Report of the Committee, p. 5.

[2] For details, see the Joint Report of the General Purposes Committee and the Town Planning Committee in the Minutes of the London County Council, 15th December, 1931, § 7. The grant was actually reduced to £700 a year, which was so inadequate that the plans could not even be kept up to date.

to terminate the appointment of Sir Raymond Unwin as technical adviser and a cessation of work even on the plans in course of preparation by the draughtsmen in his office. It is difficult to look back upon these events without a sense of shame.

In 1933 the Greater London Regional Planning Committee was reconstituted under the Town and Country Planning Act, 1932, with a membership reduced to 30 and with the relative strength of the county councils increased.[1] It remained a purely advisory body, unless and until any constituent body might see fit to delegate to the Committee its powers in connection with the preparation or adoption of any planning scheme; but the Committee was precluded from undertaking any executive function within the area of a minor authority without the consent not only of that authority but also of the county council in whose territory the area is situated. By this time there can have been few persons who entertained even the mildest expectations that any action would result, since all the important recommendations of the earlier body had been ignored. In 1936, owing to "a difference of opinion among the constituent authorities as to the best method of continuing the work,"[2] the London County Council withdrew from the Advisory Committee, which had achieved nothing whatever, either on paper or in practice. In 1937 yet another body was constituted along similar lines!

"The Minister of Health," we are solemnly told, "is of opinion that the existence of some regional planning body is of very great importance in the Greater London area, and he has come to the conclusion that a consultative body would best meet the present need—a body representative of all the local authorities in the London traffic area, but whose functions would be to consider and make recommendations on such matters as might be referred to it, rather than to initiate proposals."[3] This astounding conclusion was arrived at, be it noted, after ten years' experience had demonstrated the utter

---

[1] For details, see *London Statistics* (1934–6), vol. 39, p. 4. The area of the Committee remained unchanged.

[2] Minutes of the London County Council, 15th June, 1937. Report of the General Purposes Committee, Section 24. [3] *Ibid.*

futility of just such bodies as the Minister was now convinced would best meet the present need.

So in June 1937 a Greater London Standing Conference on Regional Planning was constituted, to consist of 24 members drawn by direct representation from the county councils and county borough councils and by collective representation from the borough and district councils. In this latest step there is not even the pretence of setting up an effective or creative body. The functions of the Conference are reduced merely to considering and making suggestions on any matter referred to it by the separate authorities. The staff is to consist of a Ministry of Health inspector working with a technical sub-committee comprising the surveyors of the county and county boroughs. The annual budget amounts to £360 a year.

All that can be said about this is that the tragedy has become a farce. It is regrettable that responsible local authorities should be willing to take part in such a burlesque performance.

In the meantime, the London County Council at long last decided to prepare a planning scheme for the whole of its area; and approval was given by the Minister of Health to the resolution in May 1935. Prior to this·some sixteen schemes had been initiated covering portions of the County of London, varying from 7,150 acres at Greenwich, Lewisham and Woolwich to 20 acres at Streatham Common.[1] Only two of these schemes, relating to Streatham Common and Hampstead Heath, have been approved by the Minister of Health. The others are in various stages of preparation.[2] It is obvious that the possibilities of planning a fully built-up area such as the County of London are infinitely smaller than the opportunities offered by an undeveloped or partly undeveloped district. For unless the land is publicly owned the cost of compensation becomes prohibitive. For this reason one should not anticipate any very important results from planning within the county. The vital place is Outer London, and here there is no comprehensive body possessing adequate power. So far there is no

[1] *London Statistics* (1934–6), vol. 39, p. 192.
[2] S. D. Adshead: *London under Statutory Town Planning*. A Chadwick lecture delivered in London on 7th May, 1937 (unpublished).

indication that we have learned anything from the mistakes of the past twenty years.

Hence we are still muddling along. But we are not muddling *through* our difficulties and problems, which continue to grow apace. That is one of the few things that cannot be done in the realm of territorial planning.

CHAPTER IV

HIGHWAYS AND BRIDGES

We may turn from the general planning of the London Region to a consideration of some of the principal services as they are administered today. Communications are closely related to town and country planning, so the question of highways and bridges may logically be dealt with first.

Within the administrative county, responsibility for the construction, maintenance and improvement of streets falls on the metropolitan borough councils and the City Corporation. The London County Council deals only with highway improvements of more than local importance. These "county improvements," as they are called, are large undertakings beyond the resources of any one district and of sufficient importance to be regarded as of general utility to the whole community.[1] They are usually authorised by special Acts of Parliament. Outside the administrative county, all classified roads are in the hands of the county councils, except in the case of urban districts and boroughs having a population in excess of 20,000. Unclassified roads are maintained by urban district councils and municipal boroughs, but arrangements may be made by agreement for these to be administered by the county council at the expense of the district authority. County borough councils are responsible for highways of all kinds in their areas. The Minister of Transport has recently taken over from the county councils the main roads of national importance, but this does not affect county borough highways, nor does it apply to the County of London.[2]

It was stated in 1926 that within the London Traffic Area there were 170 independent local authorities, all of which were highway authorities with the exception of the London County Council. The arterial roads in the same area were then vested in 90 councils and the county bridges in the hands of 11 separate

---

[1] Where a county improvement also possesses local utility, a contribution is made to the cost by the borough council of the district concerned.
[2] Trunk Roads Act, 1936.

PLATE VI. Kingsway, looking south. This is the only important highway improvement carried out in Central London during the past 50 years

(*Aerofilms*)

bodies.[1] The number of highway authorities has been somewhat reduced now as a result of the Local Government Act, 1929, but it is still very large. The Bressey Report gave the figure of 132 local authorities, excluding rural district councils and parish councils.[2]

This multiplicity of authorities exercising highway powers has many disadvantages. A single artery like the Great West Road is dealt with by 8 separate local authorities as it passes from Hammersmith to Whitechapel, each one of which has its own methods of maintenance, repair, cleaning and lighting.[3] In another stretch of modern arterial road in the environs of London, there are 27 different standards of lighting, ranging from "reasonable adequacy to none at all."[4] On a route extensively used as an exit from the centre of the metropolis 5 authorities control the lighting of 4 miles of roadway; the lamp standards vary in height from 10 feet to 22 feet and the distance between the lamp-posts varies from 15 yards to 130 yards. In some sections the posts are installed on one side of the road only while in others they are staggered on both sides. The standard of lighting is described by the Departmental Committee on Street Lighting as ranging from bad to excellent.[5]

The approaches to the river present a similar picture. The lamps on one side of Westminster Bridge are in the care of the Lambeth Borough Council; those on the bridge itself and on the Thames Embankment are under the London County Council, while Westminster City Council has charge of the illumination in Northumberland Avenue and the Strand.[6]

The Departmental Committee on Street Lighting spoke in no uncertain voice of the dangers to traffic caused by this lack of uniformity. They pointed out that the pools of darkness resulting from an uneven distribution of light on the road surface make it difficult for the motorist to judge distances

---

[1] Report of the Royal Commission on Cross-River Traffic in London (1926), p. 66.

[2] *Highway Development Survey, Greater London*, H.M.S.O. (1938), p. 7.

[3] The metropolitan boroughs of Hammersmith, Kensington, Paddington, Westminster, Marylebone, Holborn, Stepney and the City. Percy Harris: *London and its Government*, p. 65.

[4] Minister of Transport: Departmental Committee on Street Lighting, Interim Report (1935), p. 8.   [5] *Loc. cit.*

[6] A. Emil Davies: *The Story of the London County Council*, p. 13.

and for other road users to estimate the speed of approaching vehicles. Moreover, a marked variation of lighting in any one thoroughfare may also cause the driver to be uncertain regarding the need for using his headlights. The Committee concluded that, on the whole, patchy street lighting was worse than no lighting at all, from the standpoint of public safety.[1] They further declared that they did not consider the existing defects could be overcome by grouping together local authorities for lighting purposes, or by the establishment of Joint Advisory Committees on the lines of the Standing Joint Committee of Metropolitan Borough Councils,[2] because cost—and therefore finance—is a vital factor in securing consistency in lighting policy.

The methods of surfacing the roads are as unsystematic as those of lighting them. One district employs wood paving, another asphalt, another water-bound macadam and so forth,[3] More serious than this is the lack of co-ordination between the by-laws of adjacent local authorities affecting the same highway. An architect informed the Royal Commission on London Government that he was engaged in developing a county council estate in Tottenham in an area where two local authorities intersected. The Tottenham Council allowed streets to be 40 feet in width, whereas Wood Green, the adjoining Council, would not approve any street less than 50 feet wide.[4]

Passenger transport in the London Region has now been centralised in an *ad hoc* body; but no attempt has been made to set up an executive authority with power to deal with locomotion and traffic throughout the metropolis. The London County Council has never had jurisdiction over a sufficiently wide area or possessed adequate authority.[5] The principal witness for the London County Council (Mr. R. C. Norman) before the Royal Commission on London Government quoted innumerable reports issued between 1903 and 1919 in which it had been stated that an area much larger than the administrative

---

[1] Interim Report, p. 10.　　　　　　　　　　[2] *Ibid.*, p. 8.
[3] Royal Commission on London Government (1923), Minutes of Evidence, p. 641. Witnesses for Edmonton Urban District Council.
[4] *Ibid.*, Minutes of Evidence, p. 731. Evidence of W. E. Riley, witness for Royal Institute of British Architects.
[5] This was pointed out so long ago as 1905 by the Royal Commission on London Transport, Report, p. 96.

county was called for in connection with the control of main roads leading in and out of London.[1] There had been numerous suggestions for the creation of a Greater London Traffic Authority,[2] but when the London and Home Counties Traffic Committee was actually established in 1924[3] it was made a purely advisory body without executive functions, and the multitude of existing highway authorities was left undisturbed.

The cross-river traffic problem has become increasingly troublesome since 1920, and in 1926 a Royal Commission was set up to investigate this question. They reported that one of the prime causes of traffic difficulties generally, and especially of those involving cross-river communication, is the multiplicity of authorities possessing only local or partial responsibility, and the failure of Parliament to constitute for these purposes an administrative unit with financial powers.[4] The division of responsibility for Thames bridges between the City Corporation, the London County Council, and the out-county authorities in Greater London, each one of which is concerned only with its particular problem without regard to the needs of London as a whole, was declared to be one factor contributing to the unsatisfactory result. The fact that the London County Council has no direct responsibility for highways, although it is the statutory authority for bridges within its area, also affects the situation adversely, especially in regard to the provision of approaches to bridges.[5]

The Royal Commission recommended that in order to secure co-ordination of effort and rapidity of action, a single authority should be created for the London Traffic Area, armed with the necessary executive power and financial resources to deal with the cross-river traffic problem as a whole. They urged the necessity of prompt action and suggested that the existing London and Home Counties Traffic Advisory Committee was a suitable body to have conferred upon it the necessary powers.[6] No action was taken to carry out this advice.

[1] Royal Commission on London Government (1923), Minutes of Evidence, p. 37.
[2] See, for example, the Select Committee on Transport in the Metropolitan Area, 1919. [3] London Traffic Act, 1924.
[4] Royal Commission on Cross-river Traffic in London (1926), p. 66.
[5] Loc. cit. [6] Ibid., p. 67.

There has been a remarkable absence of important street improvement schemes in the County of London during the past fifty years, with the one notable exception of the Holborn to Strand clearance. Apart from this and a few small ameliorations such as the gradual widening of Piccadilly, we are struggling along for the most part with thoroughfares that were constructed before the Petrol Age. The principal routes in central London are pitifully inadequate to the existing traffic and compare unfavourably with those of almost any great city abroad, such as Paris, Chicago, Barcelona, Moscow or Stockholm, although the traffic in London is greater than that of any other metropolis in Europe. Incredible though it may sound, in at least three principal London highways, the building line is actually being brought forward at the present time![1]

Sir Henry Haward, the former Comptroller to the London County Council, remarked in 1932 that one could not fail to be struck by the comparatively small expenditure of the Council on street improvements during the previous forty years, apart from the large outlay between 1900–1910 on the Kingsway Scheme.[2] The expenditure of the Council from 1889 to 1935 (46 years) has been only about £15 millions. Its predecessor, the Metropolitan Board of Works, spent £13 millions on major improvements between 1855 and 1889, a period of only 34 years.[3] Thus the average annual expenditure has diminished from nearly £400,000 a year to £320,000 a year since the London County Council came into existence, despite the fact that traffic demands have increased enormously and the wealth of the metropolis has grown immensely.

The construction of Kingsway and the clearing away of the insanitary buildings which surrounded the area was a fine piece of work. It led to a considerable migration from the cramped quarters of the ancient City by numerous important

[1] The streets in question are Euston Road, Marylebone Road and Queensway (formerly Queen's Road, Bayswater).

[2] Sir Henry Haward: *The London County Council from Within* (1932), p. 263. *Cf. London Statistics* (1934–6), vol. 39, pp. 291, 447.

[3] Haward gives the following figures for each decade: 1889–1900, £2,051,480; 1900–10, £9,872,890 (this includes the cost of the Kingsway clearance); 1910–20, £1,703,755; 1920–30, £2,566,447. The total of £16,194,572 differs from the figure of £15 millions given in *London Statistics* which I have used above.

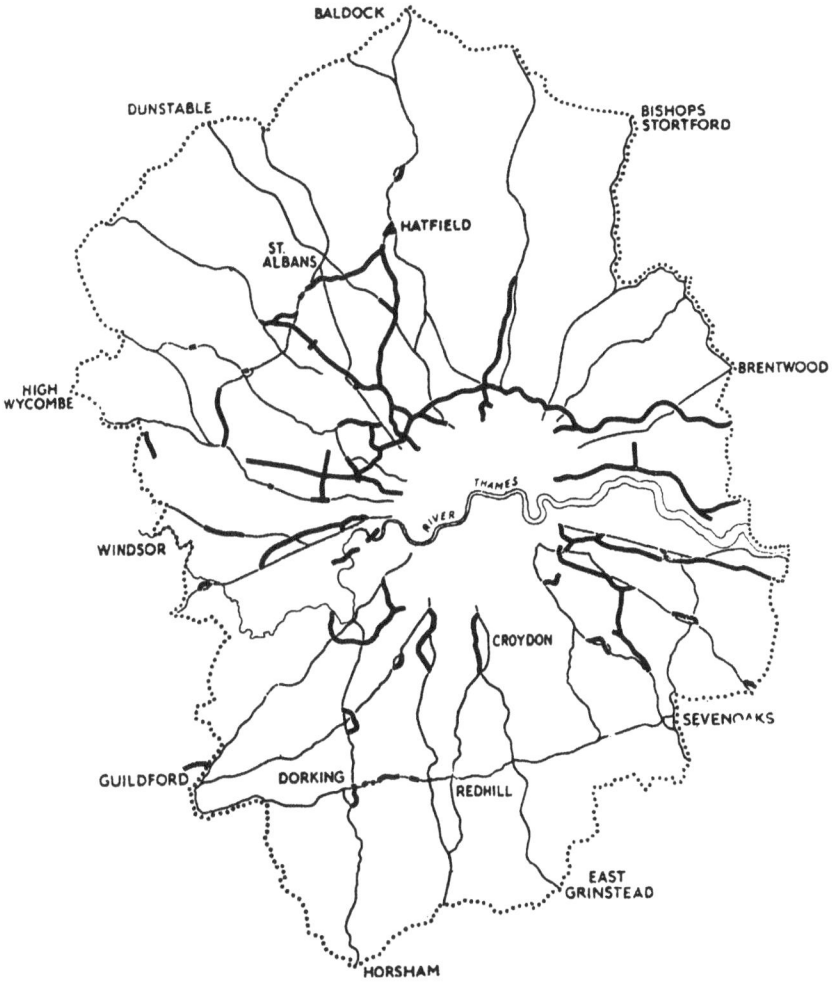

BALDOCK

DUNSTABLE

BISHOPS
STORTFORD

HATFIELD

ST.
ALBANS

BRENTWOOD

HIGH
WYCOMBE

THAMES

RIVER

WINDSOR

CROYDON

SEVENOAKS

GUILDFORD

DORKING

REDHILL

EAST
GRINSTEAD

HORSHAM

(Reproduced by kind permission of "The Economist".)

MAP OF MAIN ROADS BUILT SINCE 1918

This map was made in 1937. It shows the total absence of any main road
development in Central London.

EXPENDITURE OF LONDON LOCAL AUTHORITIES 1933-34

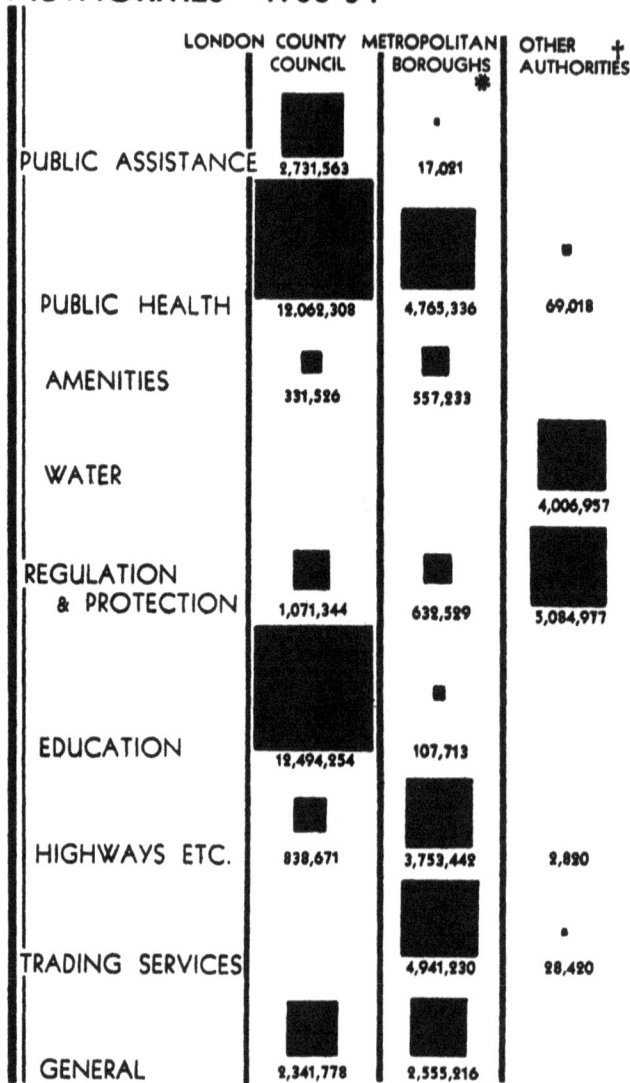

	LONDON COUNTY COUNCIL	METROPOLITAN BOROUGHS *	OTHER AUTHORITIES †
PUBLIC ASSISTANCE	2,731,563	17,021	
PUBLIC HEALTH	12,062,308	4,765,336	69,018
AMENITIES	331,526	557,233	
WATER			4,006,957
REGULATION & PROTECTION	1,071,344	632,529	5,084,977
EDUCATION	12,494,254	107,713	
HIGHWAYS ETC.	838,671	3,753,442	2,820
TRADING SERVICES		4,941,230	28,420
GENERAL	2,341,778	2,555,216	

(Figures are in £'s)

* Including the City Corporation. † e.g Metropolitan Water Board, Metropolitan Police.

(Reproduced by kind permission of "The Economist.")

The above diagram shows graphically the relatively small expenditure of the London County Council on street improvements and main highways or bridges in a typical year.

firms and the new street with the Aldwych island site at the end of it has become probably the most important business thoroughfare in London. The centre of gravity in Inner London began to move westwards from the City, and this trend was intensified by the shortsighted action of the City Corporation in failing to take advantage of the rebuilding of several great banks in their area to widen and improve the narrow City streets.

The financial aspects of the Kingsway scheme are of considerable interest. The total debt charges incurred to 31st March, 1936, were £5,209,563 and the net debt outstanding at that date £3,208,607, making a total expenditure of £8,418,170. Against this must be put the aggregate rents received to March 1936 (£2,368,887), plus the capital value of the leased sites (£3,641,044), amounting in all to £6,009,931. The difference of £2,408,239 represents the net cost to the ratepayers on the date in question. The yearly net charge on the rates, which is arrived at by deducting the sums received from rents and improvement charges from the sums needed for interest and redemption of debt, was £60,752 in 1936. This will disappear within two decades, and by 1956 a net surplus of £2,000 is anticipated, rising to £143,725 in 1966, by which time practically the whole debt will have been extinguished. The very valuable properties owned by the London County Council will thereafter be an unencumbered asset so far as existing debt is concerned. The accumulated charge on the rates is at present £2,840,676, and will rise to a peak of £3,547,336 in 1956, after which it will be gradually reduced by the yearly surpluses so that in 1986 every penny paid out of the rates will have been repaid and the Council will be in receipt of rents amounting to £146,000 a year or whatever sum is then appropriate.[1]

From these figures there emerges the highly significant fact that the one great street improvement carried out by the London County Council during its half-century of existence has not only been highly beneficial to the metropolis in terms of traffic convenience and the provision of office accommodation, but is also a good proposition from a financial point of

[1] The figures are taken from a return published by the London County Council in August 1937.

view. The public has acquired an important asset which is growing in value and in due course will pay for itself many times over. Yet despite this excellent example the London County Council has never attempted a subsequent effort of the same kind, although there are many places on both sides of the river which obviously need treatment on similar lines— Soho and Shoreditch may be mentioned as two examples.[1]

A fresh spurt of activity began to take place with the advent of a Labour majority in power at County Hall. Cromwell Road is to be extended as a main artery, Waterloo Bridge is in process of being rebuilt, and a development of the unembanked South side of the Thames is promised. These projects are all desirable ones, but it is obvious that a far more ambitious programme than this is needed if we are to get within measurable distance of solving the highway problems of the twentieth century. The Bressey Report has recently made its appearance and been received with acclamation. This Report is discussed on a later page.

Whatever the merits of the Report, it is abundantly clear that the solution of the problems with which it deals cannot and will not be attempted if the present chaotic organisation of highway and bridge authorities in the metropolis is permitted to continue. Until responsibility is concentrated instead of being dispersed there is unlikely to be available either the leadership or the resources necessary to bring to fruition the bold and imaginative plans which both the present and future require.

The indifference of the Council towards its own achievement is illustrated by the fact that no comprehensive statement of the financial results was available until Mr. Charles Latham, the present Chairman of the Finance Committee, kindly had one prepared at my request.

CHAPTER V

PUBLIC CLEANSING

We have surveyed shortly the administrative arrangements for the construction, lighting, maintenance and improvement of highways, and we may now enquire as to the manner in which they are cleansed. This scavenging function must be considered in conjunction with the two ancillary ones of refuse collection and disposal, which together constitute the public cleansing service. There are few branches of local government in which greater progress has been made in recent years, both from the point of view of technical development and administrative efficiency. No small part of the credit for this advance is due to the Ministry of Health, which has insisted on the importance of the subject and also assisted local authorities with advice and comparative statistics of costs, etc.

In the County of London all three cleansing functions are in the hands of the metropolitan borough councils and the City Corporation, except that the London County Council is responsible for cleaning and scavenging the Thames tunnels and cross-river bridges other than those in the City. We may therefore regard public cleansing as a service where the characteristics of the minor authorities are displayed to full advantage. In Outer London responsibility is distributed among the rural and urban district councils, the municipal borough and county borough councils.

As a result of some misgiving as to the state of affairs in the metropolis, Mr. Neville Chamberlain when Minister of Health appointed Mr. J. C. Dawes, the extremely able Inspector of Public Cleansing at the Ministry, to make a report on the manner in which the cleansing service was being carried out in the administrative county. Mr. Dawes's report, published in 1929, cannot aspire to the rank of a great state paper, but it deserves at least to be regarded as a notable municipal document.

It opens with the assertion that two fundamental principles must be laid down. One is that public cleansing, being an essential and indispensable sanitary service, must be organised on strictly sanitary lines. The other is that a service involving such heavy annual expenditure must be operated on sound business lines and controlled by an efficient costing system. "I have to report," declared the Inspector, "that, viewed as a whole, the London public cleansing service does not comply with the former, and that none of the separate local services complies with the latter in that there is a general lack of adequate records for checking and comparing unit costs."[1]

The cost of public cleansing in London at this time was approximately £2,200,000 a year, which amounted to one-fifth of the total annual expenditure for England and Wales, although the population of the County of London was less than an eighth of that of the whole country. The average rate charge involved in the metropolitan boroughs was 11d. in the £, equal to about a third of the average charge on the rates of all services carried out by those authorities. This burden the Inspector stated to be excessive.[2] The Report draws attention to the wastefulness caused by the absence of uniformity in methods of administration and organisation, and by the frequent overlapping between different authorities.[3]

The financial aspect of the service must, however, be regarded as of less importance than the sanitary side; and here the Report exposes the most glaring defects.

About twenty of the metropolitan boroughs adopt the discredited practice of employing private contractors to carry out their dust collection and refuse disposal services, with the inevitable result that profit-making comes before considerations of public health. The majority of the transport vehicles employed in the collection of refuse were found to be of obsolete design and giving rise to a serious dust nuisance during loading operations.[4]

The gravest feature of the system was the method of refuse disposal. No London borough tips its refuse inside the administrative county, but all of them send huge quantities to dumps situated in seven neighbouring counties. Strong complaints

---

[1] J. C. Dawes: Report of an Investigation into the Public Cleansing Service in the Administrative County of London, H.M.S.O. (1929), p. 6.
[2] Ibid., pp. 1, 15. [3] Ibid., p. 13. [4] Ibid., p. 25.

against this practice had been made to the Ministry of Health by local authorities and private individuals; and the London County Council had protested insistently against the detrimental effect on their Becontree housing estate caused by the dump at Hornchurch, one and a half miles away.[1] After a personal inspection, in the course of which he experienced on several occasions "the nauseating smell of the smouldering refuse," Mr. Dawes stated that each of these dumps constituted a potential danger to the public health.[2] "I do not hesitate to say," he observed, "that the reeking masses at South Hornchurch taken together comprise the worst refuse dump in Britain and by far the biggest. Further, I am satisfied that the complaints in respect of these dumps are reasonable and justified."[3] The large uncovered riverside dumps, of which there are several, are extensive fly-feeding belts and infested with rats. It was scarcely surprising in these circumstances that Mr. Dawes condemned in strong terms the existing dumping arrangements and advised their immediate discontinuance. Moreover, the whole attitude of about two-thirds of the metropolitan borough councils was definitely wrong because it was based on the notion that their responsibility was ended when the refuse had been handed over to contractors for disposal. Whatever its legal rights may be, wrote the Inspector, a sanitary authority should not contract out of its own sanitary responsibility in this matter, which is precisely what happens now; "I consider that they are morally responsible when they cause hundreds of thousands of tons of refuse . . . to be sent into the areas of other and smaller sanitary authorities, there to produce definitely unsatisfactory conditions."[4] The system whereby boroughs contract out of their responsibilities for the hygienic disposal of house refuse was therefore explicitly condemned and should not be allowed to continue.[5]

The 29 areas into which the county is divided for cleansing purposes have no significance either socially or technically. In 21 instances the house refuse collection vehicles of one

[1] J. C. Dawes: Report of an Investigation into the Public Cleansing Service in the Administrative County of London, H.M.S.O. (1929), p. 4.
[2] Ibid., p. 29.   [3] Ibid., p. 30.
[4] Ibid., p. 8.   [5] Ibid., p. 13.

authority pass into the district of at least one other authority during the process of collecting or disposing of garbage.[1] The disadvantages of the system are particularly noticeable in connection with street cleaning. Mr. Dawes stated that he had seen the machines clean an important road in Central London up to the boundary line of the contiguous borough and leave it in excellent condition. Five minutes later the surface of the road near this line would be covered with litter blown or carried there from the uncleansed portion of the street in the neighbouring borough. Later on, the cleaning staff of the second authority would arrive and do their part of the highway, leaving off in turn at the frontier, but of course not touching the litter which had in the meantime been transferred across the boundary. Such incidents are inevitable with the street cleansing service organised as it is at present.[2]

The local officers employed by the twenty-nine cleansing authorities take a purely parochial view of their duties; and it is nobody's business to consider the cleansing service of the metropolis as a whole in order to obtain the maximum efficiency at the lowest cost.[3] Mr. Dawes gave it as his considered opinion that with twenty-nine separate authorities an economic and efficient service cannot be provided for London. He therefore recommended the establishment of a centrally organised service[4] to deal with all branches of the subject. He took care to add, however, that although his enquiry was confined to the administrative county, it was not practicable to limit the matter to that area. If the cleansing problems of the metropolis are to be completely and not partially solved, a much larger territory would have to be considered. The refuse disposal services of many of the thickly-populated districts in Outer London are unsatisfactory and require to be improved simultaneously with those in the county itself.[5]

So outspoken a report as this could not be overlooked; but in place of the prompt action which was obviously called for the Minister of Health appointed a Departmental Committee

[1] J. C. Dawes: Report of an Investigation into the Public Cleansing Service in the Administrative County of London, H.M.S.O. (1929), p. 1.
[2] Ibid., p. 66.　　　　　　　　　　　　　　[3] Ibid., p. 7.
[4] Ibid., p. 72.　　　　　　　　　　　　　Ibid., pp. 6, 73.

to consider the Dawes report and recommend what measures should be taken on it. This Committee repeated all the main criticisms which the Inspector had made, and explicitly stated that his report was a fair and accurate statement of the facts,[1] none of which had been controverted by the evidence they had heard. They condemned the pernicious system of contracting and recommended that the local sanitary authorities and the central body should be empowered to terminate all contracts subject to compensation. They criticised the excessive number and obsolete design of the transport vehicles in use in the dust collection service and the nuisances arising therefrom;[2] they deplored the insanitary state of the refuse disposal works and recommended that many of them should be closed.[3] They drew attention to representations made by the Port of London Authority, Trinity House, and the London General Shipowners' Society of the serious danger caused to navigation on the Thames between Dagenham and Purfleet by the smoke emanating from the riverside dumps.[4] They emphasised the unsatisfactory street cleansing service provided in some of the outer districts in the metropolis and the need for improved equipment in many of the boroughs.[5] They declared that "London as a whole has not kept pace with the great advances that have been made in the science and practice of public cleansing, and that the work of the metropolitan sanitary authorities presents a great many unsatisfactory and insanitary features which should be remedied without delay."[6] They added that the existing state of affairs was not likely to be remedied while the existing organisation of the local authorities continued.[7]

The recommendations of the Committee were as follows: (1) The disposal of house, trade and street refuse should be centralised under the control of one body. (2) The central organ should also be responsible for the general co-ordination of the public cleansing service so as to secure more uniformity in equipment and organisation: for example, in regard to the

---

[1] Departmental Committee on London Cleansing Report, B.P.P., 1929–30, vol. xv, Cmd., 3613/1930, p. 4.
[2] *Ibid.*, p. 7.   [3] *Ibid.*, pp. 15, 24.   [4] *Ibid.*, p. 12
[5] *Ibid.*, p. 24.   [6] *Ibid.*, p.5.   [7] *Ibid.*, p. 21.

types of vehicles and receptacles in use, frequency of collection, the inflow of refuse to discharge points. (3) Each metropolitan borough council should be required to submit to the central organ a scheme dealing with all these matters. In case of dispute the decision should rest with the Minister of Health. (4) There should be regular conferences between local bodies and the central organ, and between their respective technical officers. (5) Costing systems should be installed throughout the metropolis.[1] There was a division of opinion as to the character of the central body. A majority of the Committee wanted a special *ad hoc* body, with the metropolitan sanitary authorities strongly represented on it "owing to the magnitude and complexity of the task which would have to be undertaken"—an incredibly foolish suggestion in view of the manner in which the metropolitan boroughs have discharged their functions. The remaining members considered that a centralised disposal service could be provided without setting up a new local governing authority—an obvious indication that the London County Council should become the central organ. There was no recommendation on this point.

The report of the Departmental Committee was presented to Parliament in June 1930. It was considered a few months later by the Metropolitan Boroughs' Standing Joint Committee, a body containing representatives of all the metropolitan borough councils. A large majority of the constituent authorities was opposed to the centralisation of refuse disposal and informed the Minister of Health in this sense. In the summer of 1932, the Minister wrote to each metropolitan Mayor stating that legislation to implement the proposals of the Departmental Committee was not practicable at that time and requesting a conference to consider how far improvement could be effected without legislation. At this conference the Minister informed the assembled delegates that he had arrived at the conclusion that evils and imperfections existed and needed attention, that the unsatisfactory conditions had become accentuated since the issue of the Dawes Report, and that the growing disinclination of outside authorities to receive London refuse had become an

[1] Departmental Committee on London Cleansing Report, B.P.P., Cmd., 3613/1930, p. 22.

urgent question. He further stated that a scheme of voluntary co-operation was preferable to compulsion and suggested a joint consultative committee of a purely advisory character to represent the metropolitan boroughs. Needless to say this feeble step was acceptable to the metropolitan sanitary authorities, since it left them in full possession of their ill-used powers and removed the threat of a central administrative organ depriving them of their autonomy.

The Metropolitan Boroughs' Standing Joint Committee then proceeded to appoint a Cleansing Sub-Committee with extremely circumscribed terms of reference. It might investigate conditions relating to the collection and disposal of refuse in London or enquire into up-to-date systems elsewhere, and prepare schemes of improvement. It might communicate with any metropolitan borough council but not with a Government department or the London County Council; and no communication or representation might be issued to the public or to the press without the authority of the full Standing Joint Committee and of the local authority concerned. Thus, the utmost secrecy is preserved over the Sub-Committee's activities and even its reports are not on sale to the public.

The situation revealed by Mr. Dawes in 1929 still exists today in all essentials and the Cleansing Sub-Committee has neither the power nor the will to change it. There have been certain improvements in detail, especially as regards the modernisation of equipment. A number of the metropolitan boroughs have purchased new fleets of vehicles for the collection service. Some of them have appointed specialist officers. The tipping dumps along the Thames have been considerably improved. But the fundamental lack of efficient organisation still obtains and there has been no administrative reorganisation, centralisation or co-ordination. The metropolitan borough councils and the City Corporation still use the method of contracting, although the contractors must now comply with the Ministry of Health precautions relating to tipping dumps along the riverside.

The cost of public cleansing in the County of London has been reduced from £1,734 per 1,000 houses in 1929–30 to £1,364 per 1,000 in the year ending March 1936, a result

entirely due to improvement in technical details.[1] But the cost
remains huge and still averages the produce of a rate of 10d.
in the £ for the metropolitan boroughs and the City.[2] It is
much higher in London than anywhere else in Great Britain.
The collection and disposal of refuse in 83 county boroughs in
England and Wales costs on an average 13s. 2d. a ton, in 131
other boroughs 12s. 8d. a ton, in 120 urban districts 9s. 3d.
a ton; but in London the cost is no less than 19s. 11d. a ton.[3]
Enormous financial savings could be effected if the service
were centralised and the garbage reclaimed scientifically at a
chain of factories erected along the Thames. About a quarter
of the material constituting the dumps consists of fine dust
which can be used for manufacturing bricks and paving blocks.
Tins and other metal objects can be re-smelted. Vegetable
matter can be used for making fertilisers, rags and old paper
pulped for making new paper. The advantages from a public
health point of view of scientific treatment of refuse in place
of the hideous, stinking dumps which are still in existence are
too obvious to need further mention.

Although the problem has so far only been tinkered with
despite expert advice of the need for fundamental reform, the
Cleansing Sub-Committee of the Metropolitan Boroughs'
Standing Joint Committee displays a degree of satisfaction
bordering on complacency. In their report dated July 1934
they stated that in general there had been a considerable and
continuous amelioration of the situation, and that whilst some
objectionable features still remain, most of the borough
councils now fully recognise the importance of the subject
and are making efforts to place their refuse disposal arrangements on an economical and efficient basis.[4] In a further report
issued in January, 1937, they reiterated this remark in almost
similar words.[5] In a later passage they state that they are of

[1] Public Cleansing Costing Returns for the year ended 31st March, 1936,
H.M.S.O. (1937), Table VIII.
[2] *London Statistics* (1934–6), vol. 39, p. 407.
[3] Public Cleansing Costing Returns for the year ended 31st March, 1936,
Table VIII.
[4] The Cleansing Sub-Committee of the Metropolitan Boroughs' Standing
Joint Committee, Interim Report on Disposal of Refuse, p. 7.
[5] Report on Collection of House and Trade Refuse, Street and Gully
Cleansing, and Costing, p. 5.

opinion that the refuse collection and disposal services of the metropolitan city and borough councils will bear favourable comparison with those services in any other part of the country. In particular the refuse disposal arrangements are alleged to be in advance of those obtaining in Outer London, and the Sub-Committee virtuously asks that the cleansing services of the outer areas should be quickly brought up to those of Inner London.[1]

Most of these remarks are highly misleading and would not stand expert examination for a moment. It is pure nonsense to suggest that the cleansing services in London could bear comparison with those of any well-run provincial city, either as regards cost or efficiency; and the Sub-Committee must know that neither Mr. Dawes nor any other independent expert would be likely to endorse their remarks.

My own experience may perhaps be typical in regard to garbage removal. I had been living for some years in the Borough of Kensington, where a collection was made twice a week in fairly modern motor vehicles. In 1935 I moved less than two miles away to the Borough of Paddington, where I found the collection was made once a week. On my informing the local authority that this was most insanitary during the hot weather, I was informed that this was the practice in the borough except in certain more or less slum quarters, where a more frequent service was instituted. A twice-weekly service appears now to have been instituted, but it is generally carried out in open carts drawn by horses.

In Paris there is a daily collection of refuse which commences at 6 a.m., and is finished by 8.30 a.m. or 9 o'clock, except in the case of the markets, which are cleared at a much earlier hour. The domestic garbage of Paris and some 31 neighbouring communes is dealt with scientifically at 4 reducing plants situated outside the cleansing zones into which the city is divided. The cleansing of the streets is also carried out in Paris with much greater efficiency than in London.[2]

[1] Report on Collection of House and Trade Refuse, Street and Gully Cleansing, and Costing, pp. 28–9.
[2] For a detailed account, see M. G. Bouly: *Services Techniques de la Voie publique de l'Éclairage et du nettoiement (Services de la Direction générale des Travaux de Paris)*, published in "Science et Industrie" of the *Ville de Paris* série, 1937 édition.

14

The cost of street cleansing in London is no less exorbitant than the expenditure on the collection and disposal of house refuse. The amount spent by the 28 metropolitan borough councils and the City Corporation on street cleansing (including street sweeping and watering, gully cleaning and snow removal) for the year ended 31st March, 1936, was £971,047 compared with a sum of £1,817,976 spent by the 83 county borough councils. In other words, the local authorities in the administrative county, dealing with an area of 74,850 acres containing 4,185,200 persons, spent more than half as much as the city councils of the 83 largest towns, comprising a territory of 825,877 acres and a population of 13,432,300! The cost per 1,000 population in London was £208 9s. as compared with £135 3s. in the provincial towns.[1] Yet the standard of street cleansing is by no means high in many parts of London, and compares unfavourably with that of the leading county boroughs.

One other point worth noticing is the effort made in recent years by the neighbouring counties to protect themselves against the dumping of refuse from London. In 1931 the Surrey County Council obtained a special Act prohibiting disposal without the consent of both the county council and the district authority concerned, which may be given subject to conditions. In 1933–4 the Middlesex County Council Bill included a similar provision, save that a right of appeal to the Ministry of Health was given against refusal of consent. Essex County Council sought powers in a Bill in the session 1932–3 to prohibit tipping unless specifically authorised but met with considerable opposition, doubtless owing to the fact that for some time past Essex has had the questionable privilege of receiving on the flat lands adjoining the Thames about one-third of the total refuse produced in London. No less than twenty-three local authorities petitioned against the Bill either jointly or severally. After negotiations between the promoters of the Bill and the petitioners the clause was amended so as to enable tipping to be continued under conditions which were practicable and not unreasonably expensive, and such

[1] Public Cleansing Costing Returns for the year ended 31st March, 1936, H.M.S.O. (1937), Table IX.

as to prevent danger to the health or amenities of the neighbourhood.

Obviously the whole situation is profoundly unsatisfactory, and will remain so until some form of central or regional control over the whole metropolitan area is instituted. But apparently it is not sufficient in London to show that a change is desirable in order to get it brought about. It seems to be necessary also to show that a state of crisis or emergency is in existence.

CHAPTER VI

HOUSING AND SLUM CLEARANCE

No one who is acquainted with the housing effort made in this country since the Great War will dispute the magnitude of the achievement or under-rate the contribution which it has made and is making to working-class welfare. It is astonishing, however, considering the enormous public expenditure involved, the many changes in policy that have been introduced, and the widespread interest in the subject, that there has been no Royal Commission appointed to enquire into the situation and advise whether the best steps are being taken to attain the desired ends.[1] If there had been a contemporary Royal Commission on Housing it is scarcely possible that the utter planlessness of the housing movement in the London Region would have escaped attention, or the tangle of areas and authorities for housing purposes allowed to remain in existence.

An attempt to state the larger issues involved was made by the Departmental Committee on Unhealthy Areas appointed by the Ministry of Health in 1920. The reports of this body were issued shortly after the first important post-war housing statute was enacted in 1919.

The Committee emphasised the vast dimensions and complexity of the housing problems to be solved in the metropolis, and insisted on the intimate connection between the location of houses, the means of transport and the distribution of factories and commercial premises. All these questions required simultaneous consideration over a wide territory.[2] In times past, the various attempts to rehouse the people had been completely disconnected from one another and unrelated to any conspectus of the situation as a whole. Sites for rehousing families displaced by clearance schemes had been acquired without any enquiry being made as to how the persons con-

---

[1] The last Royal Commission was the one on the Housing of the Working Classes in 1890.

[2] Departmental Committee on Unhealthy Areas, Interim Report (1920), p. 1.

cerned would be able to travel to and from their work. The first essential in considering the reconstruction of London was, therefore, that it should be studied and planned as a whole. "Such a plan would harmonise all future operations and correlate housing, industry and transport to their mutual advantage. Without such a plan the isolated improvement of any one of these public services may be nullified by, or may itself baulk the improvement of, any of the others."[1] As things were then proceeding, all kinds of mistakes were being made. For example, house property was being demolished to make way for business premises even in areas where there was already insufficient dwelling accommodation.[2]

The second essential, declared the Committee, was to establish a metropolitan authority with jurisdiction extending over the Home Counties. The functions conferred on this body would include the preparation of a plan for the whole area, including the built-up portions of it, the general direction of transport development, and the adjustment of local finance so that the housing burden would be fairly distributed among the various district authorities, who would retain their existing powers.[3] The possibility of such a reform had been demonstrated at a conference held at the end of 1918 composed of representatives drawn from the local authorities in Greater London, which had passed a resolution in favour of the local councils working together as a co-operative unit with a single authority for housing purposes. In 1919 another conference was held at which it was decided to ask for a new *ad hoc* authority for housing purposes. The London County Council dissented from this last-named proposal.[4] The Departmental Committee on Unhealthy Areas reverted to the matter in its final report, when it urged the Government to institute forthwith an enquiry into the scope and character of a regional organ, "the necessity of which has been recognised by almost everyone who has hitherto investigated the subject."[5]

No further action was taken to deal with the matter on these

---

[1] Departmental Committee on Unhealthy Areas, Interim Report (1920), p. 4.   [2] *Ibid.*, p. 2.   [3] *Ibid.*, p. 4.
[4] Percy Harris: *London and its Government*, p. 239.
[5] Second and final Report (1921), p. 5.

214 GOVERNMENT AND MISGOVERNMENT OF LONDON

comprehensive lines, and the question of housing was left to be discussed among a host of other subjects by the Royal Commission on London Government in 1923. It may be mentioned in passing that Mr. Sidney Webb, when giving evidence before the Royal Commission, stated that the Labour Party concurred in the report of the Departmental Committee on Unhealthy Areas that the planning of London should be considered as a whole.[1] Mr. R. C. Norman, appearing for the London County Council, informed the Commission that transport difficulties arose in connection with the Council's housing estates in Outer London. The London County Council had no power to run omnibus services to these estates and hence the provision of travelling facilities was left to chance.[2]

Despite this generally acknowledged need for a regional authority to direct policy in regard to housing, planning, transport and so forth, nothing whatever was done to bring such a body into existence. We may turn, therefore, to a consideration of the position as it actually developed.

Within the ancient City, the City Corporation is a local authority for all the purposes of the Housing Acts, 1925–35. The metropolitan borough councils are empowered to provide additional working-class accommodation within their own districts, and the London County Council may provide dwellings either inside or outside the county area to meet the needs of London generally. Both the London County Council and the metropolitan borough councils are authorities for rehousing persons displaced from unhealthy areas. The twenty-eight metropolitan borough councils are responsible for the detailed supervision of housing conditions in their areas and for carrying out repairs in default of the landlord. The London County Council, the City Corporation and the metropolitan boroughs are all local authorities for dealing with clearance areas, and any one of them may contribute towards a clearance scheme (and the consequential rehousing work) carried out by another of these councils.

As regards redevelopment schemes—that is, areas of considerable size in which overcrowding obtains or a sub-

---

[1] Royal Commission on London Government (1923), Minutes of Evidence, p. 819.　　　　[2] *Ibid.*, Minutes of Evidence, p. 44.

stantial proportion of unfit houses exists—only the London County Council and the City Corporation have jurisdiction, but the former can transfer any of its powers in this sphere to a metropolitan borough council. The intention is that the metropolitan boroughs shall be responsible for individual insanitary houses and small slum clearances while the London County Council is left to grapple with the larger slum improvement schemes. There is, however, no clear line of demarcation. "The fact that two authorities are concerned with housing should double activities," remarks Sir Percy Harris, M.P., L.C.C., "but I am afraid it often means that it falls between two stools."[1] Most students of political science would expect this result but few of them would agree with the premise.

Outside the County of London housing and slum clearance powers are possessed by county boroughs, municipal boroughs, urban districts and rural districts. There are 71 local authorities in Outer London at present engaged in providing working-class dwellings, and 30 within the administrative county, making a total of no less than 101 separate bodies for Greater London.[2]

These local authorities have been instrumental in providing nearly 140,000 houses in Greater London during the years 1920–36, which is more than a fifth of the total number built in the area during this period.[3] The above figure refers only to houses. The "dwellings" provided by local authorities, which include flats, numbered 143,500 at the end of 1935, and the total is considerably larger today. Of this, almost exactly one-half was provided by the London County Council and the other half by the remaining local authorities in Greater London.[4] About 42,000 out of the 71,634 dwellings provided by the London County Council are situated outside its own boundary in Outer London. The net outstanding debt of the London County Council in respect of housing stood at £40·6 millions on 31st March, 1935, and that of the metropolitan

[1] Percy Harris: *London and its Government*, p. 51.
[2] For details, see *London Statistics* (1934–6), vol. 39, pp. 153–4.
[3] The table on page 216 shows the shares contributed by municipal activity and private enterprise respectively to building construction in the metropolis during the period in question.   [*Continues at the foot of page 216*
[4] For details, see *London Statistics* (1934–6), vol. 39, p. 138.

borough councils at £15·2 millions, making a total of £55·8 millions.[1] The total expenditure on the service has of course been very much larger; and if we take Greater London the capital outlay incurred on municipal housing since 1920 has probably been in excess of £100 millions.

With such a heavy financial burden involved one would have thought that order and coherence would have been insisted upon as a first requisite of economy and efficiency. This would have been easy to obtain in the sphere of housing, since the state has contributed large subsidies from 1919 onwards, and Parliament and the Ministry of Health could easily have attached whatever conditions they desired to grants in respect of municipal housing. But neither Westminster nor

Continued from page 215]

HOUSES ERECTED BY VARIOUS AGENCIES IN GREATER LONDON, 1920–36

| Year | Local Authorities | | | | Private Enterprise (Houses of all Classes) | Total |
	London County Council	City Corporation and Metropolitan Borough Councils	Extra London Local Authorities	Total		
1920	268	167	712	1,147	1,495	2,642
1921	851	2,046	6,799	9,696	3,231	12,927
1922	5,497	2,085	4,465	12,047	4,860	16,907
1923	1,132	525	825	2,482	7,786	10,268
1924	1,443	360	444	2,247	15,194	17,441
1925	2,016	745	1,065	3,826	19,655	23,481
1926	4,658	862	2,833	8,353	25,176	33,529
1927	8,201	2,060	5,756	16,017	25,791	41,808
1928	9,769	1,356	3,894	15,019	27,362	42,381
1929	3,154	804	4,612	8,570	34,118	42,688
1930	3,945	664	2,922	7,531	42,652	50,183
1931	5,771	955	3,981	10,707	44,805	55,512
1932	4,506	1,688	2,131	8,325	36,288	44,613
1933	2,939	1,650	1,832	6,421	47,988	54,409
1934	3,208	2,135	2,513	7,856	72,756	80,612
1935	4,276	1,919	1,467	7,662	68,015	75,677
1936	7,076	1,673	2,385	11,134	67,704	78,838
Total	68,710	21,694	48,636	139,040	544,876	683,916

[1] For details, see *London Statistics* (1934–36), vol. 39, pp. 436–7.

Whitehall has shown any capacity to consider the housing problems of the metropolis as a whole.

Building by local authorities has been as haphazard and indiscriminate, from the standpoint of the larger issues involved, as building by private enterprise. For example, if we take the activities of the metropolitan borough councils under state-assisted schemes, we find that (on 31st December, 1935) Hammersmith had provided 2,893 rooms[1] while Shoreditch had provided only 397; Greenwich had built 4,808 rooms, Paddington none at all; Woolwich had provided 12,357 rooms and Camberwell 2,017, while Chelsea has provided only 189, Holborn 195, Southwark 228, Stoke Newington 60. Similar anomalies could be cited from out-county authorities. These enormous inequalities of performance are certainly not substantially related to differences in the housing needs of the working-class populations in those areas, nor yet to the financial resources of the districts concerned.

But inequalities of performance and the resulting injustices are not the only consequences of the metropolitan housing muddle. Actual conflict and obstruction between the various authorities are of common occurrence, especially where the London County Council is concerned. For example, the London County Council proposed to acquire some land in Chingford, Essex, which the local urban district council had zoned for building purposes. The district council heard of the proposal and hastily decided to preserve the site as an open space, solely in order to frustrate the intention of the London County Council to build in Chingford.[2] Last year the London County Council announced its decision to acquire a large site in Stoke Newington for housing purposes. This immediately arosed strenuous opposition on the part of the local borough council and certain outside interests. There are few subjects which give rise to so much friction and obstruction as a proposal by the London County Council to build a housing estate, no matter whether the site lies inside or outside its

[1] The figure for rooms is more significant than the mere number of dwellings; but the latter figure is also available. *London Statistics* (1934–6), vol. 39, p. 153.

[2] "The London County Council and the Slums," *New Statesman and Nation*, 14th March, 1936.

boundaries. The result is that the London County Council prefers to select and acquire building sites by surreptitious methods, obtaining the quiet approval of the Minister of Health with the minimum of publicity, rather than face the fuss and bother which "consultation" with the other local authority concerned nearly always implies. In no sphere are the evil effects of the disintegration of power in the metropolis more vividly illustrated than in the domain of housing and slum clearance.

CHAPTER VII

PUBLIC HEALTH

Public health is a vague and comprehensive term. It can be used to include a wide range of services, from housing to the inspection of foodstuffs, which have little connection save that they are necessary to the physical well-being of the community. It often includes the basic sanitary services of drainage and sewage disposal, street cleaning and refuse removal. I have dealt elsewhere with these matters, so I shall confine this chapter to the medical services operating directly on the individual either by way of notification, inspection or treatment.

Prior to 1929 the main items for which the London County Council was responsible in the medical field consisted of tuberculosis, venereal disease and the school medical service.[1] Of these, only the treatment of tuberculosis involved the provision of institutions by the Council.

The earliest of these in point of time was the treatment of tuberculosis, which derives from the notification of pulmonary tuberculosis initiated in 1909 which was extended to all forms of the disease in 1913. The London County Council prepared its first scheme for dealing with the illness in 1914. Under this scheme, in its amended form, the metropolitan borough councils are responsible for the provision of tuberculosis dispensaries and their management by qualified officers and visitors. This treatment is provided in some districts in municipal dispensaries, while in others there are special clinics for the purpose in voluntary hospitals. In one case a dispensary run by a voluntary association is used. Each dispensary is linked up with a hospital to which difficult cases can be sent for consultation or special out-patient treatment.[2] The London County Council is the sole authority for providing residential treatment for patients suffering from tuberculosis, and this is given mainly in its own sanatoria and T.B. hospitals or to a

---

[1] Apart from the treatment of mental disorder and deficiency.
[2] *London Statistics* (1934–6), vol. 39, p. 86.

lesser extent in voluntary hospitals with whom agreements have beeen made.

The London County Council is the supervisor and organiser in this field and pays grants to the metropolitan borough councils amounting to approximately 25 per cent of their approved expenditure.[1] In substance the powers of the metropolitan boroughs have been delegated by the public health department of the county. The advantage of assigning the dispensaries to the district authorities is that it brings the tuberculosis service into close relation with the preventive work of the local medical officer and his staff. On the other hand, it has the disadvantage of dissociating the treatment of patients by T.B. officers employed at the dispensaries maintained by the borough councils from the treatment of patients in hospitals, sanatoria or convalescent homes. The London County Council has sought to overcome this in part by giving the T.B. officer access to his patients when under observation in residential institutions or in general hospitals, even although they are not attached thereto as consultants.[2]

There may be certain defects in the existing arrangements but they reveal an intelligent attempt to organise the service as a whole and to distribute powers on a coherent basis. This is undoubtedly due to the fact that the central body (the London County Council) was authorised to prepare the scheme; and the system of grants-in-aid paid by County Hall to the metropolitan boroughs keeps their desire for autonomy within reasonable limits.

The treatment of venereal disease, which became a municipal responsibility in 1917, also shows an unusual ability to consider London "in the large." The fact that the Great War was then raging may account for the willingness of all the metropolitan authorities save one to overcome their parochial habits of mind. The law requires every county council, county borough council and the City Corporation to make arrangements for the diagnosis and treatment of venereal diseases. In London the

---

[1] Formerly the Ministry of Health also paid a grant of 25 per cent, but this has now become merged in the General Exchequer grant. *Post*, pp. 392–3.

[2] Ministry of Health: Sixteenth Annual Report, 1934–35, pp. 51, 69.

London County Council has joined with the county councils of Middlesex, Bucks, Herts, Kent, Surrey and Essex and with the county borough councils of Croydon, East Ham and West Ham, in arranging for the common utilisation of the facilities afforded by some twenty-one selected voluntary hospitals in regard to the diagnosis and treatment of both in-patients and out-patients.[1] Only the City Corporation remains outside and provides its own treatment centre at St. Bartholomew's Hospital, thus departing, as the Ministry of Health point out, from "the important principle that a patient suffering from venereal disease should not be restricted as to the locality in which he obtains treatment."[2]

The voluntary hospitals themselves had shown a sense of the magnitude and unity of the metropolis as far back as 1907, when legislation passed at their instigation to regulate King Edward's Hospital Fund for London took as its area the City and the Metropolitan Police District—that is, Greater London, with provision not only for its future extension but also for the inclusion of hospitals lying beyond the boundary and serving persons normally resident or working in Greater London.[3] This no doubt had an influence on the venereal disease scheme when it was introduced a decade later; but a point worth noting is the readiness of the voluntary hospitals to recognise the social and economic realities as compared with the reluctance of the public sanitary authorities.

The hospital and public health situation existing in London cannot be understood without some reference to the position as it grew up before 1929, when all the functions and property of the Metropolitan Asylums Board were transferred to the London County Council.

The Asylums Board was started in 1867 as a poor law

[1] *London Statistics* (1934–6), vol. 39, p. 82.
[2] Ministry of Health: Sixteenth Annual Report, pp. 52, 74.
[3] King Edward's Hospital Fund for London Act, 1907, 7 Edw. VII, c. lxx. *Cf.* Royal Commission on London Government (1923), Minutes of Evidence, p. 34. Another instance of a voluntary body which has taken a larger area than the administrative county is the London Council of Social Service, which serves the City and the Metropolitan Police District—the Greater London of the Registrar-General. The London Council of Social Service is aware of the problem presented by the size of the metropolis. See their Annual Report, 1937, p. 5.

authority superimposed on the boards of guardians for the purpose of providing for the whole metropolis "asylums" for paupers suffering from fever, smallpox or insanity. In theory and in law the Metropolitan Asylums Board was supposed to supply accommodation only for the sick poor, other persons suffering from infectious disease being the responsibility of the vestries and district boards then operating as sanitary authorities. In practice, however, the disadvantages of this division of powers were so obvious and so costly, that the Metropolitan Asylums Board gradually came to provide accommodation both for paupers and non-paupers. In 1881 a Royal Commission recommended that the function of providing hospitals for the isolation and treatment of infectious disease should be entirely divorced from the poor law and form part of the general public health arrangements of the metropolis. In pursuance of this proposal the Diseases Prevention (Metropolis) Act, 1883, was passed, whereby treatment in a hospital maintained by the Metropolitan Asylums Board was declared by itself not to be a form of poor relief; and the persons receiving it were not to suffer any civil disabilities.

In 1889 a statute dealing with the poor law further recognised the public health functions of the Asylums Board by authorising them to admit to their hospitals non-pauper patients suffering from fever, smallpox and diphtheria. After 1911, following an order issued by the Local Government Board in that year, the Metropolitan Asylums Board was empowered to receive children with measles and whooping-cough and women with puerperal fever, all without regard to whether the cases fell within the poor law or not. Other kinds of sickness and disablement were added so that by 1930 the Metropolitan Asylums Board was treating a wide variety of infectious diseases in a series of elaborately equipped general and specialised medical institutions At the date of the transfer to the London County Council they possessed *inter alia* 3 smallpox hospitals and 11 hospitals for other infectious diseases, 9 hospitals and sanatoria for tuberculosis, 3 hospitals for other specified diseases or disabilities, and 5 homes and hospitals for sick or convalescent children, at which cases were dealt with of rheumatism, orthopaedic troubles, congenital malformation, ringworm and other contagions of the skin and scalp, diseases of the eye and non-pulmonary tuberculosis.[1]

The abolition of the Metropolitan Asylums Board and the boards of guardians by the Local Government Act, 1929, led to the transfer to the London County Council of 76 hospitals and institutions, containing 42,000 beds and manned by a staff of 20,000 officers.[2] The London County Council, which had previously not been required to undertake the management of any residential institutions for medical purposes other than those appertaining to mental disease, now became responsible for the direct medical treatment of tens of thousands of patients, suffering from a multitude of divers afflictions, in a vast collection of hospitals. The London County Council now provides three out of four beds in London hospitals. The scheme formulated by the Council for carrying out its new duties announced the intention of providing assistance other than in the form of poor law relief where this could be done, as soon as circumstances permitted. The Council has engaged a corps of specialists to visit its hospitals at regular intervals and these include some of the most eminent medical practitioners and surgeons.[3]

In addition to institutional administration the London County Council also took over the domiciliary medical treatment of the sick poor in their homes, together with the ambulance services of the Metropolitan Asylums Board and the guardians.[4] As a result of all this there is now, behind the preventive services, "a comprehensive agency for the provision of skilled institutional treatment, with growing specialist facilities, available both for curative and isolation purposes and for all classes of the population."[5] It is not too much to say that since 1929 a revolution in public health work has been taking place in the metropolis.

[1] Ministry of Health Sixteenth Annual Report, pp. 75–77. This annual report contains a long section dealing with public health services in London to which I am indebted for information.

[2] Ibid., p. 55. The Council maintains (in December 1938) 44 general hospitals and institutions containing 22,309 beds and 31 special hospitals with 13,519 beds, making a total of 75 hospitals containing bed accommodation of 35,828.

[3] A. Emil Davies: The London County Council, 1889–1937. A Historical Sketch, Fabian Society, p. 28.   [4] Ibid., p. 55.   [5] Ibid., p. 96.

Despite the advantages resulting from the integration and unification of administrative power described above there still remain a number of serious defects in London's public health organisation. These are particularly noticeable in connection with maternity and child welfare work.

The metropolitan borough councils and the City Corporation are the sole authorities within the administrative county under the maternity and child welfare legislation. They provide, either directly or through voluntary associations, ante-natal clinics and infant welfare centres, health visitors and the supply of milk and meals for nursing mothers and young children. They also supply day nurseries to which medical and administrative officers are attached. Moreover they further provide, in varying degrees, midwives, maternity nursing and residential accommodation of various kinds. But here the London County Council also carries out similar functions.

Prior to the transfer in 1929, many confinements took place in the maternity wards of poor law institutions. Now that the former poor law hospitals have become appropriated as public health hospitals the maternity departments have acquired an enhanced status and are being used to an increasing extent.[1] The policy of the London County Council is to concentrate maternity work in the most suitable hospitals and to modernise the equipment and improve the staffing. At the end of 1934 there were more than 700 maternity beds available in the Council's hospitals, and 11,689 children were born in them during the year.[2] The number is now considerably larger. "The position is obviously of far-reaching importance" observes the

---

[1] The figures are as follows for the years 1931–7:

			Total births in London	Births in the L.C.C.'s general hospitals and institutions
1931	..	..	67,889	10,191
1932	..	..	64,220	11,425
1933	..	..	58,677	11,917
1934	..	..	58,697	13,253
1935	..	..	57,634	15,519
1936	..	..	58,130	18,261
1937	..	..	56,875	19,944

The totals given above include live and still-births.

[2] *Ibid.*, pp. 52–3; *London Statistics* (1934–6), vol. 39, p. 91.

Ministry of Health, "in view of the recommendation of the Departmental Committee on Maternal Mortality and Morbidity that new maternity accommodation should, where practicable, be associated with general hospitals. At the same time, the borough councils also possess the power to provide maternity homes or hospitals, either directly or by arrangement with other bodies (normally voluntary associations and voluntary hospitals). In fact, there are seven such homes containing 137 beds, which have been provided by metropolitan borough councils and in addition, considerable use has been and is made by the borough councils of beds in voluntary institutions. . . . Obviously, the proper correlation of these different sources of maternity accommodation is of the first importance to the future of the service."[1]

The full extent of this overlapping can be seen when we remember that the metropolitan boroughs are empowered to provide ante-natal clinics in their areas. The poor law authorities prior to 1929 were supplying ante-natal supervision in connection with their maternity wards, and the London County Council regard it as essential to arrange for the medical supervision of all women who intend to be confined in their hospitals, in order to ensure continuity of medical care. Hence the London County Council hospitals have largely increased their ante-natal provision and are also offering post-natal services in all cases after confinement. Where home visiting or domestic care is needed the matter is referred to the medical officer of the local borough.[2]

Overlapping also exists in regard to various infectious diseases. The metropolitan borough councils and the City Corporation are the local authorities to which notification must be made in any case of puerperal fever, puerperal pyrexia, ophthalmia neonatorum, and certain children's diseases such as measles and whooping cough. These district councils are responsible for dealing with these diseases, and they provide consultants, hospitals and convalescent homes for the purpose. But formerly the Metropolitan Asylums Board was required to provide hospital accommodation for such diseases, and the London County Council has inherited its functions in this

[1] Ministry of Health Sixteenth Annual Report, p. 52–3.   [2] Ibid., p. 54.

15

respect. Moreover, the London County Council is the supervising authority under the Midwives Acts and also registers and inspects private maternity homes. It therefore has a special interest in the infections of child-birth.[1] The fact that a copy of the notification of puerperal fever, etc., is sent to the medical officer at County Hall scarcely affects the situation.[2] The confusion resulting from the duplication of powers by concurrent authorities is not only uneconomical and inefficient, but also leads to friction.[3]

I have already mentioned the extensive series of hospitals and convalescent homes specially designed for treating children's diseases which the London County Council took over from the Metropolitan Asylums Board.[4] Here again there is overlapping with the metropolitan boroughs, who also provide homes or wards or beds in voluntary homes for children under the age of five.[5]

It is clearly desirable that maternity and child welfare work should be linked up with the school medical service, both as regards premises and staff. The Local Government Act, 1929, following a recommendation by the Royal Commission on Local Government, enables the Minister of Health, on the application of the county, to require the local education authority to take over the maternity and child welfare work in its area. In the metropolis the administration remains entirely separate, the London County Council being responsible as education authority for the medical inspection and treatment of school children, while the metropolitan borough councils and the City Corporation are in charge of maternity and child welfare functions. Only in a few cases have arrangements been made locally to house the two clinics in the same building.[6] Co-ordination of staff is presumably not considered possible.

The London County Council, again in its capacity of local education authority, attempts to control the spread of infec-

[1] *Loc. cit.*

[2] *British Medical Journal*, 7th September, 1935, supplement.

[3] Royal Commission on London Government (1923). *Cf.* Hackney, Minutes of Evidence, p. 981.   [4] *Ante*, p. 222.

[5] Ministry of Health, Sixteenth Annual Report, p. 53. The London County Council also uses beds in voluntary homes.

[6] *British Medical Journal*, 7th September, 1935, supplement.

tious diseases in the schools, while each metropolitan borough is entrusted with the control of infectious diseases in the homes and workplaces in its area, including the homes of the school children. It is scarcely surprising that Captain Warburg, one of the witnesses for the London County Council, told the Royal Commission on London Government that "the difficulties in health administration mainly arise owing to the multiplicity of authorities."[1]

One of the incidental but important results of the unscientific division of functions between the major and minor authorities in London is that the work of a medical officer in the metropolis is less extensive in scope and therefore less interesting than it is in a large provincial city, where all the public health functions are under the unified control of the county borough council. Hence it is probable that in general London does not get the pick of the men applying for these posts. In several of the counties the system of combined appointments for county and district medical officers has overcome many difficulties which otherwise exist.[2] If it were practical to introduce this method into the County of London it would certainly remedy this particular defect.

The inspection and enforcement of sanitary conditions generally is in the hands of the metropolitan boroughs and the City Corporation. These bodies are responsible not only for administering a large number of statutory provisions of a regulatory character dealing with nuisances, lodging houses, offensive trades and so forth, but also for enforcing their own by-laws. The London County Council can also make by-laws enforceable by the metropolitan borough councils, but it rests with the latter to provide an adequate and qualified sanitary staff—if they think fit to do so.[3] The variation in the standard of administration in these matters as between different parts of the metropolis is very striking. Chelsea, (population 59,031) for example, employs only 4 full-time inspectors, while Finsbury (69,888) has 9; Wandsworth (353,110) has only 17, while Stepney (225,238) has 23 and Westminster (129,579) has 18.

[1] Minutes of Evidence, p. 166.
[2] See W. A. Robson: *The Development of Local Government*, pp. 304-6.
[3] Ministry of Health, Sixteenth Annual Report, p. 47.

an analysis of the actual administration discloses remarkable variations. The 4 common lodging houses in Bethnal Green were inspected 45 times in 1935; Camberwell's 4 lodging houses were visited on 141 occasions in the same year. Fulham's lodging house was visited 75 times, the one in Greenwich 6 times, that in Hammersmith 23 times. The St. Marylebone sanitary staff inspected 224 restaurants on 426 occasions in 1935, the Southwark staff inspected 418 restaurants on 1,807 occasions, while in Paddington only 58 inspections were made of 132 restaurants.[1] Similar contrasts could be drawn in regard to the inspection of milk shops, cowsheds and slaughter-houses, offensive trades, ice-cream premises, underground rooms, smoke nuisances, houses let in lodgings and house property generally. On what grounds is it possible to justify the absence of a common sanitary standard for the capital city?

Under the Local Government Act, 1929, Section 64, the London County Council is empowered to transfer functions to the metropolitan borough councils or to appoint them to act as agents for the County Council by means of an Order made by the Minister of Health. The Minister must first receive an application from the London County Council or from any association or committee of the metropolitan boroughs; in either case he must consult the other party concerned. The Order must be laid before Parliament. An Order was made,[2] to take effect from April 1933, whereby all the duties except the making of by-laws which the London County Council formerly carried out with regard to common and seamen's lodging houses, cowhouses, infant life protection, offensive trades, slaughter-houses and knackers' yards were transferred to the metropolitan boroughs. Thus the district authorities now have responsibility for inspecting and enforcing the County Council's ordinances in regard to these matters.[3] There is doubtless a good case for decentralised administration in the regulation of local sanitary details in a great urban area, but it is remarkable that the London County Council should

[1] *London Statistics* (1934–6), vol. 39, pp. 79–81.
[2] The Transfer of Powers (London) Order, 1933. Statutory Rules and Orders, No. 114/1933.
[3] A similar power in regard to the registration and inspection of nursing homes is contained in the Public Health (London) Act, 1936, Section 249.

have been ready to surrender its powers, and the Minister willing to transfer them to the district authorities, without any assurance that the irregular standards to which I have referred would be brought up to a common minimum or at any rate subjected to some form of over-riding control.

The Public Health (London) Act, 1936, requires the London County Council to make by-laws relating to the removal and disposal of refuse, and as to the duties of occupiers for facilitating its removal by the scavengers of the sanitary authority, which is, of course, the metropolitan borough council.[1] Again, the County Council may make by-laws regulating various matters connected with the demolition of buildings, and these are enforceable by the local sanitary authority.[2] Similar arrangements exist in regard to by-laws directed to promoting hygienic conditions in the manufacture, preparation and distribution of foodstuffs.[3] There are signs here of a recognition of the inter-relation between the major and minor local authorities in the field of public health; but these provisions go no further than imposing on the local authorities the policy laid down by the London County Council, without any kind of safeguard against evasion or inefficiency in administration.

Other sections of the public health code applicable to the metropolis enable the Minister of Health, on hearing a complaint by the London County Council that a metropolitan borough council has made default in enforcing any of the provisions of the Act, to make an order calling upon the borough council to perform their duty within a specified time. If there is further default, the order may be enforced by a writ of mandamus, or the Minister may appoint the London County Council to carry out the duty in place of the defaulting authority.[4] Where the Minister thinks a local authority is making, or is likely to make, default in enforcing any epidemic regulations, he can assign those duties for a specified period to the London County Council.[5]

These enactments do not sound impressive to anyone acquainted with the long and dismal history of "default powers." The statute book of the past 50 years is littered with instances where, mistaken principles having been adopted in the organ-

[1] Section 84. (2).   [2] Section 85.   [3] Section 183.

isation of local government areas or authorities, or the distribution of functions, Parliament has attempted to salve its conscience by the insertion of default powers, whereby some other person or body can be appointed to do the job. I have yet to hear of an instance where the power of acting in default has been successfully invoked and effectively applied. It normally remains a dead letter, an empty threat; and such is likely to be the case in the instances to which reference has been made.

I have confined my remarks in this section to the position within the County of London. The situation in Outer London is profoundly unsatisfactory in regard to the organisation of public health, but it would take too long to make a detailed survey of the medical and sanitary services outside the administrative county. It is obvious that the county boundary is a line devoid of social significance, and since it has long ago been ignored by the voluntary hospitals in their co-operative activities it is difficult to see why it should be preserved for municipal functions in the field of health.

Even limiting the enquiry under this head to the administrative county, it is clear that there are grave defects in the existing organisation of authorities for health functions. Overlapping and a multiplicity of administrative organs are to be found in several vital fields, while in others there is no common standard or minimum of efficiency insisted upon throughout the area. The enormous increase in the responsibilities of the London County Council since 1929 have transformed the public health situation in the metropolis and made the solution of these problems urgent.

CHAPTER VIII

MAIN DRAINAGE

I have shown in the first part of this work[1] the way in which the drainage and sewage disposal problems of the metropolis were persistently neglected until a critical stage was reached, when, after much bickering and friction, the Metropolitan Board of Works was authorised to construct and operate the main drainage system on which the capital now relies.

The area originally defined for metropolitan drainage was the same as that which now forms the administrative county of London. But starting in 1871 the system was enlarged to take in a large part of Hornsey, and two years later part of Beckenham was included. Since then a continuous process of enlargement has taken place until at the present time the following outlying local authorities (in addition to the two already mentioned) now use the London County Council's system for the whole or part of their areas: the county boroughs of East Ham, West Ham, and Croydon; the municipal boroughs of Acton, Ealing, Brentford and Chiswick, Willesden, Wood Green, Tottenham, Walthamstow, Leyton, Ilford, Barking, Mitcham, Edmonton, Erith, Bexley and Dagenham; the urban districts of Chislehurst and Sidcup and Penge. The main drainage area changes from year to year as new districts, or parts of them, are admitted; or, as very occasionally happens, withdraw therefrom. The position is further complicated by the revision of county district areas made under the provisions of the Local Government Act, 1929. The latest figures available show the size of the area as almost 178 square miles, containing an estimated population of 5,529,500.[2]

The question of admitting neighbouring authorities to the London drainage system has hitherto been dealt with by the London County Council on public-spirited lines for the common good, on the principle that a continuous urban area

---

[1] *Ante*, p. 121 *et seq.*
[2] *London Statistics* (1934–6), vol. 39, pp. 13, 127. The total capital cost from 1856 to 1930 was £16 millions.

should, if possible, be drained as one entity, thereby avoiding separate sewage works, always a potential nuisance, and the more complete treatment of the sewage which is necessary when the effluent is discharged into a small stream.[1] The representatives of the London County Council informed the Royal Commission on London Government more than fifteen years ago that the existing London system was nearly full, owing partly to the limited size of the existing sewers and disposal system, and partly to the limited capacity of the Thames.[2] Since then some thirty square miles of built-up territory have been admitted.

Although the attitude of the London County Council has on the whole been reasonable, the negotiations have usually been long and difficult, and a special Act of Parliament is required in each case. Furthermore, in some of the recent instances the outlying authority has been left with a very considerable liability in regard to the disposal of storm water and even of sewage in time of emergency. Hence it cannot be said that the difficulties have been solved in all cases even by admission to the principal system.

Of much greater importance than these considerations is the situation created by the growth of the metropolis. The Chief Engineer to the London County Council explained to the Royal Commission on London Government as long ago as 1923 that the Thames is the ultimate drain of nearly all the territory within a radius of fifteen miles from Charing Cross,[3] and that the efficient discharge of sewage from the population within this area demanded the application of broad principles by a single authority.[4] This was far from reflecting the existing position since there were 122 local authorities operating within their separate areas.

The Royal Commission was impressed by the technical evidence, which went to show that nearly fifty separate disposal works had been established in as many localities in circumstances where "greater efficiency and economy could have been

---

[1] Royal Commission on London Government (1923): Evidence of Humphreys, Chief Engineer, London County Council. Minutes of Evidence, p. 95.   [2] Ibid., pp. 27, 95.
[3] Minutes of Evidence, Humphreys, p. 95.   [4] Ibid., p. 109.

secured in the first instance by combination between local authorities whose sewage would naturally have been taken by a single system."[1] Their attention had also been drawn to the fruitless attempts of the Metropolitan Water Board to clear away the sewage farms above the intake on the River Lee, for which purpose the Water Board had offered to make a financial contribution towards the cost of a joint scheme; but a Bill embodying these proposals had to be abandoned owing to the opposition of the local authorities concerned, who regarded the preservation of their sewage farms as more important than 'he purification of the water supply.[2] In the result, the Royal Commission unanimously recommended in its majority and two minority reports substantial changes in the drainage service and an enlargement in its area of operation. The majority suggested a small statutory committee to advise the Minister of Health in respect of an area about twenty-four miles radius from Charing Cross on main drainage and certain other questions. The first minority report proposed a central authority to administer a number of major services, including drainage and sewage, within an area at least ten miles distant from Charing Cross (about 314 square miles). The second minority report proposed that the Metropolitan Police District, which varies from between twelve to fifteen miles from Charing Cross and comprises 692 square miles of territory, should be taken as the drainage area. All these recommendations were applications in one form or another of the idea that the drainage régime ought to be under the control of one directing hand, no matter what technical methods of sewage disposal might be adopted.[3]

No action of any kind was taken to implement any of these recommendations, despite the vital character of the function. Since then the position has become greatly aggravated in every respect in which danger and inconvenience can result to the public. The reasons for this are obvious When a small town or residential settlement is surrounded by open country for several miles, a sewage disposal plant can easily be established in a place where it will create no nuisance to anyone and

[1] Report of the Royal Commission on London Government (1923), p. 64.
[2] *Ibid.*, Minutes of Evidence, p. 198.   [3] *Ibid.*, p. 103.

where the effluent can be safely discharged into the nearest stream. But where building expansion takes place on such a vast scale as that which has occurred in Outer London so as to cause these outlying towns or villages to grow together, the sewage works which were originally isolated become embedded in the midst of rapidly-growing urban communities. The sites which formerly seemed adequate in size are now found to be far too small to permit of the necessary expansion and the acquisition of other suitable land becomes increasingly difficult. Moreover, the families dwelling in the neighbourhood are troubled with noxious smells and flies; and if the works are not extended to meet the growing requirements of the neighbourhood, the water courses become polluted and offer a constant menace to the public health.[1]

These developments and their social consequences have recently been investigated by three very eminent sanitary engineers invited by the Minister of Health to report on the situation and to formulate a policy. The Report on Greater London Drainage prepared by Messrs. Taylor, Humphreys and Frank states that the Greater London drainage area consists of a territory covering 1,928 square miles, of which the London County Council main drainage area constitutes a small fragment comprising a mere 178 square miles. The area outside the London County Council's system is thus ten times larger than that inside it. The region in question lies approximately within twenty-five miles of Charing Cross.

Within this Greater London drainage region are 147 local authorities of various classes[2] serving (in 1931) a population of 9,236,978.[3] The total number of sewage disposal works contained in it is 182, omitting a few small works and those belonging to institutions and private persons. On the average there is a sewage works in every $9\frac{1}{4}$ square miles of the territory

---

[1] Taylor, Humphreys and Frank: Report on Greater London Drainage (1935), H.M.S.O., p. 25.

[2] Seven county councils (Essex, Middlesex, Herts, Bucks, Berks, Surrey and Kent), 2 county borough councils, 25 municipal borough councils, 78 urban district councils and 35 rural district councils.

[3] About 3,700,000 persons of this total live outside the area covered by the London County Council drainage system.

situated outside the London County Council's main drainage system, and the total area covered by the sites occupied by these plants amounts to nearly nine square miles.[1] The principal undertakings in the region, apart from the London main drainage system, are the West Kent Sewerage Board's main sewers and outfall works and the recently completed West Middlesex scheme. These three undertakings deal with an area of only 450 square miles, which is less than a quarter of the Greater London drainage area.[2] The Middlesex scheme was still under construction when the report was drafted. It superseded the disposal works in twenty-eight urban areas containing an aggregate of about 1,250,000 persons; the scheme cost approximately £5 millions.[3]

The three experts who prepared the report recommended unanimously that all the arrangements for sewage disposal in and around the metropolis should be co-ordinated and planned in relation to an area of about twenty-five miles radius from Charing Cross; and that for this purpose a scheme should be considered whereby the whole of this area would be served by ten, or fewer, centralised disposal works. "The outstanding conclusion is that, from an engineering point of view, co-ordination and unification over a wide area are essential in order to secure the most effective and satisfactory method of dealing with the sewage of Greater London."

The advantages resulting from the unification, co-ordination and the adoption of large specific schemes propounded by the engineers would touch on a number of different aspects. The "nuisance centres" would be reduced from nearly two hundred to ten. The difficulty of finding further sites for new disposal works would be overcome. The problem of preventing building close up to sewage plants would be solved. The amenities of numerous districts in the neighbourhood of existing sewage works would be improved by converting the works into open spaces, recreation grounds or building sites. The rateable

[1] Report on Greater London Drainage (1935), pp. 20–21.
[2] Ibid., p. 27.
[3] Eighteenth Annual Report of the Ministry of Health, [Cmd. 5516] 1937, p. 68.　　　　[4] Report on Greater London Drainage (1935), p. 44.

value of such areas would be increased. The water supply would be more adequately protected and possible sources of river pollution reduced. The sewage disposal problem which now harasses 140 local authorities would disappear. It would be possible to adopt the most satisfactory system of main sewers.[1]

The relation between drainage and other functions of city government also calls for comment. Mr. R. C. Norman, appearing as a witness for the London County Council before the Royal Commission on London Government, was asked whether he regarded the drainage area as a criterion for the boundaries of future London. He replied: "No. It is only one of the factors which you have to have regard to, but it is a very important factor in municipal life."[2] The fact that outlying authorities sent their sewage into the London County Council sewers was not in his opinion sufficient to bring their areas within the jurisdiction of County Hall.

The intervening years of undirected expansion have shown the insufficiency of this answer. Urban development necessarily depends on whether drainage facilities are available or not; hence the sewage disposal service plays a vital part in determining both town expansion and, ultimately, the extension of boundaries. The engineers' report remarks that the case for the unification of the drainage system would be definitely established if orderly building development, including the sterilisation of open spaces, were under unified control throughout the Greater London Region. "But even without such control," they add, "there is a strong case in favour of drainage unification."[3]

No steps have so far been taken to apply the advice unanimously tendered by the leading public engineers in this field of work. In the meantime, it is said that further unco-ordinated schemes are being projected by separate authorities, including a plan for the Lee valley to cost £4½ millions and another for the Colne to cost £1¼ millions. It would appear that an ancient device employed by administrative bodies whose existence is

---

[1] Report on Greater London Drainage (1935), p. 43.
[2] Royal Commission on London Government, Minutes of Evidence, p. 26.   [3] Report on Greater London Drainage, p. 26.

threatened is once more being invoked: namely, to embark on expensive but imperfect schemes within their existing powers in order to deter Parliament and the Government from ordering comprehensive reform on account of the expense involved in scrapping the schemes already adopted.

CHAPTER IX

ELECTRICITY SUPPLY

The metropolitan gas supply was appallingly bad throughout the 19th century; but when electricity came into practical use in the 1890's, the disadvantages which had been shown to result from the private ownership of public utilities in the case of gas and water did not induce Parliament to authorise the London County Council to provide a municipal service, although the Progressives at County Hall were very keen on all the utility undertakings in London being municipally owned and operated.

Instead of the London County Council or some larger authority being empowered to provide a supply for the whole metropolis, the minor municipal bodies were permitted to establish undertakings in some districts while companies were granted a franchise in others. Hence the distribution of electrical energy within the county is in the hands of some 16 metropolitan borough councils and 14 joint stock companies.[1] The allocation of districts as between public enterprise and private enterprise respectively was based mainly on political considerations rather than on technical questions. Profit-making operation managed to secure a foothold in the wealthy Central and West End portions of the town, where the local authorities were usually Conservative, while municipal ownership was established in the East End and along the south of the river and in some of the outer suburbs where Progressive majorities were in power.[2]

The results of this parcelling-out of the metropolis among parochial undertakings has been deplorable. "The current is alternating in some areas and direct in others; it is delivered at a number of different voltages and is charged for under a bewildering variety of tariffs and rates with the result that consumers living in one street may be paying 3d. a unit for

---

[1] Half a dozen of the companies have recently amalgamated.
[2] There were, of course, some exceptions, but these were the main tendencies.

lighting while those in the next street are charged 5d., and electrical apparatus may become absolutely useless when its owner removes from one district to another. More than one company is operating in the same borough; for example, the supply in the Royal Borough of Kensington is undertaken by three companies. Nowhere in the world does such a state of affairs exist."[1]

Thus wrote Sir Harry Haward, for many years Comptroller to the London County Council, in 1932. And he asks us to visualise the position which would exist today in London if the great unified schemes which were prepared at County Hall in 1905 and 1907 had been carried out by or under the control of the Council. Electricity would have been generated at a few large stations situated well out of the way along the lower reaches of the river. Hence most of the plant used at the seventy stations then established in the area would have been scrapped and written off with a consequent saving of the millions of pounds spent in extending these stations. The authorised distributors, both municipal and company, would have become retailers only and been able to devote the whole of their efforts to developing the sales and consumption of electricity. Consumers would have been supplied at far lower prices and under more consistent tariffs. In short, the objects sought to be achieved by the Electricity Supply Act, 1926, would have been attained twenty-five years earlier, to a much greater extent so far as the metropolis is concerned, and at a much lower cost of money. The organisation of a public service essential both to domestic life and industrial efficiency, would have been formed on a basis to challenge comparison with that found in any of the other great cities of the world.[2]

Unfortunately, electricity supply shared the same fate as that to which the other public utilities in the metropolis were doomed; and municipal administration on a parochial scale or operation by profit-making companies was allowed to continue unchecked until our own day. Another great opportunity was available in 1931, when the powers of the London supply companies were due to expire and the undertakings became

[1] Sir Harry Haward: *The London County Council from Within* (1932), p. 358.   [2] *Ibid.*, p. 368.

liable to compulsory purchase by the London County Council. But the London County Council, with a majority at that time averse to municipal ownership, failed to accept the opportunity and agreed to postpone the date of compulsory purchase until 1971, on condition that a quite inadequate measure of re-organisation was undertaken by the units concerned. The London and Home Counties Joint Electricity Authority was established on the passage of the London Electricity (Nos. 1 and 2) Acts, 1925, promoted by two groups of companies to confirm agreements between the London County Council and the companies with regard to future working.

The joint electricity district covered by the Authority contains 1,841 square miles with a population of 9,088,600. Its boundaries almost coincide with those of the London and Home Counties traffic area. Within this district are no less than 82 separate authorities distributing electricity. These consist of 16 metropolitan borough councils and 14 companies inside the County of London; and in Outer London of 22 municipal corporations, 6 urban district councils and 24 companies.[1] About 42·7 of the total supplies in the London district are given by local authorities and 57·3 by the companies.[2] More than 40 of the generating undertakers are giving supplies in bulk to other authorised distributors.

The Acts of 1925 empowered the companies in each group to amalgamate; provided for the regulation of dividends by sliding scales; required the companies to set up sinking funds to secure the transfer of their undertakings in 1971 to the joint electricity Authority; make it obligatory to notify the Authority of any proposal to expend a capital sum exceeding £5,000 on assets which may be purchasable by the Authority in 1971; to maintain transferable equipment to the satisfaction of the Authority; to dispose in accordance with the Authority's reasonable directions of any electricity generated in excess of requirements; and to carry out in their own districts the relevant part of the technical scheme.[3]

[1] For details, see *London Statistics* (1934–6), vol. 39, pp. 347–52.
[2] Graeme Haldane: *The Central Electricity Board and other Electricity Authorities* in *Public Enterprise*, edited by W. A. Robson, p. 147.
[3] *London Statistics* (1934–6), vol. 39, p. 346.

The London and Home Counties Joint Electricity Authority consists of thirty-six members appointed or elected by various groups of companies and of local authority undertakers inside and outside the county; by the London County Council, by the City Corporation, and by other county councils; by the workers in the industry; and by the Railway Companies Association.[1] It has a most unwieldy constitution.[2]

The Joint Authority has few duties so far as generation and main line transmission are concerned. It has, however, begun to acquire distribution powers in the southern fringes of Outer London, such as Dorking, Leatherhead, Surbiton and Weybridge, and now supplies directly about 100,000 consumers. But these areas amount to only about 190 square miles out of a total area of 1,850 square miles covered by the electricity region.[3]

The reorganisation effected by the legislation of 1925 was not drastic, and the provisions relating to the co-ordination of generation and main line transmission have been largely superseded by the South-East Electricity Scheme prepared by the Electricity Commission under the Electricity (Supply) Act, 1926, and carried out by the Central Electricity Board, a national body set up by that Act. The functions of the London Power Company, which was formed by the amalgamation of ten companies to supply the West End of London, now overlap those of the Central Electricity Board.[4] It appears that production costs of electricity have been considerably reduced but prices are still far too high, although the sliding scale requirements have had an appreciable effect in reducing charges. The number of different tariffs in operation within the metropolis is very large.[5]

The position in regard to electricity supply and distribution within the metropolis is clearly unsatisfactory, and its defects are obvious to any competent engineer or administrator. The

[1] *London Statistics* (1934–6), vol. 36, p. 3.
[2] Haldane: *op. cit.*, p. 144. [3] *Ibid.*, pp. 146–7.
[4] Mr. Haldane's essay, to which reference has been made, gives an excellent account of the Central Electricity Board and other electricity authorities.
[5] See the *Statistical Review* (1936), No. 9, published by the Joint Electricity Authority.

16

multitude of authorised undertakers remain in operation under separate management until 1971; the division of the territory between public and private enterprise is unscientific and undesirable; the London and Home Counties Joint Electricity Authority is a weak federal body, cumbersome in structure and feeble in function, unable to exercise effective control over the vital aspects of electricity distribution.

CHAPTER X

WHOLESALE FOOD MARKETS

The ownership and management of markets is one of the oldest and most widely recognised municipal functions. All over the world the provision of markets is regarded as one of the most essential tasks of a city council. In London this matter, like so many others of public concern, has been left to the vagaries of ancient privilege and private profit.[1]

In the year 1326 a royal charter was granted to the City Corporation confirming the members in their liberties, customs and rights to hold markets, and forbidding the holding of others within seven miles of the City boundaries. At that time, and for long afterwards, the citizens both lived and worked inside the square mile of the ancient City, and the effect of this monopoly was merely to prevent the establishment of rival markets under separate authority on the outskirts of London. With the passing of six hundred years since the reign of Edward III the situation has entirely changed; and the vast metropolis which has grown up around the City is now subjected to arrangements whereby most of the wholesale food markets are under the control of the 40,000 or so persons who nominally constitute the body of local government electors within the square mile.

The City Corporation owns and operates the Billingsgate fish market, Smithfield meat market, the Leadenhall market for poultry, game and eggs, the Islington market for cattle, sheep and horses, and the Spitalfields market for fruit and vegetables. Both the Islington and Spitalfields markets are situated outside the area of the City. The Corporation has always been extremely zealous to protect its charter privileges but on more than one occasion it was unable to resist an encroachment on its monopolistic powers. Charles II granted to the fourth Earl of Bedford the right to hold a fruit and vegetable market at Covent Garden and this remained a

---

[1] A good account of London markets in the 19th century is contained in Sir G. Laurence Gomme: *London in the Reign of Victoria*, pp. 94-125.

substantial source of income to the Bedford family until 1913, when the market was sold. There are two other wholesale food markets in public ownership: the Borough market in Southwark for fruit, vegetables, dairy produce, eggs and poultry (owned by Southwark Borough Market Trustees), and a market for fruit, vegetables, corn, etc., at Brentford, owned by the Municipal Borough of Brentford and Chiswick. The London and North Eastern Railway owns a potato market at King's Cross and a market for fruit, vegetables, corn, forage and eggs at Stratford; while the London, Midland and Scottish Railway owns a potato market at Somers Town, St. Pancras.[1]

The London County Council has from time to time attempted to obtain powers to acquire or establish markets. In 1890 it asked Parliament for the necessary powers, but the clause was deleted by the House of Lords. In 1899 the London County Council promoted a Bill to acquire Spitalfields Market, which was then under private ownership. In 1919 it persuaded the Government to appoint a Departmental Committee to consider the whole question of market facilities and ownership in London, with a view to reducing the high price of food.[2] Nothing came of any of these efforts. But the County Council failed to take advantage of two golden opportunities to assert its claim to represent the majority of Londoners in the vital matter of wholesale food markets. In 1913, when Covent Garden was sold, the purchaser was willing to resell the property to the London County Council, but the Municipal Reform majority refused to consider the project on political grounds, and so the market has always remained in private hands.[3] In 1911, the Metropolitan Borough of Stepney, which had obtained Parliamentary powers to purchase or lease Spitalfields, decided to offer its rights to the London County Council, which in 1899 had been so anxious to acquire the market. But once again political considerations intervened and the County Council stood aside in favour of the City Corporation.[4]

The Covent Garden market has been severely criticised in recent times. The Departmental Committee on Wholesale

[1] For details, see *London Statistics* (1934–6), vol. 39, pp. 326–7.
[2] Percy Harris: *London and its Government*, pp. 218, 221.
[3] *Ibid.*, p. 216. [4] *Ibid.*, p. 221.

Food Markets in 1921 described the buildings as obsolete, inconvenient and badly lighted; and condemned the internal conditions for causing delay in the delivery of goods both to and from the market.[1] Sir Percy Harris, who has given much attention to the subject, emphasises the extreme inconvenience of its situation, which is far from railway termini or goods yards, the river and the port of London. It lacks the advantage of proximity to the residential quarters of the metropolis and is nowhere near a cheap shopping centre. The passage of horse and motor vehicles conveying the million tons of produce which are dealt with annually at the market through the narrow streets of Central London adds greatly to the unnecessary traffic congestion.[2] The highly perishable nature of some of the fruit dealt in at the Garden results in frequent economic loss through delay in transit. The Departmental Committee drew attention to the fact that the owners of Covent Garden were then making active efforts to dispose of it, and expressed the opinion, "in the strongest possible terms, that this market, the largest of its kind in the Kingdom, should be placed under a public authority with a view to its development in the interests of the trade and the consumer."[3]

The fish market at Billingsgate was also severely criticised by the Committee on the ground that the area was inadequate —it covers only 1·2 acres—and the position unsuitable. The market, they stated, compares very unfavourably with the wholesale fish markets in the leading provincial cities, and the situation is highly inconvenient for handling the 225,000 tons of fish which are now bought and sold there each year.[4] The City Corporation has effected a number of improvements since the publication of the report, but the fundamental objections to the site remain.

The evidence adduced at the inquiry convinced the Committee that improvements and extensions are urgently needed at some of the markets which cannot be carried out under their present ownership and management; and that in the case

[1] Percy Harris: *London and its Government*, p. 217.   [2] *Ibid.*, p. 216.
[3] Departmental Committee on the Wholesale Food Markets of London (1921), B.P.P., vol. xii, Fourth Report, p. 8–9.
[4] Percy Harris, *op. cit.*, p. 218.

of other markets the question of removal ought at once to be considered. But, they pointed out no existing authority has the necessary powers to deal with this question.[1] Hence they recommended that a new market authority should be established in the metropolis.

In formulating this part of their advice the Departmental Committee laid stress on the obvious and indisputable fact that the functions of the wholesale food markets are not confined to the area controlled by any existing local authority. The produce is distributed over a region extending far beyond the present boundaries of London. "We are of opinion that the London wholesale market problem should be dealt with as a whole, and in this connection would point out that there is at present no one authority which has, so far as Greater London is concerned, power to create new markets or to improve, remove, or co-ordinate those now in existence."[2]

The Committee considered three possible alternatives in suggesting the appropriate authority: the City Corporation, the London County Council and a specially constituted body. They recommended in favour of an *ad hoc* body on which the first two councils would be adequately represented.

The duty of the new authority would be to apply the principle that market facilities ought to be administered in the public interest of the whole metropolis and not in the interests of private owners or of separate municipal councils. It would have power to acquire land and other property compulsorily; to raise funds for the purchase of existing rights and new markets; to close markets or remove them to more suitable sites; to fix tolls, rents and charges; to construct subsidiary buildings, storage plants and other facilities; to make by-laws regulating porterage and other matters; and generally to provide and maintain a modernised and well-planned system of markets suitable to meet the needs of a great metropolitan community.[3]

The fruit of these labours fell on deaf ears. Nothing whatever has been done to rationalise the wholesale food markets in Greater London or, for that matter, in the administrative

[1] Departmental Committee on the Wholesale Food Markets of London (1920), First Report, p. 3.   [2] *Ibid.*, p. 4.   [3] *Ibid.*, pp. 4, 5.

county or the City of London. The City Corporation remains entrenched in its ancient privileges and continues to make a handsome profit out of markets serving the whole metropolitan region. The history of the City markets, wrote Sir Laurence Gomme, is not a pleasing one. "It is the history of a struggle for private rights, not for municipal duties."[1] The owners of Covent Garden and the other private markets continue to levy tribute on the foodstuffs which sustain London's teeming millions. Unnecessary inconvenience continues to prevail, prices are higher than they need to be, the streets more congested, traders less prosperous, the poorer families less well nourished, than they would be under a coherent and unified system of wholesale food markets.

[1] Sir G. Laurence Gomme: *London in the Reign of Victoria*, p. 94.

CHAPTER XI

FINANCE

Finance is the key to local government activity, and, indeed, to all forms of public administration. The system of municipal finance is somewhat involved throughout Great Britain but it is specially complicated in London, partly owing to the peculiar organisation of local authorities and partly because special arrangements have been introduced in the metropolis which do not exist elsewhere. I shall not attempt a comprehensive survey but merely deal with a few matters of outstanding importance.

The total gross expenditure of the local authorities within the administrative county for 1933–4 was £59·3 millions.[1] Of this, some £18·7 millions was met from receipts in aid of services, including receipts in the nature of indirect taxes, such as water assessments made by the Metropolitan Water Board, and rents from municipal houses. This left £40·6 millions to be defrayed from rates and taxes. Rather more than £12 millions was paid by grants in aid from the central Government, of which £7·8 millions was for specific services, such as education, and £4·2 millions by way of the General Exchequer contribution. This left a sum of £28·3 millions to be raised out of the rates.

These figures relate only to the Administrative County. Table I on page 249[2] shows the totals of population, rateable value and rates raised for Greater London.

It will be observed that the London County Council, which spends more than twice as much as all the other local authorities in the Administrative County put together—including the Commissioner of Police for the metropolis and the Water Board—is not mentioned in the table. This is because the London County Council is not authorised to levy rates itself but is required to issue precepts or demands on the

[1] This figure was arrived at after deducting certain payments included twice in the gross disbursements which totalled £61 millions. See *London Statistics* (1934–6), vol. 39, p. 397.

[2] Reproduced from H. Morrison: *How Greater London is Governed*, p. 151.

borough councils within its area. These minor bodies are responsible for the levying and collecting of their own taxes and those embodied in the London County Council's precept.

TABLE I

Rating Authority	Population (1931)	Rateable Value (1933)	Approximate Amount of Rates raised (1933-4)	Per cent. of Rateable Value
		£	£	
City of London ..	10,999	8,937,765	3,545,000	39·7
Metropolitan boroughs	4,386,004	51,446,827	24,800,000	48·2
Administrative County of London	4,397,003	60,384,592	28,345,000	46·9
County boroughs ..	669,780	4,165,515	2,860,000	68·6
Municipal boroughs ..	1,869,596	16,567,307	7,490,000	45·2
Urban districts ..	1,173,149	10,777,023	6,200,000	57·5
Rural districts ..	94,414	1,133,053	505,000	44·6
TOTAL—Greater London (Metropolitan and City Police districts)	8,203,942	93,027,490	45,400,000	48·8

The allocation of rates in 1933–4 among the various categories of local authorities in Greater London was as follows:[1]

```
For services of—                                    £
    London County Council      ..      ..      18,785,000
    Metropolitan borough councils ..    ..       6,405,000
    City Corporation        ..      ..      ..     905,000

        Total for Administrative County  ..    26,095,000
    Outer London authorities—
        County councils (parts)      ..    ..    7,120,000
        County borough councils    ..    ..      2,680,000
        Municipal borough councils ..    ..      3,515,000
        Urban district councils      ..    ..    2,200,000
        Rural district councils      ..    ..      115,000
    Metropolitan Police    ..    ..    ..        3,675,000

        Total for Greater London    ..  £45,400,000
```

[1] H. Morrison: *How Greater London is Governed*, p. 152. The apparent discrepancy between the figure of £26,095,000 given in this table and the figure of £28·3 millions previously stated as being the sum raised by rates during 1933–4 is explained by the fact that the latter includes a proportion of the sum raised for the Metropolitan Police (i.e. £2,248,160).

One of the results of the present system is that the London County Council has no control over the efficiency with which rates are collected or not collected, as the case may be. The procedure and the practice of the twenty-nine rating authorities within the administrative county vary from district to district, with large differences in what are termed leakages—that is, losses in collection. These losses arise from unoccupied premises, abatements to owners accepting liability instead of the occupier, absconding and bankrupt ratepayers, and poverty. In the year ending March, 1935, a sum of £1,615,698 was written off as irrecoverable within the administrative county, of which £1,481,961 was attributed to empties (i.e. unoccupied property), £81,640 to allowances to owners for compounding, £46,624 to absconding and bankrupt occupiers, and £5,473 excused by justices on the ground of poverty. The distribution of these aggregates among the various districts reveals some remarkable discrepancies which cannot be explained by differences in the social and economic conditions of the neighbourhoods in question.[1]

There is, however, no reason to expect any considerable degree of consistency in these matters in view of the fact that Parliament has thought fit to regard the subject as one properly left to the minor authorities. But it should be realised that the law permits large variations in practice. For example, under the Poor Rate Assessment and Collection Act, 1869, the landlord may be granted an abatement up to 30 per cent in return for his accepting responsibility for payment of rates regardless of whether the property is occupied or not; subject to this maximum the local authority is free to determine the figure in agreement with him. Holborn makes the largest allowance at $17\frac{1}{2}$ to 20 per cent and Chelsea comes next with 16 to 20 per cent; while other high figures are $17\frac{1}{2}$ per cent in Hampstead and 16 per cent in Westminster. Most areas allow considerably less than this. Lambeth and Deptford offer 5 per cent, Hackney $7\frac{1}{2}$ per cent, and Wandsworth $2\frac{1}{2}$ to 5 per cent.[2]

It is almost certain that losses in collection would be lower and more justly distributed on consistent principles if the task

[1] *London Statistics* (1934–6), vol. 39, pp. 426–7.
[2] *Ibid.*, p. 413.

of rate collection was subjected to a common policy throughout the metropolis.

But this is a small matter compared with the extremely important question of inequality in rating.

The aggregate rateable value and the rateable value per head varies enormously in the administrative county. The rateable value of the whole county stood at £61 millions in 1935, of which nearly £9 millions lay in the square mile of the City and more than £11 millions within the City of Westminster.[1] So that these two small areas together contained nearly a third of the whole rateable wealth of the administrative county. At the other end of the scale is Stoke Newington, whose rateable value is only £415,372, and Bethnal Green which has only £520,526. A rate of a penny in the £ produces £36,921 in the City of London, £46,181 in Westminster, £13,348 in Marylebone, £13,865 in Kensington and £12,611 in Wandsworth, but only £3,524 in Shoreditch, £3,112 in Poplar, £4,459 in Woolwich and £3,638 in Bermondsey.[2]

It is, however, the rateable value per head of the local population which provides the most significant indication of taxable wealth in relation to administrative responsibility. Here again we find very large differences between the metropolitan districts. The rateable value per head of the estimated population is (1934–5) £879 19s. in the City, £88 13s. in Westminster, £34 11s. in St. Marylebone and £18 9s. in Kensington, while in Poplar and in Bethnal Green it is only £5 in each case and £7 5s. in Battersea.[3] The inevitable consequence is that we have high rates in the poorer areas and low rates in the richer ones. In 1935–6 the poundage rates varied from 9s. 8d. in the City, 9s. 6d. in Westminster and 9s. 9d. in St. Marylebone to 17s. 4d. in Poplar, 16s. 6d. in Bermondsey and 15s. 2d. in Bethnal Green.[4] The following table[5] collects

[1] *London Statistics* (1934–6), vol. 39, p. 412.
[2] *Ibid.*, p. 430. These figures are for central rates; the amount raised locally by a rate of a penny is slightly smaller owing to certain differences in the method of assessment.
[3] Ministry of Health Statement, Rates and Rateable Values, 1935.
[4] *London Statistics* (1934–6), vol. 39, p. 412.
[5] Ministry of Health Statement, Rates and Rateable Value, England and Wales, 1934–5, H.M.S.O., 1935.

together this information relating to those areas which represent the extremes of wealth and poverty.

TABLE II

RATES AND RATEABLE VALUES, METROPOLITAN BOROUGHS AND CITY

	Rate in £ (1934–5)	Rateable Value	Average Amount per Head of Estimated Population	
			Of Rateable Value	Of Rates collected
	s. d.	£	£ s. d.	£ s. d.
City of London	8 8	8,975,428	879 19 0	348 10 0
Battersea ..	9 2	1,120,245	7 5 0	3 6 0
Bermondsey ..	15 6	874,983	8 3 0	6 0 0
Bethnal Green	13 6	520,447	5 0 0	3 3 0
Poplar ..	16 0	747,829	5 0 0	3 15 0
St. Marylebone	8 8	3,250,554	34 11 0	13 17 0
Kensington ..	8 11	3,343,364	18 9 0	7 16 0
Westminster ..	8 6	11,195,294	88 13 0	34 11 0
Woolwich ..	12 5	1,089,801	7 9 0	4 13 0

There are very large rating inequalities in Outer London: that is, the parts of Greater London situated outside the administrative county. Broadly speaking, the thickly populated districts lying north of the Thames in Essex are the most heavily rated while the well-to-do residential suburbs in Surrey, Herts and Middlesex have to bear the lightest burden. Thus, the rates in West Ham are 19s. 4d., in East Ham 16s. 8d., in Dagenham 14s. 6d., in Leyton 13s., while in Wimbledon they are only 8s. 6d., in Surbiton 8s. 8d., Epsom and Ewell 9s., Coulsdon and Purley 8s. 8d., Harrow 8s. 10d. The average for the Essex portions of Greater London is 14s. 7d. (1935–6), for the Herts portions 10s. 4d., for the Kent portions 9s. 10d., for Middlesex 9s. 11d., and for the Surrey portions 9s. 2d.[1]

It is clear that while the rateable value of property depends mainly though not entirely on economic and social factors lying outside the sphere of local government, the aggregate rateable value of an area and its rateable value per head depend largely on how the area is defined. The enormously high rateable value per head of the square mile of the City arises

[1] *London Statistics*, vol. 39, pp. 417–25.

from the concentration of extremely valuable business property within a narrow territory containing very few residents. If one were to draw boundary lines with that consideration in view, it would be quite possible to divide up London, or any other place for that matter, so as to arrive at a wide range of areas in regard to aggregate rateable values or rateable values per head.

We may now consider the extent to which the administration of services by the London County Council equalises the great differences in wealth between the boroughs within the County of London. The rates levied within the metropolis are divided into what are known as central rates and local rates. The central rates are levied by the London County Council and the Receiver for the Metropolitan Police. The Metropolitan Water Board is also authorised to levy a deficiency rate on its "contributory" areas[1] to make good any losses which arise when the water charges do not meet the Board's annual expenditure. An Act of 1921 permitted the Water Board to increase its charges, since when no deficiency rate has been required.

The London County Council's rates consist of the county rate, which is levied throughout the administrative county, and the special county rate. The special county rate (amounting to about 3d. in 1935–6), and the Metropolitan Police rate are not levied upon the City of London nor upon the Inner and Middle Temples. The City Corporation maintains its own police force and also administers within its area the services elsewhere provided by the London County Council and charged on the special county rate. These include the provision of a coroner and his court, an ambulance service, buildings for mental hospitals, deficiencies arising under some of the Housing Acts, and the majority of the regulatory functions such as the inspection of weights and measures, dangerous structures, massage establishments, shops, and the employment of children and young persons.[2]

The principal items in the precepts of the London County Council (for the year ending March 1936) are education 2s. 6d.,

[1] For details, see *London Statistics* (1934–6), vol. 39, pp. 335, 403.
[2] Herbert Morrison: *How Greater London is Governed*, p. 25.

public assistance 2s. 1d. and public health 1s. 10d. Other London County Council services came to 1s. 7d., making a total of 8s. After deducting Exchequer grants, the proceeds of local taxation, licence duties and various balances, the total precept issued by the London County Council was reduced to 7s. The Metropolitan Police precept amounted to 11d., making 7s. 11d. in all, compared with an average of 3s. 1d. raised by the metropolitan boroughs for their own purposes. Thus, after allowing for the adjustment grant paid under the Local Government Act, 1929, approximately 72 per cent of the total average rates of 11s. levied within the county in 1935–6 was required for "central" services.[1]

Shortly after the London County Council was constituted a vigorous demand arose for a more equitable distribution of the cost of local government over the metropolis. A number of expedients were introduced for this purpose, such as the common poor fund and the equalisation fund, the former assuming substantial proportions. But the demand continued, and much of the time of the Royal Commission on London Government and other inquiries was occupied with this question. The majority of the Ullswater Commission concluded that although about 70 per cent of the total expenditure from rates was then being equalised, "some further equalisation of rates between the richer and poorer districts within and surrounding the administrative county of London is fair in principle and ought to be brought into operation with the least possible delay."[2] The report recommended that an equalisation fund should be established to cover the administrative county of London together with all urban areas—fifty-five in number—wholly or partly within a radius of ten miles from Charing Cross. Rural districts were to be excluded. The fund was to be fed by the produce of a rate of 1s. 6d. in the £ levied uniformly throughout the area together with the produce of a

---

[1] *London Statistics* (1934–6), vol. 39, p. 414. The figure of 11s. 0d. excludes the City. If the City is included, the average rate is 10s. 9d., as previously stated.

[2] Report of the Royal Commission on London Government (1923), p. 105. *Cf.* p. 87 and Minutes. Evidence, Webb, pp. 828–9, 838, 840–1, 943. The Committee on Local Taxation (1894) and the Royal Commission on Mental Deficiency also made similar proposals.

further rate of 1s. levied in similar manner. The former was to be used for general services, the latter to defray expenditure on the poor law.[1] It was estimated that the total amount produced by these equalisation rates would be about £8 millions a year.

No action was taken on these recommendations, but in 1929 the Local Government Act, introduced by Mr. Neville Chamberlain as Minister of Health, transferred the poor law from the local boards of guardians to the London County Council and thereby made the cost of administering public assistance a charge on the administrative county as a whole. The common poor fund and the pre-existing equalisation fund were abolished.

This was no doubt a step in the right direction but it fell far short of what the Royal Commission had proposed. In the first place, none of the urban areas lying outside the county were brought into the scheme. Second, the figure of £8 millions considered necessary was abandoned. The total expenditure of the London County Council on public assistance (1934–5) was £6·6 millions and of this less than £6 millions was charged on the county rate.[2] No further equalisation took place in respect of the remaining £2 millions. Third, the rate for the common poor fund was actually higher in 1928 and 1929 (when it stood at 2s. 2d.) than the precept of the London County Council for public assistance in the years between 1933 and 1936 inclusive.[3] Fourth, the proportion of total rates allocated to central purposes has scarcely risen since 1929, as the following table shows:

			Per cent				Per cent
1925	..	..	74·51	1931	..	..	74·49
1926	..	..	73·15	1932	..	..	72·81
1927	..	..	66·73	1933	..	..	73·16
1928	..	..	69·48	1934	..	..	71·34
1929	..	..	72·73	1935	..	..	71·58
1930	..	..	74·07	1936	..	..	72·16

It cannot be said, therefore, that the inequalities of wealth either within or without the administrative county of London have been remedied effectively by the transfer of the poor law.

[1] Report of the Royal Commission on London Government (1923), pp. 107–8.
[2] *London Statistics* (1934–6), vol. 39, p. 381.   [3] *Ibid.* (1933), p. 414.

We are now in a position to see what are the residuary inequalities remaining when the precepts from County Hall and the Metropolitan Police have been met. The rates in respect of purely local expenditure range from 1s. in Westminster and 1s. 1d. in the City, to 7s. 8d. in Bermondsey and 7s. 4d. in Poplar.[1] Some part of the difference is due to variations in efficiency between the various metropolitan boroughs and to the amount of service provided by them in relation to the size of the area. Chelsea Borough Council, for example, which levies the third lowest rate for local purposes in the administrative county (1s. 5d.), spends less than it should on street cleansing, sanitary inspection and certain other public health services.[2] Bermondsey has provided far more housing than any other metropolitan borough, and spends the proceeds of a 7d. rate on housing purposes. This is one of the reasons why the rates in Bermondsey are higher than they are in Fulham or Hampstead, where a rate of less than 1d. is devoted to housing.[3] There are large variations in the excellence and extent of the public libraries, the baths and wash-houses, in the standard of street cleansing, lighting and repair, and indeed in all the services provided by the district authorities; and these variations are usually reflected in variations of cost, and hence in the rates which have to be levied. But when due allowance has been made for these factors, it is abundantly clear that such influences are of minor importance compared with the inequality of wealth between the different areas. The rateable value per head in an area is easily the predominant factor in determining the amount of the rates in the £ which the local authority is compelled to raise.

The question arises, therefore, whether the existing financial arrangements within the administrative county can be regarded as satisfactory or equitable. In my opinion they cannot. But it is impracticable to consider the problem of finance apart from the reorganisation of authorities within the London Region.

The majority report of the Ullswater Commission envisaged

[1] These figures are prior to adjustment under Section 100, Local Government Act, 1929.   [2] *Ante*, p. 227.
[3] For details, see *London Statistics* (1934–6), vol. 39, pp. 404–7.

that the equalisation fund which they recommended to be set up should be distributed as regards that part of the fund to be devoted to general purposes among the local government units on the basis of their day population; while the public assistance part of it was to be divided on the basis of the night population in the various poor law unions which then existed, with a special weighting in favour of the population living in overcrowded conditions. The fund was to be administered by the Minister of Health.[1]

The principle of considering the day population which a local authority may have to serve was an important innovation which had hitherto never been recognised in connection with municipal finance. In the case of a great urban area such as London, in which some districts are mainly residential and others mainly industrial or commercial, it is obviously wrong to consider the responsibilities of local authorities solely in terms of their resident inhabitants. I have already referred to the burden thrown on the local authorities of Inner London by the huge army of workers living in the dormitory settlements beyond the frontier which on six days of each week invades the administrative county. This problem is fundamentally one of local government structure; but the majority Commissioners were unwilling to contemplate reorganisation. They accordingly recommended financial arrangements which would not involve a reform of municipal areas or a redistribution of functions, either in the administrative county or within Greater London. Their financial proposals imply, however, the existence of a larger problem than the one they are designed to solve.

The Donald-Walsh minority report criticised the majority proposal to include, for rating purposes, the urban communities lying outside the administrative county as being contrary to the principles of English local government, since it transgresses the maxim of no taxation without representation. If the common interests of the area are so extensive as to involve the mutual interchange of service, then, said the minority Commissioners, the proper course is to widen the metropolitan area. Every argument used by the majority in favour of their equalisation scheme is an argument for the unity of London

17   [1] Report, pp. 108–10.

government.[1] They suggested as an alternative an equalisation fund based on a proportion of the estimated current expenditure on services remaining under local administration. This was far less thorough-going than the scheme put forward by Mr. Sidney Webb on behalf of the Labour Party whereby an equalisation fund would be based on assessable value per head of the population. Under the Webb plan areas whose rateable value per head was above the average for the metropolis would be rated in proportion to their excess, while those below it would receive payment in proportion to their deficit. The payment would be unconditional but could be refused to a local authority which neglected or failed to carry out any of its duties with regard to sanitation or other matters.[2]

The inequalities of wealth among the metropolitan boroughs and the outlying suburbs is no new phenomenon. So long ago as 1908 the London Reform Union[3] was reminding the public that the separation between rich and poor is even more obtrusive in the outer suburbs than it is within the administrative county. On the north and east, the Union pointed out, great agglomerates of population have grown up consisting almost entirely of the poor. The rateable value per head is low, and the need for municipal expenditure high. Conditions in the south and the west are with a few exceptions quite the opposite. The communities are composed of comparatively well-to-do residents; rateable value per head is high, while the expenditure required for such services as education or public assistance is small.[4]

This contrast between needs and resources has been steadily growing more pronounced in the thirty years which have elapsed since 1908, until today it may be said that the desirability of securing a substantial measure of equalisation is one of the principal grounds on which the reform of London government can be urged. The question is not that of ironing out the differences between the administrative county and Outer London, but of diminishing the inequalities within

[1] Report, pp. 200–3.
[2] Royal Commission on London Government (1923), Minutes of Evidence, p. 823.
[3] *London To-day and To-morrow* (pamphlet, 1908), p. 12.
[4] See the rating map at end of volume for the present position.

Greater London as a whole. This is shown by the fact that the average rates for the administrative county, for Outer London and for Greater London are almost the same. In 1935–6 the first was 10s. 9d. and the two latter 10s. 8d.[1] Or, to put the matter in a somewhat different form, the inequalities between districts within the administrative county tend to be continued over the border into Outer London.

A word may be said concerning the new system of grants-in-aid introduced by the Local Government Act, 1929, in case the reader should think that this has materially affected the situation. The methods applied to London differ entirely from those adopted in the remainder of the country. The City Corporation and each metropolitan borough council is entitled to receive from the Ministry of Health out of the county apportionment of the administrative county a sum equal to one-third of the amount which would have been payable to it if it had been a county borough council. Thus, during the first four fixed grant periods it gets an appropriate percentage, decreasing from 75 per cent to 25 per cent, of its losses from derating and discontinued grants, together with one-third of the sum it would have received as a county borough council on its weighted population. Ultimately, in the fifth and all succeeding grant periods, when the whole of the General Exchequer Contribution is to be divided in accordance with the formula contained in the Act, the metropolitan councils will receive one-third of the sum to which they would be entitled if they were county boroughs. One other difference is that in calculating their weighted populations no account is taken of the percentage of unemployed persons in the locality.[2]

The residue of the county apportionment of the County of London is paid to the London County Council and constitutes its General Exchequer Grant.

Additional Exchequer grants are also payable in London, as elsewhere, but the sum involved is small. There is also a

[1] *London Statistics* (1934–6), vol. 39, p. 417. *Cf. post* p. 454 *et seq.*
[2] Local Government Act, 1929, Sections 88–98; Hart and Hart: *An Introduction to the Law of Local Government and Administration*, pp. 238–9; *London Statistics*, vol. 39, pp. 400–1.

supplementary grant available in London to enable the metropolitan boroughs to recoup any losses they may suffer through the operation of the Local Government Act, 1929, partly from the Exchequer and partly from a charge on those areas where the Act produces a gain. The following table shows the grants distributed in London under these provisions:[1]

	Years 1933–4, 1934–5, 1935–6 £
Amount of county apportionment	3,941,542
Amount of additional Exchequer grant	38,533
Total	3,980,075
Amount payable to the City Corporation and the metropolitan borough councils	1,481,563
Residue payable to the London County Council ..	2,498,512
Amount of supplementary Exchequer grant for London	172,760

There is no provision in these arrangements for the equalisation of any pre-existing differences of wealth within the administrative county except to the small extent to which rateable value is allowed to weight the estimated population for the county and district apportionments under the formula.[2] The very slight importance given to this factor is shown by the statement opposite.

From this it will be seen that the entire sum which is manipulated by the factor of rateable value per head amounts to only £53,000, which is less than 3 per cent of the total annual grant payable to the metropolitan borough councils and the City Corporation during the years 1933–6 under the Local Government Act, 1929. In no case does the adjustment made by the weighting in respect of rateable value produce a substantial rectification of the immense inequalities of taxable

[1] *London Statistics* (1934–6), vol. 39, p. 402.
[2] One of the four factors for weighting the population in the formula used to distribute the General Exchequer Contribution is a rule whereby if the rateable value per head of the estimated population is less than £10, the estimated population is increased by the percentage represented by the proportion which the deficiency bears to £10. Local Government Act, 1929, Fourth Schedule, Part III.

STATEMENT GIVING CERTAIN PARTICULARS RELATING TO THE RATEABLE VALUES AND THE EXCHEQUER GRANTS PAYABLE UNDER THE LOCAL GOVERNMENT ACT, 1929, FOR THE THIRD FIXED GRANT PERIOD (I.E. THE FIVE YEARS BEGINNING ON 1ST APRIL, 1937) OF THE CITY OF LONDON AND THE METROPOLITAN BOROUGHS.

Name of Area	Product of Rate of 1d. in the £ for the Year 1938-9	Rateable Value per Head of Population based on Registrar-General's Estimate of Population in 1936 and Provisional Figures of Rateable Value on 6th April, 1936	Portion of the Exchequer Grant payable to each Council in the Third Grant Period which is calculated in accordance with the Provisions of Section 98 (1) (b) of the 1929 Act, i.e. based on a Formula	Sum by which each Amount in Column (4) would be reduced if the Weighting for Rateable Value were excluded from the Calculation
(1)	(2)	(3)	(4)	(5)
	£	£	£	£
City of London	32,500	836·065	845	—
Battersea ..	4,697	7·871	19,835	2,755
Bermondsey ..	3,440	8·735	13,630	1,130
Bethnal Green ..	1,973	5·427	16,005	3,941
Camberwell ..	6,976	7·566	30,250	4,964
Chelsea	5,072	21·849	5,025	—
Deptford ..	2,678	6·857	14,902	2,764
Finsbury ..	4,894	19·384	7,786	—
Fulham ..	5,365	8·931	16,925	1,343
Greenwich ..	4,030	10·279	10,917	—
Hackney ..	6,077	7·157	29,461	5,287
Hammersmith ..	5,025	9·579	15,015	479
Hampstead ..	6,248	16·753	8,066	—
Holborn ..	6,602	47·275	3,099	—
Islington ..	8,925	7·320	44,115	7,174
Kensington ..	13,435	18·881	16,914	—
Lambeth ..	9,476	8·408	35,518	3,929
Lewisham ..	7,338	7·971	28,744	4,042
Paddington ..	7,105	13·100	13,370	—
Poplar	3,038	5·530	24,293	5,577
St. Marylebone	14,095	36·634	8,217	—
St. Pancras ..	9,030	11·828	19,625	—
Shoreditch ..	3,133	9·558	11,880	336
Southwark ..	5,212	8·741	20,654	1,704
Stepney ..	6,861	8·340	29,309	3,067
Stoke Newington	1,751	8·484	5,963	679
Wandsworth ..	13,655	9·492	33,883	1,549
Westminster ..	42,305	84·741	11,036	—
Woolwich ..	5,050	7·972	19,058	2,642
	245,986	—	514,340	53,362

capacity between the various districts within the administrative county.

The whole question of the Exchequer contribution has recently been reconsidered by the Minister of Health and a further statute passed to regulate the grant; yet nothing whatever has been done to modify the position in regard to this supremely important aspect of the matter.[1]

Thus, the financial changes introduced by the Local Government Act, 1929, have left unsolved the two great problems in the municipal finance of London government: namely, the enlargement of the boundaries of London for financial purposes, and the more equitable distribution in the incidence of rating burdens to correspond with rateable wealth.

The methods of authorising capital expenditure applicable to the metropolis are peculiar. The London County Council has inherited the system relating to the Metropolitan Board of Works laid down by the Metropolis Management Act, 1855, the Metropolitan Board of Works (Loans) Act, 1869, and later statutes. These Acts confer borrowing powers on the London County Council but a special Money Act obtained annually limits the amount of capital expenditure under various heads which may be incurred by the Council during the current financial year and the following six months until the next Act is passed.[2]

The London County Council does not have to obtain the sanction of a central government department in order to borrow money, but it is the only local authority in Great Britain which is controlled by Parliament in this manner. If Manchester or Liverpool obtains Parliamentary powers to carry out a large scheme of municipal construction requiring capital expenditure over several years, authority to raise the whole sum required is given at the same time; and a similar practice is adopted where the sanctioning authority for a loan is the Ministry of Health or one of the other departments. But the London County Council must go cap in hand to Parliament

---

[1] Report on Result of Investigation under Section 110 of the Local Government Act, 1929. A.C. 42/1937, H.M.S.O.; Local Government (Finance Provisions) Act, 1937.

[2] Sir Harry Haward: *The London County Council from Within*, p. 116.

each year. "This exceptional position," writes Sir Harry Haward, "arose from the desire of Parliament to take a comprehensive annual survey of the large expenditures which sixty years ago were being undertaken by the Metropolitan Board of Works under various borrowing powers and it has continued up to the present time."[1]

The principal defect of the present system is, however, that it fails to ensure this "comprehensive annual survey" of capital expenditure by local authorities in London which it was intended to provide. The reason for this is twofold. In the first place, since the middle of last century, when the procedure was first laid down, a series of administrative organs has been established whose financial commitments on capital account are not considered at all in relation to the London County Council's annual programme. Parliament has no cognisance of the intentions of the Port of London Authority, the Metropolitan Water Board, the London Passenger Transport Board, the Commissioner of Police, the City Corporation or the metropolitan borough councils, or at any rate they are not brought into the discussions on the annual Money Bill promoted by the London County Council. The sanctioning authority or the procedure varies in almost every case.[2] These several public authorities therefore embark on schemes involving large capital outlay without regard being paid to each

[1] Sir Harry Haward: *The London County Council from Within*, p. 117.
[2] The following table shows the sanctioning authorities for most of the London authorities:

Borrowing Authority	Sanctioning Authority
London County Council	Parliament and Treasury.
Metropolitan Water Board.	Parliament and Ministry of Health.
Receiver for Metropolitan Police District.	Home Office.
Port of London Authority.	Parliament and Ministry of Transport.
London Passenger Transport Board.	Parliament and Ministry of Transport.
Metropolitan borough councils:	
For electricity supply.	Electricity Commissioners.
For some services.	Ministry of Health.
For other services.	London County Council.

other's commitments or the effect of them all on the money market, the trade cycle or the volume of unemployment. "They may each propose to do large capital expenditure in the same year," Mr. Sidney Webb observed to the Ullswater Commission, "when probably it would be desirable that the capital expenditure should be taken one year after another."[1] Mr. and Mrs. Webb, in the famous minority report of the Poor Law Commission, were the first social investigators in this country to point out the important part which capital expenditure by local and national authorities can and should play in regularising the demand for labour and diminishing the effects of slump and boom.[2] Their views on this matter are now generally accepted both in Great Britain and in other countries, but no steps have so far been taken to co-ordinate the borrowing projects of the local authorities in the metropolis. When we recollect that the total gross debt still outstanding of the London County Council, Metropolitan Water Board, the metropolitan boroughs and the City Corporation is about £212 millions and the net debt £153 millions (in 1935), with a further sum of about £112 millions for the London Passenger Transport Board, and that this represents only a small fraction of the capital expenditure undertaken by these bodies since their inception,[3] or the current value of their assets, the magnitude of the matter becomes evident.

But, in the second place, these organs comprise only a part of the local authorities engaged in capital works within Greater

[1] Royal Commission on London Government (1923), Minutes of Evidence, p. 841.

[2] Minority Report of the Poor Law Commission (1909), Part II, Chapter V (c).

[3] To take one example, the London County Council and its predecessors since 1856 raised £198 millions. Of this, £70 millions has been repaid. Of the remaining gross debt of £128 millions, assets applicable to the redemption of the debt amount to £55 millions, leaving only £73 millions as the net debt outstanding on 31st March, 1935. *London Statistics* (1934-6), vol. 39, p. 448. The corresponding figure had increased to £77 millions by 31st March, 1936. £44 millions of this was in respect of block dwellings and cottage estates, which represent valuable financial assets; the remainder of the debt is secured against schools, hospitals, fire brigade stations, bridges, the County Hall and other municipal property. See the speech by Mr. C. Latham when submitting the Annual Estimates, 1936-7, to the Council, Publication No. 3179.

London. The capital commitments of the multitude of local authorities of all classes in Outer London are not easily available, and there is no particular point in making the calculation, since I am concerned only to suggest that they are of considerable dimensions. Some indication of their probable magnitude can be obtained by comparing the rateable value of the administrative county of London, which stood at £61 millions in 1935, with that of Outer London, which was £38 millions in the same year.[1]

The sanctioning authorities and procedure for borrowing by Outer London municipalities are the same as for the rest of the country and differ entirely from those applicable to the in-county authorities. There is something to be said for "a comprehensive annual survey" of the borrowing operations of all the public bodies within the metropolitan region, having regard to the immense economic and social importance of this area. There is nothing to be said in favour of the disorderly arrangements which now exist. Their main result is to subject the London County Council to a special procedure as regards its capital expenditure without applying a coherent financial policy throughout the region. In any case the question arises whether the legislature is the right body to exercise control of this kind. If so, Parliament should assume responsibility in all cases; but here again no clear idea seems to have been evolved as to whether control over municipal borrowing is best carried out by Parliament, or a central government department, or by a combination of the two.[2]

When we come to the metropolitan boroughs we find further signs of confusion. The borrowing powers of the district authorities are regulated by two distinct codes. One of these is contained in the Metropolis Management Act, 1855, and relates to the purchase of land, street lighting, housing, local highway improvements, paving works, sewerage and drainage, small parks and recreation grounds, refuse destructors, town halls, municipal offices and similar activities. Under its provisions the London County Council is the sanctioning authority,

---

[1] *London Statistics* (1934–6), vol. 39, p. 417. I have referred on pages 235–6 to drainage schemes in Outer London running into many millions of pounds.
[2] See footnote (2) to p. 263.

subject to a right of appeal to the Minister of Health if the London County Council refuses permission, or fails to grant it within six months, or attaches conditions to the sanction. This system was inherited from the epoch of the Metropolitan Board of Works, which supervised the borrowings of the vestries and district boards.

The other code was laid down by the Public Health (London) Act, 1891, and provides that the sanction shall be given by the Ministry of Health direct without any intervention or consent by the London County Council. Subsequent statutes conferring borrowing powers on the metropolitan boroughs have always incorporated either one system or the other.[1] The only exception was the Electricity (Supply) Act, 1919, which transferred from the London County Council to the Electricity Commission the duty of sanctioning loans. The Minister of Health has become the approving body for loans required in connection with the construction of baths and wash-houses, public lavatories, coroners' courts, disinfecting stations, cemeteries, maternity and child welfare clinics, public libraries and certain other undertakings.

There is no intelligible principle to be discerned in this division of control. Either the London County Council or the central department is the appropriate body for supervising the district councils in this matter. It cannot be right that concurrent powers should exist. The explanation appears to lie in the opposition of the metropolitan boroughs to any form of control by County Hall,[2] and their ability to persuade Parliaments and Governments hostile to the London County Council to allow a second and unco-ordinated method to grow up.

The London County Council does not appear to have exercised its powers in such a manner as to justify resentment by the local bodies. During the forty-eight years from 1889 to 1937 some 4,731 loans sanctions were granted by County Hall for an aggregate amount of £56 millions. In this period of nearly half a century there have been only 17 cases in which

[1] Hart and Hart: *An Introduction to the Law of Local Government and Administration*, p. 234.
[2] Sir Harry Haward: *The London County Council from Within*, pp. 162-3.

a vestry or metropolitan borough council has appealed to the Minister of Health (or his predecessor the Local Government Board) against a refusal by the London County Council to sanction a loan. In 12 instances the appeal was dismissed; in 3 instances the appeals were allowed in full in special circumstances; in one case the appeal was partly allowed; and in the remaining case the original scheme was modified and subsequently approved.[1] The practice of the relative Committee of the Council entrusted with the scrutiny of applications to

[1] The details of the appeals are as follows:

Number of cases	Amount	Reason for Refusal	Result of Appeal
5	£ 1,280,505	Borough Council had no legal power to expend for desired purposes.	Appeal dismissed.
2	7,448	Expenditure not of a capital nature; should be defrayed out of revenue.	Appeal dismissed.
1	14,665	Sanction given to amount, but period of repayment reduced.	Appeal dismissed.
3	7,300	Excessive cost.	Appeal dismissed.
1	500	Sanction given to amount, but with a condition as to repayment.	Appeal dismissed.
1	23,000	Expenditure not of a capital nature; should be defrayed out of revenue.	Minister agreed in principle with L.C.C., but allowed appeal in special circumstances.
1	15,312	Sanction given to amount, but period reduced.	Appeal as to period allowed.
1	888,423	Buildings (housing schemes) contravened London Building Act.	Appeal allowed.
1	18,563	Insufficient information given by Borough Council.	Appeal allowed in part.
1	11,100	Excessive cost.	Scheme subsequently modified and sanctioned and appeal dropped.

I am indebted to the Deputy Controller of the L.C.C. for kindly supplying me with this information.

borrow has been to seek to remove differences of opinion by friendly and informal discussions prior to referring the matter to the Council with a recommendation for formal decision.

The total net debt of the metropolitan boroughs and the City Corporation outstanding in 1936 was £39 millions, amounting to 62 per cent of their rateable value.

The London County Council has, indeed, gone beyond its strict duty of sanctioning loans by the local bodies and advances money to those of them which desire to take advantage of the unrivalled facilities for raising capital at low rates of interest possessed by the principal municipal organ. Sir Harry Haward, the late Comptroller to the Council, has stated that during his long official career the London County Council never refused to lend money to a local authority authorised to borrow and to whom the Council had power to lend, either on account of the financial position of the borrowing body or for any other reason.[1] By 1932 the London County Council had advanced £55 millions to the metropolitan borough councils (or their predecessors, the vestries), the former boards of guardians, Metropolitan Asylums Board and the London School Board.[2] In 1935, out of a total net debt outstanding by the metropolitan borough councils of £34·3 millions, 51 per cent (£17·5 millions) was in the form of mortgage loans advanced by the London County Council.[3] The metropolitan boroughs are, of course, entirely free to choose the method by which they shall raise money. If they prefer to do so they can apply to the Public Works Loan Board, or issue bonds or debenture stock, or obtain loans from banks, insurance companies or similar institutions. All these methods are used in varying degrees.

The chief points dealt with in this chapter may now be summarised. The principal municipal body in London is not responsible for rating or assessment, which is left in the hands of the City Corporation and the twenty-eight metropolitan borough councils, whose practices and procedures vary considerably. The inequalities in the burden of rating between the districts vary enormously and are chiefly due to differences in the rateable value per head of the population, though to a

---

[1] Haward: *op. cit.*, p. 177. [2] *Ibid.*, p. 172.
[3] *London Statistics* (1934–6), vol. 39, p. 445.

lesser degree to differences in quantity or quality of service provided. There is no budgetary control over the metropolitan boroughs or the City Corporation, nor any arrangements for consultation or co-ordination with the London County Council in regard to rating or expenditure. There is no provision for equalising the substantial lack of correspondence in financial resources and municipal responsibilities between rich and poor districts which exists both within the administrative county and in Outer London. There has been no attempt to widen the area of charge for local government services so as to include the homes of hundreds of thousands of persons who work inside the County of London but live outside it. The system by which the Exchequer Contribution is distributed in London gives the maximum degree of independence to the district councils and scarcely affects the existing inequalities. The methods of controlling capital expenditure are diverse and incoherent, and fail to ensure a comprehensive survey of borrowing by metropolitan authorities.

CHAPTER XII

FIRE PROTECTION

Defence against fire is one of the most elementary forms of protection with which a great city may be expected to equip itself. This applies with special force to London, in whose long history the Great Fire of 1666 still stands out as a major event. Yet the fire-fighting organisation of the metropolis is hampered, like practically all the other services, by the obsolete ideas which hang like millstones round the neck of the capital, and the persistent refusal of the governmental authorities to recognise the obvious facts of life and growth.

Within the administrative county the duties of fire prevention and protection are in the hands of the London County Council; and not even the City Corporation claims the right to operate a separate service within its area. The London Fire Brigade is a body of the highest efficiency and repute. It is not only the largest but also the best fire-fighting force in Great Britain. It has a total staff of 2,142 and manages 60 land fire stations and 3 river stations.[1] The fine new headquarters which were recently opened on the Albert Embankment represent the last word in modern scientific practice. The Brigade deals with about 6,000 fires a year and requires an annual expenditure of £886,000.[2] Towards this a grant of £10,000 a year is made by the Treasury (in view of the large amount of Government property in the area), and the fire insurance companies and underwriters contribute in proportion to the insured value of property in the area.[3] There is thus a clear recognition that the efficient conduct of this service is of material importance both to the central government and to business interests. Yet the course of events which I am about to describe would not lead one to suppose that there were any major outside interests concerned in promoting the administrative improvements which are necessary.

---

[1] *London Statistics* (1935-7), vol. 40, pp. 238-9.   [2] *Ibid.*, p. 393.
[3] The sum contributed in 1936 by the companies amounted to nearly £80,000.

In Outer London very little is known about the size or equipment of the fire brigades. The principal reason for this obscurity is that hitherto no central department has been responsible for supervising or inspecting the fire fighting or fire prevention work of local authorities, and no grant-in-aid has been payable in respect of it. Hence there are no general statistics or annual reports covering the whole country, and we are in consequence dependent for information on the dim light thrown by an occasional Royal Commission or departmental committee.

A recent departmental committee stated in 1936 that in the Greater London area there are about 64 brigades possessing 264 pumping appliances.[1] From these figures we must deduct 112 pump-carrying appliances maintained by the London Fire Brigade[2] which leaves a total of 63 brigades with 152 appliances for the whole of Outer London. Thus, an area six times as large as the Administrative County and containing a slightly larger population, is served by only about 35 per cent more appliances. The strength of the personnel is unknown. We do know, however, that a number of local authorities in Outer London are not exercising their fire protection powers at all, since there are 82 county boroughs, municipal boroughs, urban and rural districts possessing such powers and only 63 brigades in operation. Hence, 19 areas are either without any protection at all against fire or else dependent on neighbouring brigades. Only the London County Council and the City Corporation has an obligation to make provision for fire brigades.[3]

A sidelight on the efficiency of the brigade in a particularly well-to-do area in Outer London was given by the report of the Commissioner appointed to hold a public enquiry into the circumstances in which a child lost her life in a fire at Golder's Green in March 1938. The Commissioner's report made severe reflections on the conduct and management of the Hendon Borough Council's fire brigade. The evidence showed that the

[1] Departmental Committee on Fire Brigade services, 1936, p. 13.
[2] L.C.C. Minutes of Proceedings. Report of the General Purposes Committee, 11th February, 1936.
[3] Departmental Committee on Fire Brigade Services, 1936 Report, p. 11.

instructions requiring emergency calls to be immediately recorded were not always obeyed or precisely carried out. The superintendent was absent during the critical period of the fire and a station officer was called upon to carry responsibilities beyond his powers. The superintendent had for long not been in a physically fit state to take active command at a serious fire. The Commissioner in conclusion raised the question whether the Hendon fire brigade's standard of training and leadership are such that it can be regarded as competent to deal efficiently and effectively with fires of any serious magnitude. He disagreed with the principle enunciated by the chairman of the Works and Fire Brigade Committee of the Council that it must be left to the chief officer to see that training is efficient.[1]

There is apparently a small degree of co-operation between some of the fire brigades in Greater London. Thus, there is a standing arrangement for mutual assistance within half a mile of the county boundary between the London County Council and the Willesden Borough Council. The London County Council entered into this arrangement in view of the exceptional length of the boundary and the peculiar manner in which Willesden juts into the county. Moreover, one side of Edgware Road is in the Administrative County and the other side in Willesden.[2] Apart from formal agreements of this kind, the London Brigade usually endeavours to give assistance in response to any urgent call outside its boundaries. But to ask for such assistance involves an outlying authority in what they would perhaps regard as a humiliating dependence on the London County Council; and this may be one of the principal reasons for the few instances in which it occurs.[3] It would require a very serious conflagration to overcome the petty parochialism of the Outer London authorities.

The absurdity of the administrative boundaries from the point of view of fire prevention and fighting is so obvious as to need no emphasis. The Royal Commission on Fire Brigades

[1] *The Times*, 14th June, 1938.
[2] Royal Commission on Fire Brigades and Fire Prevention, [Cmd. 1945] 1923, B.P.P., vol. xi, p. 136.
[3] The London Fire Brigade is stated to attend "a few fires outside the County boundary." *London Statistics* (1935–7), vol. 40, p. 238.

and Fire Prevention pointed out in 1923 that in the industrial portions of the East End a factory may be situated partly in East Ham and partly in West Ham.[1] The Port of London Authority have complained of the extreme inconvenience resulting from the administrative boundaries passing right through warehouses. When a fire occurs, it is necessary to determine in which part of the warehouse it originates before the Port Authority knows whether it has to summon the London Fire Brigade, or the East Ham Brigade, or the West Ham Brigade.[2] The Port of London Authority requested in vain that some action should be taken to rectify this confusion.

During the Great War of 1914–18 a scheme was brought into operation, by a Home Office order under the Defence of the Realm Act, which introduced a limited amount of pooling of resources within Greater London. Under this scheme the Home Secretary constituted the City of London, the Metropolitan Police District and the urban districts of Watford, Dartford and Egham into the Metropolitan Fire Brigade area under the executive command of the Chief Officer of the London Fire Brigade.[3] The arrangements provided for assistance within the Administrative County being rendered by the suburban brigades where help was needed in case of air raids or the resulting fires.[4] But the sending of pumps was subject to their not being locally engaged and in the last resort depended on the willingness of the local authorities to send their appliances outside their own areas.[5] Outer London authorities could in their turn obtain the assistance of the London Fire Brigade if they invited it. The arrangements were said to have worked satisfactorily during the War; but they were allowed to lapse at the end of it.

The subject of fire defence was considered by the Ullswater Commission on London Government, 1921–3. The London

[1] Report, [Cmd. 1945] 1923, p. 136.
[2] Royal Commission on London Government, 1923. Norman: Minutes of Evidence, p. 56.   [3] *Ibid.*, pp. 31–2; Report, p. 9.
[4] Royal Commission on Fire Brigades and Fire Prevention, [Cmd. 1945] 1923, p. 136.
[5] Departmental Committee on Fire Brigade Services, [Cmd. 5224] 1936, p. 23

County Council in their evidence proposed a larger authority whose powers should include those of fire protection and prevention. But this did not commend itself to the majority of the Commission, who reported that in their view the evidence showed that outlying districts were able to receive from the London Fire Brigade any help they might require. The Commissioners were therefore prepared to accept the extraordinary assortment of arguments which the Outer London authorities employed to resist any organic unification. The Essex County Council pleaded that a "centralised" system would cost more; Middlesex County Council objected on the ground that it would mean the replacement of voluntary or semi-voluntary brigades by a professional staff with consequent loss of local interest; the Wimbledon Borough Council contended that the local brigades are accustomed to co-operate in order to provide increased security; while the County Borough of Croydon added that they would feel less secure if protected by the central authority.[1] The picture of the worthy inhabitants of Croydon, dependent during their working hours within the Administrative County on the London Fire Brigade, but unwilling to expose themselves at night to the horrid danger of its alleged incompetence, is a touching one which calls for no comment. What all these arguments amounted to was that the outlying authorities preferred to remain parasitic on the London Fire Brigade in case of need rather than to accept administrative reform.

The subject was not quite disposed of by the Ullswater Report. Still small voices in the Home Office kept on asking awkward questions until a Royal Commission on Fire Brigades and Fire Prevention was appointed in 1923 to investigate the question throughout the country. The Royal Commission advised that while the arrangements in Greater London are no doubt susceptible of improvement in both efficiency and economy, this should be sought through co-operation rather than consolidation. They agreed with the majority of the Ullswater Commission that a case for altering the existing system had not been made out.[2] Generally speaking, no action

[1] Royal Commission on Fire Brigades and Fire Prevention, Report (1923), p. 137.   [2] *Loc. cit.*

has been taken on the recommendations of the Royal Commission.[1]

The arrangements in the metropolis in regard to defence against fire are thus unchanged from what they were 15 or 25 years ago. There is a multitude of separate fire brigades operating independently, there is an immense variation in size and efficiency of the forces, a great diversity in equipment, technique and methods of training, there is a negligible amount of co-operation, and no over-all planning or direction. But the international situation has undergone an almost unimaginable change since 1923; and we are now faced with the possibility of another European war in which London would certainly be the principal object of aerial attack so far as this country is concerned. Its capacity for dealing with incendiary bombs might easily be a decisive factor in deciding the issue. In these circumstances the question of fire-fighting services has become a matter of crucial importance.

Official recognition of this was indicated by the appointment in 1936 of a Departmental Committee on Fire Brigade services under Home Office auspices. The Committee stated explicitly that the degree of co-operation introduced during the Great War of 1914–18 would certainly be insufficient in the event of a future conflict, since it merely provided that some of the brigades surrounding the Administrative County should in an air raid send motor pumps to stand by at certain fire stations in Outer London, while the London Fire Brigade despatched an escape to stand by at the local station.[2]

Mr. Herbert Morrison, in his evidence on behalf of the London County Council, stated the indisputable fact that from the point of view of efficient fire defence a case can be made out for a Greater London fire brigade organised under a single command. He pointed out, however, that equally good grounds for unity of administration throughout Greater London exist in the case of several other services for which the London County Council is responsible. The creation of an *ad hoc* authority for a Greater London fire brigade would, he em-

[1] London County Council: Minutes, 11th February, 1936. Report of the General Purposes Committee.
[2] Report, [Cmd. 5224] 1936, pp. 35, 58.

phasised, be undesirable in that it would have grave repercussions on the structure, stability and comprehensiveness of the normal machinery of local government. Difficulties already exist in regard to the water supply for fire-fighting purposes. The public utility undertakings are under no obligation to provide an adequate supply and pressure of water for use with fire extinguishers, and the London County Council has no power to expend money in providing, extending or enlarging water mains. The Fire Brigade and Main Drainage Committee of the London County Council is of the opinion that in certain parts of London the water supply is not satisfactory for fire protection purposes; and in the event of air raids this might give rise to grave danger.[1]

The disadvantages resulting from the fact that the fire brigade is in the hands of one authority and the water supply in the hands of another, would be followed by further drawbacks if an *ad hoc* body were set up in charge of fire brigades throughout the metropolis. For example, the inspection and regulation of places of public entertainment, such as cinemas and theatres, and the administration of the London Building Act, needs to be carefully correlated with the requirements of the fire brigade; and this could not be done if the duties were in separate hands. The London County Council therefore rightly resist the creation of an independent fire brigade authority for Greater London. At the same time they advised the Committee that the scheme of aid raid precautions for the metropolis should embody a comprehensive plan for protection against fires arising from incendiary bombing, including unified control, a special system of communications, advance training of reserves, and augmented personnel and appliances. In conclusion, the Council placed on record its belief in the efficacy of a large administrative unit.[2]

The Committee accepted the general thesis put forward by Mr. Morrison and advised against the formation of an *ad hoc* fire brigade area involving the creation of a new local government organ. Nor did they recommend an enlargement of the

---

[1] London County Council Minutes, 11th February, 1936. Report of the General Purposes Committee; Report of the Departmental Committee on Fire Brigade services, [Cmd. 5224] 1936, pp. 23–4. [2] *Ibid.*

London County Council's own area.[1] All that they proposed, in fact, was the setting up of a joint committee containing representatives of all the local fire brigade authorities within the Greater London area. This joint Committee is to formulate a scheme for effecting necessary improvements in personnel and equipment, and to plan a regional organisation which will come into operation in the event of war under the control of a regional mobilising officer.[2]

The obstacles to the most efficient organisation of the fire-fighting service, even in the face of a vital national interest, is a good illustration of the insuperable difficulties which exist in many branches of municipal administration in the present state of London government. If there is good reason for requiring the organisation of fire protection services on a regional basis in time of war, it follows that there must also be good reason for reforming the organisation at the present time. Even the peace demands on the fire brigades are less adequately and efficiently met with the 64 independent and unco-ordinated brigades which now exist than they would be with a single unified force of the unrivalled standard of the London Fire Brigade. From the point of view of preparation against air attack, there is everything to be said for establishing and perfecting during the years of peace a single brigade with uniform methods, training and appliances, accustomed to unified command, rather than waiting until the outbreak of hostilities to bring into operation an untried scheme imposed on 64 independent bodies with the inevitable friction and delay which that involves.

Yet what else could the Departmental Committee propose? It was limited by its terms of reference to a consideration of fire brigade services alone, and could not examine the general question of London government. It saw the indubitable force of the London County Council's objections to an *ad hoc* fire brigade authority; and thus was left with no other alternative than the feeble makeshift arrangements which it actually recommended.

We are thus driven to conclude that the problem of improving the administration of a particular service cannot be tackled

[1] Report, p. 58. [2] *Ibid.*, pp. 31-2.

separately, but must form part of a comprehensive reorganisation of the metropolis as a whole. One wonders how many air raids it would take to drive this elementary truth into the obtuse minds of those who cling so obstinately to forms of government which are manifestly obsolete.

CHAPTER XIII

EDUCATION

It is often said that education is the local government service about which the local elector cares most deeply; and that the flame of civic patriotism burns more brightly round the schools than round any other kind of municipal institution. It is, therefore, of particular interest to observe the situation in London in regard to education.

Within the administrative county the London County Council is responsible for all forms of public education: elementary, higher and technical. In Outer London there are 32 local education authorities, of which 23 are responsible for elementary education only, while 8 possess powers in respect of higher education. All the 33 local education authorities in Greater London may and do act independently. There are 33 education rates levied in the metropolis. The financial burden varies greatly as between one part of London and another. The average education rate within the administrative county is 2s. 8d. (1936–7).[1] In the outlying areas the figure depends chiefly on the rateable value per head of the district. In West Ham the education rate is 5s. 8d., in Tottenham 3s. 10d., in Walthamstow 4s. 2d., in Ealing 2s. 6d., in Richmond 1s. 9d.[2]

A certain number of difficulties occur in the provision of elementary education. They are of less importance than those arising in connection with higher and technical education, but they are by no means negligible.

Owing to the spread of London and the migration of residents to the outlying areas, the school population in the inner districts is diminishing rapidly. For example, the total roll in elementary schools within the administrative county fell from 650,000 in 1926–7 to 450,000 in 1937–8. The average attendance in the coming year (1939–40) is expected to fall by nearly 30,000. In consequence, hundreds of teachers are becoming redundant every year, no new ones are being engaged, and the age of

[1] *London Statistics*, vol. 40, p. 421.   [2] *Ibid.*, p. 432

existing teachers is rising substantially[1] This is undesirable in every branch of school work, but in some schools in the administrative county there are actually no teachers young enough to give the children physical training. In Outer London, by contrast, the school population is increasing, and local education authorities are therefore taking on considerable numbers of young teachers.[2] The growing disparity between the age of teachers within and without the county cannot be to the advantage of the metropolis as a whole. It cannot be overcome while the prevailing organisation continues. The Board of Education is unable to deal with the matter.

In regard to special schools for backward, deaf or dumb children the balance of advantage lies the other way. The larger the area, the more specialised the provision can be. Hence, the London County Council is undoubtedly able to cater more efficiently than the Outer London authorities for the special needs of minor groups such as these. Here again a pooling of resources would improve the position.

One of the few instances where a satisfactory degree of co-operation has been achieved is in regard to the operation of the Education Act, 1936. This statute raises the school-leaving age to 15 but permits children to be released from school between the ages of 14 and 15 in order to enter "beneficial employment" or on account of difficult home circumstances. Great apprehension was felt by educationists at the disadvantages which would result from local education authorities in the London Region adopting inconsistent policies in working this Act. Moreover, while in the administrative county the London County Council could require children to attend day continuation classes for a specified number of hours each week as a condition of releasing them from school,[3] there was

[1] The average age of elementary teachers in the administrative county is probably well over 40; while that in an expanding borough in Outer London such as Hendon is about 28.

[2] For example, the population of Middlesex increased from 1,253,002 in 1921 to 1,638,728 in 1931.

[3] Section 2 (4) of the Act provides that the authority which issues the employment certificate, in determining whether any employment will be beneficial, shall have regard (inter alia) to "the opportunities to be afforded to the child for further education," and may require as a condition precedent to the grant of a certificate, such undertakings as they think necessary from the employer.

no certainty of this being done in Outer London, because day continuation work falls under the heading of higher education and many local education authorities are responsible for elementary education only. Furthermore, there would be no medical supervision of children released for employment unless they are in attendance at a day continuation school, in which case they are inspected by the school medical officer.

Fortunately, these difficulties have been overcome by an agreement arrived at from a conference of 34 local education authorities held at County Hall in November, 1937, and March, 1938. The conclusion reached was that while it was not possible to draw up a comprehensive list of "non-beneficial" employments and much discretion would have to be left to individual local authorities, several important types of undesirable employment could be definitely ruled out. These include occupations connected with the sale of intoxicating liquor, the collection or sorting of rags and refuse, employment on any racecourse or speedway where betting is allowed, in dance halls and amusement saloons, and so forth. The conference also agreed that the weekly hours of employment should not exceed 44, of which 8 must be devoted to attendance at classes, and that no overtime should be allowed. Principles were also formulated as to the hours of the day during which work might be permitted, wages, holidays and other related questions.[1]

This desirable result may be regarded as a triumph of goodwill over malorganisation. There are some problems, however, which cannot be solved by mere co-operation because they arise out of the physical planlessness from which the metropolis has suffered so acutely in the past.

The urban elementary school is essentially a local institution in the sense that the children almost always live not only in the area of the education authority providing or supporting the school but also in the vicinity of the school itself.[2] In the congested conditions of London, however, it is not possible

[1] A statement containing full information was issued by the education officer of the London County Council on 22nd October, 1938, and was reproduced in part in *The Times* on 24th October, 1938.
[2] There were only 2,050 children resident in Outer London attending L.C.C. schools and institutes in 1934-5 and 250 attending special schools.

to continue this genuine localism into the sphere of recreational and health-promoting activities. The insurmountable difficulties of providing children with adequate facilities for playing games under suitable conditions in the thickly built-up parts of London, except those which have access to pitches available in the public parks, led the London County Council to adopt an entirely new policy in this matter. Sites have been acquired in open fields outside the administrative county, playing pitches have been made and classroom accommodation provided in the form of wooden huts. Groups of children are sent for a whole day's outing during which the time is divided between work and play. Between July, 1931, and May, 1937, 8 fields were acquired for this purpose, comprising a total area of 70 acres and containing 31 classrooms in which 5,840 children could be accommodated each week. Further plans envisage a site of $22\frac{1}{2}$ acres at Becontree to accommodate 1,600 children weekly and the acquisition of 70 acres near Upminster on which 30 classrooms will be constructed capable of receiving 6,000 children each week. Improvements will be made in regard to the amenities provided, and shower baths and changing rooms are being introduced.[1]

The usual practice is for each school participating in the scheme to send one class each week in charge of a teacher. This means that a party of about 40 children leave their school at 9 a.m. and arrive at the playing field between 9.45 and 10.15 a.m. according to the distance. They remain there about five hours. Ultimately, when the scheme is complete, 150,000 will be paying a weekly visit to a playing field.[2] Most of the children bring food with them and supplement it with a hot drink prepared on the site. Sometimes a hot meal can be purchased locally but improved provision for the mid-day dinner is under consideration. Much attention is paid to proper coaching in team games and athletics by the physical education staff; and an attempt is made to relate the classroom work to the rural environment.

The scheme represents a humane and enlightened policy

[1] Annual Report of the London County Council, 1936, vol. v, Education, No. 3347, p. 6.
[2] The total cost will amount to £1,500,000.

for which the London County Council deserves great credit. But recognition of the Council's courageous and imaginative effort to overcome the "insurmountable difficulties"[1] with which it was faced should not blind us to the tremendous mistake which was made (by all the central and local authorities concerned) of permitting those difficulties to arise by refusing to face in proper time the essential question of planning the growth of London so that the vast majority of its children (and their families too) should not be cut off from easy access to playgrounds and country surroundings. Nor is it possible to believe that an adequate substitute has been found by means of these lengthy journeys to distant places and the elaborate arrangements which they involve. In general, the children appear to derive substantial benefit to their health, but some of the attendant drawbacks are referred to by the district inspector in his official report. He states that a considerable number of school children prove to be "bad sailors" on the trams—and also presumably on the trains and buses —in some cases too bad to take part in the scheme at all.[2] The loss of time resulting from the long journeys, amounting often to 2–2½ hours in the day, and the discomfort arising from wet weather, are serious where the field is situated a considerable distance from the nearest omnibus or tram stop.

In the sphere of higher education a substantial number of children living outside the area are in attendance at secondary schools provided or maintained by the London County Council. In 1919 the number was stated to be 6,000 out of a total of 36,000 pupils enrolled in public secondary schools within the County of London.[3] At that time there were no financial arrangements in force for recovering the net cost other than fees, and these out-county families from which the children came were in consequence being heavily subsidised by the London ratepayers.[4] Shortly afterwards the London County Council introduced a scheme whereby an out-county authority is required to pay the net cost (after deducting fees) of a child living in their area and attending an L.C.C. secondary school.

[1] Op. cit., p. 6.   [2] Ibid., p. 7.
[3] Royal Commission on London Government, 1923: Minutes of Evidence, p. 116, Sir Robert Blair.   [4] Ibid., Norman, p. 45.

If the local authority refuses, the parent must pay the entire cost.[1]

The figures show that (in 1934-5) 3,824 pupils living outside the administrative county were attending secondary schools aided and maintained by the London County Council. The total number of children attending such schools was (in 1936) about 34,000, so that the proportion coming from over the border exceeded 11 per cent. By 1938 the number had dwindled to 2,650.

The diminution in the proportion of out-county children from nearly 17 per cent in 1919 to 11 per cent in 1936 and a still lower proportion in 1938 is no doubt partly due to the improved provision of secondary schools by the outlying local education authorities. But it is also due to some extent to the practice which the London County Council was almost bound to adopt of charging the net cost on the possibly reluctant outside local authority, which is under no obligation in the matter whatever. "It is obvious," said Sir Robert Blair (then Education Officer at County Hall), to the Royal Commission on London Government, "that the new policy of charging the net cost to the outside authority or to the student, while inevitable on the part of the London County Council, raises grave difficulties for a large number of thoroughly good outside-county students, difficulties which would not exist were the one economic area also one education area. The difficulties are not merely those attendant on transfer from an inside- to an outside-county institution. For many no suitable provision exists outside; and it has already been remarked that it would be financially unwise to provide it."[2]

At the present time arrangements have been made between the London County Council and the County Councils of Middlesex, Essex and Kent whereby what are called "vouchered" pupils from those counties are admitted to London schools at the ordinary fee. The London County Council receives from

---

[1] This arrangement dates from 1st August, 1921. The total annual cost for each child was £42, of which £14 was payable as fees by the parent and £28 by the local authority. Half the latter sum is recoverable from the Board of Education.

[2] Royal Commission on London Government, 1923: Minutes of Evidence, p. 118.

the appropriate county council the difference between the ordinary fee and the full cost of the education provided. In addition to these "vouchered" pupils there are a few out-county pupils paying much larger fees which represent the whole cost of the education. There are no similar arrangements in force with the county councils of Surrey or Herts, or with the county boroughs in Outer London.

It is not suggested that secondary education is always provided on either better or more generous lines inside the administrative county than in the Outer London areas. That is far from being invariably true. The Education Officer for Kent (Mr. Salter Davies) was able to show the Royal Commission on London Government that the Kent County Council was making more extensive provision for secondary education than the London County Council, since in 1921 in the former area secondary school pupils numbered ten per thousand, whereas in London the proportion was only seven per thousand. Moreover, Kent was not only already in advance of London in this respect but had also been progressing more rapidly. Furthermore, in the part of Kent falling within the Greater London area, 34·8 per cent of the pupils in secondary schools were admitted free of charge, whereas in the administrative county the proportion was only 28·1 per cent. In the same part of Kent the percentage of secondary pupils holding University Exhibitions was 2·4 as compared with only 0·6 in London. The output of trained teachers in Kent was 17 per 100,000 population; in the L.C.C. area it was 13·6.[1] In presenting this information to the Royal Commission, Mr. Salter Davies disavowed any desire to criticise the educational provision made by the London County Council. "I am only suggesting," he said, "that Sir Robert Blair's picture of the denizens of outer London shivering at the gates of Paradise is somewhat highly coloured."[2]

Middlesex, again, has certain superior virtues in regard to secondary education. Thus, the Middlesex County Council holds a common examination for all candidates seeking entry to its secondary schools and accepts the top children regardless

[1] Royal Commission on London Government, 1923: Minutes of Evidence, Salter Davies, p. 360.   [2] *Ibid.*, p. 360.

of whether their parents pay fees or not. The number of scholarships is determined by the number of poor children on the list—a system possessing obvious advantages. The London County Council, on the other hand, predetermines the number of free places in secondary schools and then holds a special examination for poor children to fill those places separate from the examination for fee-paying children—a far less admirable method.

Hence no simple case for organising higher education on a regional basis can be based on the argument that the standards and methods of the London County Council are invariably superior and ought to be extended to the outlying authorities. As we have seen, the practices of some of those authorities are considerably in advance of the London County Council. Indeed, one of the main contentions which may be urged in favour of greater unity in London education is that it might tend to promote a raising of the standards and methods of the London County Council to the level of the best outlying local education authorities. The differential treatment of children living in different parts of the metropolis by the several educational authorities is irrational and unjust.

In the field of technical education we find an entirely different situation. Here the London County Council holds an undisputed primacy, and the obsolete boundaries of the administrative county are so completely unrelated to contemporary needs that local authorities have been virtually compelled to find ways of overcoming them.

It is obvious that the efficient provision of technical and vocational education demands a high degree of specialisation; and this in turn involves large financial resources and an extensive population from which a sufficient number of students can be drawn for each specialised course. The London County Council satisfies both these requirements in the highest degree and has in consequence been able to make extensive and excellent provision over a wide range of subjects.

The Council maintains thirteen day technical schools for boys, including the Central School of Arts and Crafts, the Hammersmith School of Building, the Shoreditch Technical Institute for instruction in the furniture trades, the Smithfield

Meat Trades Institute, the Westminster Technical Institute for training in hotel and restaurant work, the School of Retail Distribution and Salesmanship, the School of Photo-Engraving and Lithography, and several schools of engineering and navigation. There are eight day technical schools for girls who wish to receive instruction in such occupations as ladies' tailoring, dressmaking, lingerie, millinery, upholstery, hairdressing, photography and cookery. There are seven art schools offering both day and evening courses. Evening classes in technical subjects are also held in eighteen technical institutes maintained by the Council. In addition to these "maintained" schools, the London County Council also aids by money grants many other educational institutions offering instruction in art, science and technology. The grants are given subject to certain conditions, including inspection and the appointment of London County Council representatives on the managing body.[1] The institutions thus aided comprise ten polytechnics and some eighteen miscellaneous institutions including Toynbee Hall, the Royal School of Needlework, Morley College, the School of Woodcarving, Goldsmiths' College, the Bermondsey Settlement, the Mary Ward Settlement, King Edward VII Nautical School and the Cordwainers' Technical Institute.[2] Lastly, there are eleven voluntary day continuation schools for young persons between the ages of fourteen and eighteen offering courses mainly of a vocational nature in such subjects as technical drawing, workshop calculations, woodwork, shorthand, typewriting, book-keeping, metal work, dressmaking and practical science.

Mention should also be made of a large number of evening institutes which the London County Council provides for various classes of students. These are usually held in elementary school premises and include courses in preparatory and advanced commercial education, and instruction in practical subjects as well as in cultural studies of an elementary kind.[3]

The work of the London County Council in the sphere of higher technical instruction is a very fine achievement. It is

[1] *London Statistics* (1935–7), vol. 40, p. 269.
[2] *Ibid.*, p. 272.   [3] *Ibid.*, p. 275.

unique both in magnitude and quality, taken as a whole. For many years it has attracted an increasing body of students living both inside and outside the county. Prior to about 1920, students residing in outlying areas were admitted as freely and on the same terms as those whose homes were inside the boundary. As a result of the post-war economy campaign, the London County Council introduced a scheme whereby outlying authorities were required to pay the balance of the cost for students coming from their areas, after deducting the amount paid in fees. This arrangement remained in force for more than a decade. In 1931, another campaign for economy in public expenditure led the local authorities in Outer London to put a severe check on the amount of money they were willing to spend on their students receiving technical education from the London County Council. In consequence there was a sharp drop in the attendance figures. This naturally caused considerable difficulty at County Hall, since the elaborate system of technical education had been developed in the anticipation of a steady flow of out-county students. The unexpected diminution in numbers and revenue jeopardised specialised courses and introduced an element of uncertainty and confusion into the situation.

In 1934 a revised scheme was introduced under which (after deducting fees) the cost of educating out-county students is shared equally between the outlying local education authority and the London County Council.

The number of out-county students admitted in 1935–6 to technical institutions, schools of art and evening institutes aided and maintained by the London County Council was approximately 28,000, and the local education authorities contributed in respect of them a sum of £84,200.[1] If fees are added, a total revenue of £145,000 was received in respect of out-county students. If we include the numerous out-county students who are admitted to technical and similar institutions either free of charge or without charge to their own local authorities, by virtue of concessions granted by the London

---

[1] I am indebted to the Education Officer of the L.C.C. for this information. The sum mentioned excludes £2,000 paid direct to the Northern Polytechnic by the Middlesex County Council.

County Council, the total number in attendance within the administrative county was 36,400 in 1935-6.[1]

Under the arrangements now in operation, the London ratepayer is subsidising the out-county student attending technical, art and evening schools maintained or aided by the London County Council by an amount approximately equal to that paid by the local education authority of the area from which the pupil comes. An annual sum of nearly £85,000 falls on the London rates in respect of these students.

The proportion of out-county students at London County Council technical schools varies considerably between the different grades and categories of institutions. There are no detailed figures available, but the following aggregates give a rough indication of the position.

In 1934-5, 82,545 students were in attendance at technical institutions, schools of art, polytechnics and trade schools aided and maintained by the London County Council; of these, 20,658 were out-county students. The attendance in the same year at L.C.C. evening institutes was 139,458, of which only 10,518 students came from over the border. As one would expect, it is to the higher and more specialised centres of instruction that the majority of the out-county students go. The great monotechnics provided by the London County Council, such as the schools of printing and book production, photo-engraving and lithography, enjoy a supremacy which is unchallenged; whereas lower down the scale, in the realm of polytechnics and evening institutes, it is far easier for the outlying education authorities to make their own local provision.

It will be seen that the London County Council has fulfilled its responsibility for technical education in the broadest possible spirit, offering facilities and opportunities to the student population of all London without regard to the narrow confines of the administrative boundaries. It has been financially generous almost to a fault, subsidising the out-county student with funds derived from the rates. It is, indeed, not easy to see how this burdening of the London ratepayer can be justified,

---

[1] L.C.C. Education Committee Minutes, 21st July, 1937, p. 343. Report of the Higher Education Sub-Committee, par. 26; see also Circular T2 relating to out-county students.

19

except on the ground that it is necessary to attract the out-county students in order to maintain the most specialised technical schools on an efficient basis. The authorities in Outer London, like the eager students who flock to the schools in Inner London, have not permitted the administrative boundaries to form an insuperable obstacle in the students' path.

In spite of all this co-operation, it cannot be said that the problem of technical education in London has been satisfactorily solved in all respects.

In the first place, the present procedure is unnecessarily complicated. A student living beyond the border who desires to attend an L.C.C. institute must apply to his own local education authority for permission, unless he is able to pay the full cost of the fees himself. The local education authority often prefers him to go to their own institution, if there is one, and hence they are likely to approve his attendance at the London County Council school only if special cause can be shown either on the ground of lack of local provision, inaccessibility, etc. Second, the out-county students are not eligible for the London scholarship system, which is much more favourable in this branch of education than the scholarships offered by most outlying authorities. Third, there is no proper planning of technical education for the whole metropolis. A conference between the local education authorities concerned with technical education is held once a year, but questions of planning and development are not usually discussed on these occasions.

The London County Council, strictly speaking, can only consider its own population when making provision for technical institutions and cannot properly take account the requirements of the students it may and does receive from outside its border, although these students need the L.C.C.'s technical education and the London County Council needs them. There is no organised machinery for considering the industrial and technological requirements of the whole metropolitan region. Out-county students have to be content with facilities at existing institutions and are not deliberately catered for. No account is taken by the London County Council of the technical training required for special industries situated

outside the boundary—e.g. motor engineering at the Ford
works at Dagenham. There are few instances, if any, on record
where an outlying local authority has requested the London
County Council to make provision for a particular form of
technical education.

Nevertheless, despite their shortcomings, these arrangements
do show that the local patriotism of which so much is heard
whenever a reorganisation of areas is under consideration,
disappears with remarkable speed whenever an opportunity is
given to the inhabitants of outlying areas to take advantage of
the services, institutions or money provided by the London
County Council. We have seen this in connection with main
drainage, the green belt, and other services no less than in
regard to technical education. One sometimes wonders whether
the readiness of the London County Council to offer its
services to persons and communities outside its boundaries
may not have had the effect of encouraging a parasitic outlook
on the part of the Outer London local authorities, and of
postponing much-needed reform by making things easier for
the out-county inhabitants than they really ought to be.

A similar state of affairs exists in regard to London University. The University serves the whole metropolis, in
addition to which many of its students are drawn from other
parts of Great Britain and oversea countries. For the year
1935–6 the University received by way of grants from local
authorities £142,170, of which no less than £129,000 was paid
by the London County Council. Thus, all the Outer London
authorities, containing a population larger than that of the
administrative county, contributed only £16,000 to the University, while the London County Council paid nearly eight
times as much. Yet the residents of Outer London attend the
University probably to as great an extent as do the inhabitants
of the administrative county.[1]

[1] From 1936–7 onwards the annual grants payable by local authorities
to the University were as follows: Bucks C.C., £200; Essex C.C., £1,500;
Herts C.C., £500; Kent C.C., £2,500; Middlesex C.C., £5,000; Surrey C.C.,
£2,500; Croydon C.B.C., £500; Southend C.B.C., £200; East Ham C.B.C.,
£100 (approximately); West Ham C.B.C., £270. The annual maintenance
grant of the London County Council for the quinquennium 1935–40 is
fixed at £129,000. Contributions to the General Building Fund raised by

"I hope to show," said Sir Robert Blair to the Ullswater Commission on London Government, "that (1) the use of London educational institutions (above the elementary stage) by workers within the Greater London area is founded on the unconscious belief on the part of the pupils, parents and students that the area is already one educational area; and (2) that there are great educational advantages to be obtained from making administratively true what is already a not unnatural conception or belief."[1] His argument was that Greater London is one economic area; that pupils, parents and students regard it and use it as a single educational area; that the existence of a large number of separate local education areas within the metropolis creates obstacles both to its use in this way and to the appropriate organisation and location of educational institutions, as well as to the solution of the wider problems of educational policy and to the effective and economic use of the area's resources; and that as the educational needs of the area are not static but in active movement they can only be dealt with appropriately by a combination of local and central management within Greater London.[2]

The London County Council proposed that the principal London authority should have responsibility in matters of educational policy, such as the fixing of salaries and wages, the adoption of scholarship schemes, the control of training colleges and higher technical schools, the standards of provision for higher education, relations with the University and so forth. The local authorities within the metropolis would

the University to pay for the new building on the Bloomsbury site are much more nearly equal as between the local authorities in Inner and Outer London. The London County Council gave £250,000, the City Corporation £100,000, Middlesex County Council £100,000, and the councils of the five neighbouring county councils and the three county borough councils together £151,100. The London County Council have also granted a further sum of £200,000 for new buildings for the Institute of Education and Birkbeck College. The London County Council has made substantial grants for the capital requirements of schools of the University not connected with the Bloomsbury site (two sums of £150,000 each having been voted for this purpose during the decade 1930–40), while so far none of the outlying local authorities have made any contributions of a similar nature. For details, see the University of London Calendar, 1937, pp. 32–5, 37–8.

[1] Royal Commission on London Government (1923), Minutes of Evidence, p. 116.　　　　　　　　　　　　　　　　　[2] *Ibid.*, p. 120.

have the power of initiative and administration over other educational institutions; they would provide and manage school sites and buildings, award scholarships and exhibitions, look after the curricula, appoint and promote teachers, etc. In regard to finance, the estimates of the local authorities would be scrutinised by the principal authority and the expense defrayed up to a prescribed standard of cost by a single rate levied over the whole area. Any excess cost would be borne by the respective local authorities. By this means a common standard of education would be maintained for the entire metropolis and the cost spread evenly throughout its parts by a common rate.[1]

Much of the opposition to this proposal was based on the difficulty of dealing with the residual parts of the outlying counties if the portions of them in Greater London were abstracted. Thus, the witnesses for Surrey pointed out that teachers' salaries throughout the county were fixed on the Burnham Scale III. They were apparently able to do this because assistants in urban schools had good chances of becoming heads of rural schools. With the abstraction of the portion of Surrey falling within Greater London, it was urged, the county would become mainly rural, and the county council might no longer be justified in retaining Scale III. During the examination of these witnesses it appeared, however, that the county was then saving £20,000 a year in elementary teachers' salaries within the Greater London part of their area by not granting them the higher salaries laid down by Scale IV of the Burnham award.[2] This particular claim could therefore scarcely be substantiated.

Nevertheless, there is a problem raised by the effects on residual areas of treating the metropolis as it needs to be treated. But that aspect of the matter belongs to Part III of the present work.

[1] Royal Commission on London Government (1923), Minutes of Evidence, p. 181–2. Appendix IV, Memorandum by R. C. Norman and Captain Warburg.
[2] *Ibid.*, pp. 285–6 (Evidence of W. A. Powell, J.P.).

CHAPTER XIV

THE ULLSWATER COMMISSION ON LONDON GOVERNMENT

Shortly after the war of 1914–18 there were symptoms of a renewed interest in the administrative problems of the metropolis. This was stimulated partly by the brief enthusiasm for social reform which accompanied the peace and the demobilisation; and partly by an appreciation of the growing urgency of the problems confronting London. Mention has already been made of the enquiries set on foot to deal with the subjects of housing, wholesale food markets, water supply, the Thames Conservancy and other aspects of local government in London. The most comprehensive event in this movement was the appointment in 1921 of the Royal Commission on London Government, with Lord Ullswater, a former Speaker of the House of Commons, as chairman. The Commission arose out of a report issued by the London County Council in 1919, containing a resolution asking the Government to investigate the desirability or undesirability of a reform of London government. The terms of reference of the Commission charged it "to enquire and report what, if any, alterations are needed in the local government of the administrative county of London and the surrounding districts, with a view to securing greater efficiency and economy in the administration of local government services and to reducing any inequalities which may exist in the distribution of local burdens as between different parts of the whole area."

It may be said at the outset that the Ullswater Commission was an unmitigated fiasco, despite the fact that it included among its members persons of extensive knowledge of local government and wide experience of public affairs.[1] The responsibility for its failure lay partly on the London County

---

[1] The members were Viscount Ullswater (chairman), Sir Richard Vassar-Smith, Sir Horace Munro, Sir Albert Gray, K.C., Mr Neville Chamberlain, Mr. E. H. Hiley, Mr. G. J. Talbot, K.C., Sir Robert Donald, Mr. E. R. Turton and Mr. Stephen Walsh. Sir Richard Vassar-Smith resigned in 1921 and Mr. Chamberlain in 1922.

Council, which presented its case in a feeble and unconvincing manner; partly on the Minister of Health, who made no effort to place a plan of reform before the Commission or even to invite attention to the prevailing difficulties; and partly on the Commissioners themselves, most of whom were unable to grasp the fundamental nature of the problems involved.

Some of the chief officers of the London County Council gave admirable evidence showing the difficulties which existed in their respective departments and the best manner of meeting them.[1] But the official policy of the Council was to point out that the existing system was inadequate and to leave to the Commission the task of formulating a better one. Thus, the memorandum of evidence presented by Mr. R. C. Norman, the principal witness for the London County Council, stated by way of general conclusion that "few, if any, of the more important powers of local government can now be exercised with complete efficiency if the county boundary is regarded as the administrative limit. As the inconvenience of the existing boundary has made itself felt in one service after another various more or less satisfactory expedients have been adopted for enlarging the area of administration. . . . The situation has again become acute in the case of several other services." These problems were pressing for a solution, which must be either piecemeal, as in the past, or based on a single considered plan. The time is ripe for a careful enquiry into the whole question of London government. "But," the statement continued, "we do not consider that it falls within the province of the Council to present to the Royal Commission a scheme purporting to be a complete and satisfactory solution of the present complicated problem." Hence the witness stated, "We are not prepared with proposals for defining a new administrative area or for establishing a suitable administrative authority and determining its constitution and powers and its relations to existing authorities."[2] This diffidence was due to the fear that a definite scheme emerging from County Hall would arouse

[1] Notably Sir Robert Blair, the Education Officer, and Mr. Humphreys, the Chief Engineer.
[2] Royal Commission on London Government (1923), Minutes of Evidence, pp. 47, 50.

the suspicions of the neighbouring authorities;[1] but since their suspicions and hostility were in any case inevitably aroused, there was nothing to gain and much to lose by treading softly. The decision to say virtually nothing positive was both important and unfortunate: the Commission found the attitude of the London County Council unhelpful and did not hesitate to say so.

The Commissioners asked the London County Council to describe the administrative difficulties with which they were faced, but instead of complying with this request the representatives from County Hall assumed the unsuitability of the existing area and attempted to support the assumption by citing instances where Parliament and other authorities had ignored the existing boundaries for various purposes, such as main drainage, parks and open spaces, water supply, fire protection (during the War), allotments and small holdings. Thus, the fact that the Metropolitan Police District had been defined in 1829 on much more extensive lines than the administrative county was taken to show that the present county area was found to be wrong a century ago.[2] But the Commission was not prepared to accept inferences of this kind or to base its findings on unverified assumptions.[3]

The procedure of the London County Council, declared the Commission, was prejudicial to the usefulness of the enquiry, because it invited a maximum of criticism directed against the proposals put forward by the Council itself and of objection to any change whatever in the *status quo*. At the same time it failed to encourage the ready disclosure of difficulties in administration by other local authorities, or the discussion with those bodies of possible remedies. A practicable scheme could only be formulated in the light of the fullest information concerning the prevailing situation, followed by full and frank discussion between the authorities concerned with a view to devising an improved organisation. Nothing of this kind was even attempted by Mr. Norman and his colleagues.

An essential weakness of the London County Council's

[1] Percy Harris: *London and its Government*, p. 237.
[2] Minutes of Evidence, p. 36; Report, pp. 7-10.
[3] Report, pp. 7-10, 13.

handling of the situation was that they made no attempt to demonstrate the disadvantages resulting from the existing arrangements outside their own area. The principal witnesses from County Hall expressly disclaimed any knowledge of the places beyond their frontier, and the Council made no attempt to ascertain whether or not they were efficiently administered, or what difficulties, if any, were caused by the present organisation.[1] Moreover, as the Commission complained, the London County Council had made no endeavour to ascertain the sentiments of neighbouring authorities, or to explain to them why they considered the existing state of affairs to be unsatisfactory. There had, in fact, been no consultation whatever on these matters of vital common interest.[2] It was scarcely to be expected that the outlying authorities would reveal their own shortcomings, and since the Royal Commission showed no signs of investigating matters independently, there was little likelihood of the defects being brought to light at all without the active intervention of the London County Council. Incidentally, one of the Minority Reports quite justly criticised the procedure of taking evidence only from the interested parties, on the ground that it tended to make the enquiry resemble a form of litigation, save that no opportunity was given for testing the statements of witnesses by cross-examination.[3] The Commission made no attempt to go into the field for the purposes of making observations directly or through trained investigators attached to the Commission. They were in consequence entirely dependent for their information on the evidence brought before them.

The statement submitted by the London County Council was intended to show, first, that the area laid down by Parliament was too restricted for certain purposes; and second, that the Council lacked the necessary powers to deal with some services which ought to be administered by a so-called "central authority"—i.e. an authority for the whole metropolis. In particular, the Council contended that a larger area was

---

[1] Report, p. 57.
[2] *Ibid.*, p. 58. "London said that there were difficulties, but London, as far as I am aware, never approached the counties as to the methods of meeting those difficulties." Holland, Surrey County Council, Minutes of Evidence, p. 270.   [3] *Ibid.*, pp. 141-2.

required for education, electricity supply, housing and the regulation of wholesale food markets.[1]

On the question of the proper area which ought to be selected for the administration of these functions, the representatives of the London County Council were either evasive or ignorant. Mr. Norman contented himself with pointing out that the principle followed hitherto—both in 1855 and on most other occasions—in defining urban areas for local government purposes was to include "the whole continuous urban area . . . together with such a surrounding belt as was likely to become of an urban character within a short time."[2] This principle, he suggested, should be followed as far as possible in reforming the structure. He was careful to add that while no part of the continuous urban area should be omitted, the London County Council would not ask for the inclusion in the enlarged boundary of any considerable territory of a rural character. Regard might be paid, however, to the territories employed for such services as the police forces, water supply and electricity distribution, in order that one Greater London Region should if possible be used for all local government purposes. On the other hand, when Mr. Norman was asked whether it would be desirable to take one of the existing areas for the proposed new authority, he answered in the negative. "It seems to me that the dominant factor is where the town is," and the witness reverted to his statement concerning the inclusion of the continuous urban area.[3] In reply to another question by Mr. Neville Chamberlain, the witness insisted that the London County Council was disinclined to include exclusively rural areas with the built-up portion unless they were threatened with building development in the near future. They had not even considered the desirability of preserving a belt of rural country round London.[4]

It would be difficult to find a more unimaginative and feeble outlook than that shown by the London County Council's chosen representatives. Their conception of the metropolis was unable to rise above the idea of an endless vista of cement streets and built-up districts, despite the fact that they con-

[1] Report, pp. 10–12.
[2] Minutes of Evidence, p. 51.
[3] Ibid., pp. 64–5.
[4] Ibid., p. 48.

tended the area should be one in which the new authority could exercise town planning powers.[1] It was left for the Royal Commission to point out that the lines of continuous building had followed the main transport routes and did not in themselves indicate an area well adapted to local government.[2] The Council's suggestion was cursorily dismissed as offering no solution of the problem of defining the area of optimum efficiency.

The major authority which was to operate in the new area should be a directly elected council consisting of 124 members elected by an equal number of constituencies. The London County Council was opposed both to the creation of further *ad hoc* bodies to deal with particular services[3] and to the establishment of joint committees representative of all the independent authorities concerned in the administration of municipal functions.[4]

On the other hand, Mr. Norman went to great lengths in upholding the so-called dual system, whereby certain services are reserved to a major authority and others entrusted to minor bodies elected for particular districts. The objects of his plan were, indeed, not merely to concentrate power in the new central organ but also to relieve it of much detailed work now carried out by the London County Council, and to increase the powers of the proposed district authorities in comparison with those possessed by the metropolitan borough councils.[5] Whether this emphasis on the proposed aggrandisement of the minor councils was introduced in the hope of conciliating potential opposition, or from a genuine belief in localism within the metropolis, is not known. But part of the ineffectiveness of the London County Council's proposals arose from the fact that although the evidence showed there was a considerable amount of friction and conflict in the working of the dual system within existing boundaries, the representatives from County Hall demanded an extension of that system without explaining the causes of the friction or offering any remedy designed to overcome it. There was, however, something almost ludicrous in the spectacle of the principal witness for the

London County Council entreating the Royal Commission "not to destroy local patriotism"; telling them that civic patriotism was more local than general inside London, and that on such local patriotism and upon a feeling of community depends in a large measure good local administration.[1]

Considering that throughout the past century the good government and proper planning of London have been persistently sacrificed to narrow parochial interests and the prejudices of a hundred self-regarding districts, one would have thought that local patriotism might well have been left to look after itself, as it has so far managed to do only too efficiently. The real problem in London is how to arouse a civic consciousness for the whole metropolis, and it was with this question that the London County Council might more usefully have concerned itself.

The functions of local government were to be divided into three groups: (1) services administered by the new principal authority, (2) services to be supervised by this organ, (3) services to be administered independently by the district bodies. The London County Council were not prepared to submit detailed proposals concerning the powers to be allocated to the principal authority, nor had they thought out its relationship to the district councils. But it appeared that the first two categories would include fire protection, town planning, housing, drainage, parks and open spaces, public health, education, the regulation of building—for all of which the London County Council is already responsible within its own area, either acting alone or in conjunction with the metropolitan boroughs—together with such additional functions as water supply, transport, roads, poor law and wholesale food markets.[2]

The district authorities would be reduced in number and enlarged in size. They would have wider powers than the metropolitan borough councils in regard to education, public assistance, public health and the administration of the London Building Acts. They would share with the principal organ the responsibility for drainage, housing and the provision of parks and open spaces.[3]

[1] Minutes of Evidence, p. 53, Q. 638.
[2] Report, pp. 16–17. [3] *Ibid.*, pp. 20–1.

These proposals from County Hall met with strenuous opposition from all the larger and more powerful local authorities outside the County of London. The City Corporation, which was left untouched, opposed the plan on the grounds that the London County Council had not shown that it would secure greater efficiency and economy,[1] though why the City should suddenly become interested in efficiency and economy was not explained. The metropolitan borough councils, needless to say, eagerly besought the Commission to give them powers at present exercised by the London County Council. The county councils outside the administrative county directed their efforts to proving the efficiency of their administration and the necessity for maintaining the *status quo*, to criticising the London County Council and denying the existence of any problems arising from the present system. The only exception to this appeared to be Surrey County Council, which agreed that a wider area was needed for dealing with a few services such as transport and main drainage, housing and town planning, higher education and vocational training.[2] But this admission did not imply for a moment that the County Council was prepared to "surrender" any portion of the County of Surrey to a Greater London Council. Indeed, their witness expressly stated that the integrity of Surrey must be preserved at all costs.[3] Hertfordshire took a similar view, and was willing to co-operate only on the understanding that there should be no attempt either directly or indirectly to extend the boundaries of London. "The Hertfordshire County Council," declared their representative, "are of the opinion that whether their County or any portion of it is placed under the existing London County Council or joined with London under a new central body it will equally be the annexation of Hertfordshire by London, and to this the inhabitants of Hertfordshire strongly object."[4] How the views of the inhabitants of Hertfordshire or anywhere else on the subject were ascertained was not stated; so far as I know no local authority made any attempt to discover the state of local opinion on the question.

[1] Minutes of Evidence, p. 939.
[2] *Ibid.* (Holland), p. 302.
[3] *Ibid.*, p. 270.
[4] *Ibid.* (Barnard), p. 190.

The frequent use of such terms as "annexation" to describe the orderly process of extension of boundaries was by no means the only attempt to import prejudice into the discussion. Another method was to seek to appeal to the Conservative members of the Royal Commission by emphasising the radical and "political" character of the London County Council as compared with its neighbours, although as it happened a Conservative majority had been in power for some years at County Hall. "The London County Council are admittedly a political body," declared the honourable representative for the Herts County Council, "the Hertfordshire County Council are precisely the opposite, being entirely non-political." The following interrogation then took place on this statement:[1]

Q. "I suppose in your own County you have Moderates and Progressives, have you not?"
A. "No, we have not."
Q. "Are you all Progressives, then?"
A. "I think as one understood these things a few years back, of the entire number of our County Authority there were 67 of one sort and 13 of another—I was one of the 13."
Q. "Then there were sheep and goats in those days? I congratulate you on having converted the others. You do not like the London County Council; that is evident?"
A. "I do not like their political actions."

The implication underlying this statement was that only the Labour Party, which was then beginning to gather strength at County Hall, could be called "political"; and the attempt was made to discredit the London County Council by this means. A similar attitude to municipal politics was shown by the witnesses for the Essex and Surrey County Councils. Sir Herbert Nield, K.C., M.P., for the Essex County Council, was asked, "Why should you deprecate the fact that men who have been on the London County Council should stand for Parliament?" "I do not deprecate it at all," he replied, "but I say it is a strong nursery for Radical Members of Parliament."[2]

[1] Minutes of Evidence, p. 191. [2] Ibid., p. 373, Q. 5553.

The county borough councils based their opposition to the London County Council scheme on the simple maxim that there is nothing like leather. Hence both Croydon and West Ham demanded that instead of creating a larger principal authority, the metropolis outside the administrative county should be divided up between a number of substantial county borough councils. A Greater London Council, contended the Croydon witness, would not be able to administer so large an area without leaving the bulk of the work to the paid officers— a state of affairs, incidentally, which prevails among all local authorities in Britain above the status of a parish council. Further, if the local authorities in Outer London were shorn of their powers in favour of a larger organ, they would not attract the services of the men and women who now serve on them.[1] And thirdly, minor authorities in the metropolis would no longer be in direct touch with the Ministry of Health and other Government departments but would have to work through the principal London Council.[2] The officials of the Ministry of Health were doubtless flattered and surprised to learn of this yearning for direct contact with Whitehall.

The West Ham County Borough Council argued in favour of a belt of entirely autonomous county boroughs circling the administrative county—graphically but inaccurately described as—"a sort of ball-bearings as between the County of London on the one hand and the Home Counties on the other." (A member of the Commission reminded the witness that the object of ball bearings is to diminish friction, whereas his proposals might increase friction.) The grounds on which this proposal was put forward was, first, that conditions inside the county were not good enough to warrant an extension of boundaries. Second, amalgamation of units into county boroughs and boroughs would reduce the number of contracting parties to reasonable units. Third, the poor would be better

---

[1] One of the points made by Mr. Norman for the London County Council was that the creation of a larger principal authority and reduction of minor authorities might be helpful in securing good candidates for election in view of the enhanced importance of the work. It would also permit of higher salaries being paid to officers, with a consequent improvement in the quality of the professional staff. Minutes of Evidence, p. 73, Q. 1158.

[2] *Ibid.*, p. 330.

cared for and their interests more fully safeguarded by smaller authorities than by a great metropolitan council.[1] Moreover, "If you get this great London you are talking about, you may conceivably get a stick-in-the-mud London which will do nothing."[2] The contrast between the views of the poor East End dormitory area and the apprehensions of the county councils on this point is interesting. But in the result they stood shoulder to shoulder.

Most of the smaller local authorities in Greater London outside the administrative county were opposed to the London County Council proposals. Divers reasons were put forward as the grounds of opposition, but the underlying objection appears to have been the fear that the establishment of a larger body would lead to an increase in the rates.[3] The witnesses for Barnes and Richmond (called together) pointed out that in existing circumstances the municipal work in a particular area was divided up among eight persons. If the London County Council scheme were adopted they would be replaced by only one or two representatives. This would mean finding someone able to devote nearly all his time to the job—an elderly man past his prime who had made money and found time hanging heavily on his hands. They regarded this as a disadvantage and preferred younger men.[4] All these aspects of the subject were discussed on a basis of pure guesswork without any attempt to elicit the facts. They were put forward as mere obstacles to reform.

The boroughs or urban districts of Ealing, Acton, Chiswick and Willesden were in favour of amalgamating the smaller areas with common interests into county boroughs; and they agreed to collaborate for this purpose. Hornsey took a similar view, but could not come to terms with her neighbours.[5] Ilford was willing to combine with the urban districts of Barking and Dagenham and the rural district of Romford.[6]

Almost the only local authorities supporting the London County Council proposals were the urban district councils

[1] Minutes of Evidence, pp. 562–4.   [2] Ibid., p. 568, Q. 8737.
[3] See, for example, the witnesses of Bromley, Penge and Richmond.
[4] Minutes of Evidence, p. 499, Q. 7683.
[5] Ibid., pp. 770–1.   [6] Ibid., p. 877.

of Edmonton and Southall-Norwood. When asked to explain this strange phenomenon, the witness for Southall-Norwood ascribed it to the enlightened opinion of the authority he was representing. He said the opposition was based entirely or mainly on monetary issues and that the "local patriotism" argument was not to be taken seriously.[1]

The Labour Party gave evidence in support of the London County Council's proposal, although, as Mr. Sidney Webb stated on their behalf, they had nothing to gain politically by enlarging the municipal boundaries of the metropolis so as to include the home county constituencies,[2] and despite the fact that the Labour Group at County Hall was in a minority and therefore had not formulated the policy put forward. At the same time the Labour Party pursued a line of its own in suggesting a suitable area for the new authority differing both from that proposed by the London Labour Party and the nebulous area indicated by Mr. Norman.[3] Mr. Webb dealt at length with the need for equalisation of finance throughout the area, and a common scheme of recruitment, promotion and superannuation of local government officers.[4]

Mr. Herbert Morrison appeared as a witness both for the London Labour Party and for the Metropolitan Borough Council of Hackney. In his former capacity he criticised the conception put forward by the London County Council of a metropolitan authority comprising only "the continuous urban area" referred to by Mr. Norman. The London Labour Party proposed instead to establish in the Home Counties a number of satellite towns around which there should be spread a wide agricultural belt. For such a plan the principal London authority must cover a wide region and be able to command considerable financial resources. The area designated should be not less than the London and Home Counties Electricity District together with those parts of Surrey excluded from the scheme. The Greater London Council should have jurisdiction over the Thames and its tributaries within the area thus defined and also down the river to its mouth.[5]

The regional authority should take over the powers and

[1] Minutes of Evidence, pp. 639, 796–7.   [2] Ibid., p. 816.
[3] Ibid., p. 821.   [4] Ibid., p. 823–4.   [5] Ibid., pp. 681–2.

duties of the Port of London Authority, the Metropolitan Water Board, the Joint Electricity Authority, the Metropolitan and other police forces, the Thames and Lee Conservancy Boards within its area, and the Metropolitan Asylums Board. It was to have control of large general hospitals, specialist hospitals and sanatoria; to undertake housing and slum clearance schemes beyond the capacity of minor authorities; to control highway and traffic functions; to manage the fire brigades and wholesale food markets; to be the education authority for such matters as teachers' salaries and university grants; to supervise the operation of an equalisation fund in regard to financial expenditure. It was to have power to acquire land compulsorily for public purposes, to make by-laws enforceable throughout the region, and to be able to delegate any of its functions and powers to the minor authorities.[1] It would have devolved upon it the task of supervising the district organs, thereby relieving the over-burdened Government departments.

As a witness for the Hackney Borough Council Mr. Morrison indicated the numerous powers which should be transferred to the district authorities. These were mainly of a regulative character and included the enforcement of the Shop Acts, the licensing and inspection of massage establishments, the protection of infants, the licensing and inspection of common lodging-houses, the licensing or supervision of dairies, cowsheds and milkshops, slaughterhouses and knackers' yards, fried fish vendors, fish curers, rag and bone dealers and ice cream merchants, the administration of the weights and measures legislation, the registration of lying-in homes and midwives, and the regulation of outdoor advertisements.[2] On behalf of Hackney Mr. Morrison reiterated his previous statement that the principal authority should have a boundary which would include the continuous built-up area "with a margin sufficiently large to prevent, if necessary, the further growth of Greater London outwards, except on town planning lines."[3]

After hearing this and much other evidence, the Ullswater Commission issued three separate reports. The majority report rejected the proposals for a Greater London authority put

[1] Minutes of Evidence, p. 683.   [2] *Ibid.*, p. 982.   [3] *Ibid.*, p. 980.

forward by the London County Council, the Labour Party, Mr. Herbert Morrison and various other witnesses. Their conclusions in regard to the County Hall evidence were that the difficulties disclosed were no greater than those existing in other parts of the country between contiguous authorities;[1] that the London County Council had failed to prove any shortcomings in the administration of Outer London authorities which would be remedied by their inclusion in a Greater London area; and that so far as centralisation was necessary the territory to be covered should extend far beyond the limits within which the proposed principal organ could operate efficiently.[2] They illustrated this by reference to the transport problem.

The dual system of local government by major and minor authorities was then examined by the Commissioners. It has the advantage of making possible the centralised administration of large-scale services in which there is a common interest over the whole area while leaving a substantial residue of functions in the hands of local bodies in more direct touch with the ratepayers. But it also has the disadvantage, they said, of generating friction between the major and minor organs over the distribution of powers. The hostility of the metropolitan boroughs towards the London County Council had come to light in much of the evidence, and some of the borough councils are apt to regard the London County Council as an alien authority, although it is an elected body representing the whole County of London.[3] An extension of the dual system would involve curtailing the powers of 3 county borough councils and 8 municipal borough councils, and would be strenuously opposed by them, no less than by the county councils bordering London.[4] Such a catastrophe could not be contemplated with equanimity by the majority Commissioners, and they therefore recommended that the existing area of the London County Council should be retained and the county areas left unaltered.[5]

[1] For an account of the general position throughout the country, see W. A. Robson: *The Development of Local Government*, Part I.
[2] Report, p. 67.
[3] *Ibid.*, p. 68.
[4] *Ibid.*, p. 69.
[5] *Ibid.*, p. 70.

It may be noted that the Commissioners made no attempt to analyse the causes of the friction to which they adverted, and they made no suggestions as to how it could be overcome. Moreover, they overlooked the fact that a dual or even treble system of local government exists throughout the entire country outside the county boroughs. They were, however, obsessed by the thought of the political opposition which would be raised by any attempt to establish a Greater London Council. "Even if the dual system of London had greater merits and fewer difficulties than it has," they wrote, "we view its extension as being impracticable in face of the opposition it would encounter." And they quoted Mr. Lloyd George's warning to the deputation from the London County Council which waited on him in December 1920, when he remarked: "A report that would get 128 local authorities up in arms against you is not a report that any Government would face with equanimity."[1]

As a result of these preoccupations the recommendations of the majority report left the municipal structure within the metropolitan region completely intact and made virtually no attempt to solve the local government problems within the area. Or, perhaps one should say, the problems were scarcely recognised to exist. A statutory committee of a purely advisory kind was to be set up consisting of not more than twenty members nominated by the London County Council, the City Corporation, the metropolitan borough councils, the county and county borough councils, the Commissioner of Police, the transport undertakings, traffic interests and labour organisations. The area it was to consider in its deliberations was to be the electricity district, but provision was to be made for subsequent enlargement.

This committee was not to have any executive power whatever; nor was it even to advise the constituent local authorities. It was to assist the respective Minister responsible to Parliament for housing, town planning, transport and main drainage. In regard to transport, the committee was to replace both the existing statutory Roads Committee and the non-statutory technical committee attached to the Ministry of Transport. Town planning was conceded to be inseparable

---

[1] *Loc. cit.*

from transport, and in this sphere the committee would assist the Ministry of Health in examining the various town planning schemes proposed within the area and seeing that they harmonised with one another. In the sphere of housing the committee would "exert a useful influence" in helping local authorities to decide the number and kind of houses required, and the places where they should be situated. It would seek to prevent sites for municipal housing estates being purchased without consultation with the district authority concerned. In regard to main drainage, the committee would advise the Minister on the provision of new facilities throughout the area, and also on the admission of sewage from outlying authorities into the main drainage system.[1] The staff was to be provided by the appropriate department.

The advice tendered by the Ullswater Commission has been carried out in regard to only one service—namely, traffic and transport. The constitution and functions of the London and Home Counties Traffic Advisory Committee have already been described.[2] Otherwise the report fell stillborn from the printing-press. It is worth noting that it is precisely in this sphere of activity that the greatest and most decisive inroads on local autonomy have been made since 1924: not in the direction of larger or more scientifically designed areas of local government, but rather towards the supersession of local government by central departments or *ad hoc* bodies appointed by Whitehall. The establishment of the London Passenger Transport Board and the Metropolitan Traffic Commissioner, the designation by the Minister of Transport of a central official (Sir Charles Bressey) to be responsible for planning London highway development, the taking over by the Ministry of Transport in 1937 of almost the whole mass of main roads in the counties: these events disclose an unmistakable trend towards centralisation or at least towards detachment from the system of local government.

This is exactly what a study of social development might lead us to expect. Where administrative institutions fail to adapt themselves to contemporary needs or deliberately resist necessary change, an entirely different solution of an apparently

---

[1] Report, pp. 75–9.   [2] *Ante*, pp. 149–55.

irrelevant kind is likely to be adopted. Such a solution, often applied with successive variations, seems to make no attack on the existing institutions and therefore fails at first to excite their opposition or even notice. But what was at first a tiny cloud on the horizon no bigger than a man's hand gradually assumes larger proportions and is later seen to be threatening the importance, if not the existence, of the older institutions, which eventually sink into a position of declining importance. When this position of decadence is reached, the obsolete organs seek to reassure themselves of their status by an exaggerated display of ornamental ritual or by an obstinate clinging to antiquated procedure.

This process has occurred in the case of the City Corporation and the City of London Companies. It has overtaken the Inns of Court and parts of the judicial system. It took place with the old municipal corporations prior to 1835. It is now in course of occurring again in regard to certain portions of the contemporary local government system, particularly in the case of London and certain other centres obviously demanding regional treatment. The process can only be arrested by a willingness on the part of the organs concerned to adapt themselves to changing social needs. Up to the present the local authorities in the metropolitan region, apart from the London County Council and one or two minor exceptions, have shown no disposition to submit to necessary changes, as the resistance they displayed in the evidence given to the Ullswater Commission amply demonstrated. It is unlikely that the members of these authorities will grasp the significance of what has occurred in regard to highways and transport; but if they are capable of drawing the inescapable conclusion, it might not be too late even now to save the situation.

It is in the light of these considerations that we must interpret the majority report as a document of centralising import.

We may now turn to the two minority reports, one signed by Messrs. Hiley and Talbot, the other by Sir Robert Donald and Mr. Walsh. The brief Hiley-Talbot report proposed the division of the metropolis into a number of substantial units having a status equal to that of county boroughs. In addition there should be a principal authority to administer throughout the area certain large-scale services such as water supply, main drainage, tramways, and the making of by-laws. The area should be limited to those places where building was practically continuous and the prevailing conditions more or less identical with the rest of London—an area lying approximately within a distance of ten miles from Charing Cross.[1]

The Donald-Walsh report was a document of a much more ambitious kind. It was based on a critical analysis of the evidence, a keen sense of the historical development of London government, and insight into the needs which had been revealed by previous Committees and Commissions.[2]

The report dismisses at an early stage the claim made by so many of the witnesses that the local autonomy of the existing units should be preserved at all costs. It points out that there are in fact only three autonomous councils—i.e. the County Boroughs of Croydon, East Ham and West Ham—within the whole of Greater London. Everywhere else power is shared. It points out also that the ideal unit alleged to be required for local administration varied in size between 15,000 and 200,000 persons according to the size of the place from which the witness came.[3] "A representative of a borough of over 200,000 population presented it as a model which should be lived up to. Other witnesses from communities of 100,000 considered that the size of their districts provided conditions guaranteeing the greatest efficiency in local government, and so on down to communities of less than 20,000, which were represented by witnesses who considered that the best plan was to allow them to develop on their own lines. In one instance a district with a population of 1,500 was represented as an ideal area for local government efficiency."

Messrs. Donald and Walsh then examined the various proposals made to the Commission. The idea of an indirectly elected body to carry out a limited number of large-scale services in Greater London, which was favoured by several important local authorities, they considered to be unsound, having regard to the proved failure of the Metropolitan Board of Works and the unsatisfactory results, from a democratic

[1] Report, pp. 138–40.   [2] *Ibid.*, p. 149.   [3] *Ibid.*, p. 154.

point of view, of such bodies as the Metropolitan Water Board and the Metropolitan Asylums Board.

The advisory committee recommended by the majority report was also in their opinion open to grave objection, since it tended to withdraw responsibility from the elected representatives of the people. If a committee of this kind went beyond traffic and dealt with questions of housing and town planning it would be encroaching on the sphere of municipal government and many difficulties would arise.[1]

The only practical and desirable reform would consist in the creation of a new directly elected authority having jurisdiction over an enlarged area, which the Commissioners thought should be that of Greater London. This body would replace within its territory all the county councils now exercising functions there. In terms of the population represented by witnesses, the Commissioners estimated that the greatest body of evidence supported a solution of this kind, though there might be differences in matters of detail.[2]

The new Greater London Council should have responsibility in regard to transport and town planning, fever hospitals and mental hospitals, main drainage and sewage disposal, water supply and the provision of small holdings. It should share responsibility with the district authorities in regard to housing, education, poor law, fire brigades, parks and open spaces, wholesale food markets, public health, the river conservancies, the supervision of the London Building Acts and the administration of by-laws.[3]

The Donald-Walsh report then dealt in some detail with certain of these services. It proposed a scheme of co-partnership or "mixed enterprise" in order to secure a unified system of transport undertakings. It emphasised the need for the principal authority not only to construct housing estates but also to be in charge of traffic, drainage and educational services required by the persons migrating to these estates, in order to avoid throwing unfair burdens on other local authorities and the disadvantages of unco-ordinated administration.[4] A large housing scheme can only be properly planned and executed

---

[1] Report, pp. 159–61.
[3] Ibid., pp. 164–5.
[2] Ibid., pp. 160–2.
[4] Ibid., p. 180.

when all the major services are under one control. In regard to main drainage it was obvious that the problem should be dealt with in terms of health, geography, engineering and economic considerations. It was generally agreed by the witnesses that if a larger authority were set up it should deal with main drainage and sewage disposal.[1] In the field of education, the Greater London Council should be in supreme command throughout the area; and it should be directly responsible for the municipal aspects of university, trade, technical and other specialised forms of education. It would fix salary rates and organise the training of teachers. The metropolitan borough councils should have their functions enlarged so as to correspond with municipal boroughs elsewhere, which would mean that they would administer elementary education and also be able to contribute in aid of higher education. The powers of the metropolitan borough councils would also be greatly increased in the sphere of public health. The merging of small areas in Outer London and the consequent replacement of part-time medical officers by full-time officers was a badly needed reform.[2] The projected transfer of the poor law and the medical institutions of the Metropolitan Asylums Board would not yield its fullest advantage unless the area of the metropolitan authority receiving the new functions were enlarged so as to cover Greater London. The Metropolitan Water Board and the Thames and Lee Conservancy Boards should be abolished, and their duties transferred to the Greater London Council.[3]

The whole tenor of the majority report, declared Messrs. Donald and Walsh, indicates "a suspicion of progress and a fear of local government development."[4] Their own proposals, they contended, sought to establish one homogeneous area which would increase the sense of common citizenship.[5] "We have in view in all our recommendations the unity of London. We hold that Greater London is one and indivisible in all the essentials which constitute one great civic and urban community. It differs only from other large urban communities by its immensity. The problem of size as it affects local self-

[1] Report, pp. 186–7.   [2] *Ibid.*, p. 182.
[3] *Ibid.*, pp. 190, 193.   [4] *Ibid.*, p. 203.   [5] *Ibid.*, p. 163.

government is met by the dual municipal system. . . . Our scheme combines simplicity with respect for the dignity of local institutions. It should enable the citizen to take a keener interest in local affairs. He will more easily understand the system and conditions under which he is governed."[1]

But eloquence and reason were alike in vain when all the London authorities were at sixes and sevens with each other and the members of the Royal Commission divided between three reports and an additional note. And so it came about that, apart from the one measure of centralisation involved in the creation of the Traffic Advisory Committee, the labours of the Ullswater Commission were completely abortive. Once again the possibility of imposing a rational system of municipal government on the metropolis was sacrificed to the discordant voices of a hundred vested interests: the interests not of profit-making companies but of the stubborn, selfish and narrow-minded local government organs whose resistance intimidated the majority and defeated the minority of the Royal Commission.

But the problems which the Commission failed to solve still persist. Much water has flowed under the bridges since 1923, the bridges themselves have in several cases been rebuilt in the intervening years, but no solution of the chaos of London government has been attempted. The dangers and disadvantages of the situation have immeasurably increased in the fifteen years which have elapsed since the Ullswater Commission reported; and they are much harder to deal with now than they were then. The difficulties press for a remedy with increasing urgency; and the time is approaching when it will no longer be possible to bury our heads in the sand.

[1] Report, p. 207.

Part III

THE FUTURE

CHAPTER I

THE GREAT LEVIATHAN

We have already seen that London is absolutely larger in terms of population than any other metropolitan centre in the world.[1] It is more populous even than the City of New York.[2]

To grasp the true significance of London we must, however, see it in relation to the rest of the country. Thus, Greater London consists of a territory which comprises $\frac{1}{137}$th part of the area of Great Britain, and within that territory is one-fifth of the population of Great Britain and almost a quarter of its rateable value.[3] If we take the London Passenger Transport area as our unit of measurement, we have an area slightly more than $\frac{1}{50}$th part of Great Britain containing a third of the entire rateable value of the country. The value of insured property in London reaches the enormous figure of £2,275,211,661.[4] It is abundantly clear that London contains a disproportionately large share of the national wealth and population. It includes relatively a far larger proportion of the nation than is to be found in any other city in the world.

The importance of London is not, however, to be judged solely in terms of its size nor even of its wealth. It has to be considered also in terms of the managerial institutions or directing organs of one kind or another which are concentrated within its territory.[5]

The size of London has in the past often given rise to anxiety on the part of Governments. This was due to a variety of causes, such as the difficulty of providing supplies of food

---

[1] *Ante*, p. 163.

[2] Unless we take the whole metropolitan area of Greater New York City, within a radius of 50 miles from the City Hall in Manhattan, which is said to comprise 11,500,000 persons (1930), but this covers thickly populated parts lying some distance away in the states of New Jersey and Connecticut. So the generalisation stands.

[3] Third Report of the Commissioner for the Special Areas (England and Wales) [Cmd. 5303], 1931, pars. 20–24 and Appendix I; *Rates and Rateable Value*: Ministry of Health Statement, 1934–5, p. 5; *London Statistics* (1934–6), vol. 39, p. 417.

[4] *London Statistics* (1934–6), vol. 39, p. 236.   [5] *Ante*, p. 161.

and water, the danger of fire in the days of wooden houses, the fear of plagues and epidemics spreading among a congested populace. In consequence there were several attempts in the Tudor and Stuart reigns to restrict the size of the capital. Proclamations were issued forbidding the erection of new houses in London, and enjoining people from the country to return to their homes.[1] None of these regulations or appeals were effective for any length of time, for the simple reason that no adequate machinery was in existence to enforce the policy of the Government. No relevant conclusions can therefore properly be drawn from these early failures which have any bearing on the problems that confront us at the present time.

An entirely new factor has recently become a cause of grave disquiet to Government and people alike—the menace of attack from the air in the event of war. We do not know for certain how deadly modern methods of aerial warfare are likely to be when opposed by up-to-date methods of defence; but the indications of probable damage and danger are of a highly disturbing character. Great Britain is clearly running enormous risks in having so great a proportion of her population, wealth and manufacturing resources concentrated in one centre situated within a few minutes' flying distance of the coast. It is one of the ironies of history that while in the past many of the great cities owed their position and growth partly to the strength of their military defences, the invention of aircraft has made many of these same cities the danger spots of our age. In former times men flocked to great cities in search of shelter behind their walls. Nowadays women and children have to flee from these cities for reasons of safety.

The growth of the metropolis has in recent years been proceeding at a breathless rate. An increase of nearly a million and a quarter persons has taken place in Greater London since 1921.[2] In Outer London alone the population increased by

---

[1] W. Cunningham: *Growth of English Industry and Commerce*, vol ii, p. 172; A. F. Weber: *Growth of Cities in the 19th Century*, p. 454.

[2] *Cf.* the opening remarks of Sir Montague Barlow at the first public session of the Royal Commission on the Geographical Distribution of the Industrial Population; Third Report of the Commissioner for the Special Areas (England and Wales) [Cmd. 5303], 1931, pars. 20–4 and Appendix I.

628,000 during the quinquennium 1931–6: the equivalent to a town nearly half as big again as Edinburgh. The future increase in the population within a thirty mile circle of Charing Cross is expected to be 5½ per cent between 1934 and 1941 (540,000 persons) and 4 per cent between 1941 and 1951 (a further 410,000). Thus, the forecast gives a total population for the thirty mile circle of 10,350,000 by 1941 and 10,760,000 by 1951. The population of Greater London will presumably rise from 9 to 9½ millions between 1941 and 1951 according to the present trends. The migration from other parts of the country is expected to continue for a considerable time to come unless something is done to check the trend.[1]

The central core of London exerts a centripetal force by day and a centrifugal force at night. A huge army of workers invades the offices and shops, the factories and docks, on each weekday morning; and retreats every evening to the ever-expanding mass of suburban settlements. According to one estimate (which may be excessive), the daily tidal wave of population is composed of more than 2,500,000 persons, surging in to work, ebbing out to sleep.[2] So great is this diurnal movement that 250,000 workers in London are engaged in transport undertakings. Put in more homely words, one in every ten of London's workers lives by carting the others about, at an annual cost of £40 millions. A London family spends on the average £16 a year or 6s. 3d. a week in travelling in buses, tubes, trains and trams. This amounts to something like 10 per cent of the average Londoner's income.[3] Taxi-cabs and private cars have been excluded. If they are taken into account, the average cost is much higher.

Great anxiety has recently been expressed lest this elephantine growth of the metropolis should be checked or the feverish rushing to and fro of its inhabitants be diminished. Mr. Pick, the Vice-Chairman of the London Passenger Transport Board, declared to the Royal Commission on the Geographical Distribution of the Industrial Population that "a living organism

[1] Bressey and Lutyens: *Highway Development Survey* (1937), p. 10.
[2] E. C. Willatts: *The Land of Britain. Middlesex and the London Region*, p. 166 (Report of the Land Utilisation Survey of Britain).
[3] F. J. Osborn: *London's Dilemma* (a pamphlet published by the Garden Cities and Town Planning Association).

cannot be static and survive," by which he meant that London should be permitted to grow indefinitely. "We are concerned that London should remain a healthy organism," he continued, "and to remain a healthy organism it cannot be allowed to decline. Once it begins to decline there is no controlling that decline. It may crash in and cause a much more acute problem than you have today with the distressed areas."[1] Asked what would be the limiting factor in the further expansion of London, Mr. Pick agreed that "the limit of the patience of the straphanger" would be the ultimate determinant. The London Passenger Transport Board estimates that this point will be reached when there are 12 million persons within a radius of 12 to 15 miles from Charing Cross; but Mr. Pick was careful to explain "It is our limit, not his. It may be he will think he can do even more than that, I do not know."[2] A member of the Commission then asked the witness whether he was putting the capacity of transport as the real limit to the growth of London quite irrespective of other conditions, to which he replied: "I can imagine at some time or other a problem of water supply, and a problem of sewage disposal. There are other factors, but mainly at the moment it is transport."[3]

This deliberate and cool-headed discussion concerning the patience of the straphanger between a responsible executive and the trusty and well-beloved Commissioners recalls forcibly le Corbusier's violent denunciation of the modern giant city as a place in which the growth and tempo of urban development have got beyond control. "Now that the machine age has let loose the consequences attaching to it, progress has seized on a new set of implements with which to quicken its rhythm; this it has done with such an intensification of speed and output that events have moved beyond our capacity to appreciate them; and whereas mind has hitherto generally been in advance of accomplished fact, it is now, on the contrary, left behind by new facts whose acceleration continues without cease; only similes can adequately describe the situation: submersion, cataclysm, invasion. This rhythm has been accelerated to such

---

[1] Royal Commission on the Geographical Distribution of the Industrial Population, Minutes of Evidence, Qs. 3403–4.
[2] Ibid., Qs. 3391–3.　　　　　　　　　　　　　[3] Ibid., Q. 3413.

a point that man (who has after all created it with his small individual inventions, just as an immense conflagration can be started with a few pints of petrol and one little match)—man lives in a perpetual state of instability, insecurity, fatigue and accumulating delusions."[1] The centres of the great cities, he says, are like an engine which is seized. The centres are in a state of mortal sickness, their boundaries are gnawed at as though by vermin.[2]

New and rapid means of transport, new demands for air and space, new forms of power, have led the metropolis to burst through the invisible bonds which formerly restrained it. We see it now flooding the countryside, creating new land values, exchanging new slums for old, sprawling far and wide without direction or control, devoid of coherence or integrity. London has undergone a revolution, and we have not realised it or taken steps to guide it.[3] The capital is thrusting outward and upward with increasing force and diminishing social advantage.

[1] Le Corbusier: *The City of Tomorrow*, pp. 85–6.
[2] *Ibid.*, pp. 94, 96.
[3] A. G. Gardiner: *John Benn and the Progressive Movement*, p. 85.

CHAPTER II

WHAT IS LONDON?

I propose in this part of my study to deal with two questions
of outstanding importance: namely, the proper organisation of
London government, and the territorial planning of London.
But before proceeding to these problems, it is necessary to
enquire into the character of the area to which they relate.
Many of the defects in the structure of London government
are due to the fact that no effort has ever been made to
comprehend the nature of the metropolis. Hence we have the
absurdity of its being partitioned among county councils,
borough councils, county borough councils, metropolitan
borough councils, urban district councils, rural district councils,
parish councils, and several other kinds of authority, none of
which are suited to its needs. Hence also the establishment of
the London County Council as an incident in the creation of
county councils for the rural shires.

A town, observes Professor Mess, can be regarded from
three points of view. First, there is the aggregation of buildings
and persons. Second, there is the sense of community. Third,
there is the machinery of local government and of social
organisation.[1] Greater London obviously has the first of these
components, but lacks as a whole both the second and the
third. It would, however, in any case be impossible to regard
the vast sprawling mass of the metropolis as a town in the
ordinary sense of the word. It falls more appropriately under
the heading of a conurbation, a term used to designate a number
of urban communities, large or small towns, which were
formerly separated from one another by expanses of open
country and each of which had its own community life. When
these towns expand until they become contiguous, and the
population moves freely throughout the whole area for the
purposes of work, business, education and pleasure, a conurbation has been formed. "Has the conurbation ceased to be

[1] H. A. Mess: "The Growth and Decay of Towns," *Political Quarterly*,
July–September 1938, p. 406.

many towns and become one town? We cannot say that, but it may be on the way to becoming a single town or city. Should we call that area a town which has its own local authority? That would be to fly in the face of facts, the facts of change, of inter-penetration. . . . A town, like a nation, must be conceived partly in subjective terms, in terms of the consciousness and sentiments of a population; but, again like a modern nation, it normally requires some kind of geographical unity and separateness, and its own organ of government. Geographical definition, civic sentiment, municipal status: all three seem to be needed for the existence of what can be described without reservation as a town—at least in modern England."[1]

I have already expressed the opinion that Greater London cannot be regarded as a town without a misuse of language. The size and scale of the metropolis are beyond anything normally connoted by the word town or city. In my view London is a region or province.

With this idea in mind we can return to the question mentioned above and ask: Has the conurbation of Greater London ceased to be many towns and become a single entity? The answer is equivocal. As regards the aggregation of buildings and persons, it is abundantly clear that the various villages, districts, towns and parishes have become completely contiguous and agglomerated. But as regards the outer edge of the region, the process of expansion is so rapid and so haphazard that it is difficult to accord the quality of geographical coherence to its blurred and moving line.

In regard to the second criterion, there is again a certain ambiguity in the answer. Most of the inhabitants of the London Region are in greater or less degree conscious of belonging both to the small local area in which they live and to the larger metropolis of which it forms a part. If one asks a Londoner in London where he lives, he will answer Chelsea, or Richmond, or Forest Hill or Highgate. If he is asked the same question when he is abroad, or in a provincial city, or in the country, he will simply answer London. In many parts of the region local sentiment is very strong and regional patriotism

[1] H. A. Mess: "The Growth and Decay of Towns," *Political Quarterly*, July–September 1938, pp. 391–2.

very weak, and this tendency increases as one leaves the administrative county and moves either inward to the City (where it approaches 100 per cent) or outward towards the periphery. Professor Mess mentions the county borough of West Ham as an instance of a place where allegiance to the locality is strongly felt. Yet most inhabitants of West Ham, he adds, would say that they were Londoners and have some feeling that St. Paul's and Westminster Abbey and other historic buildings were in their city. But on the whole, he suggests, the effect of the spread of London outside the administrative county has been to weaken the civic consciousness previously prevailing without creating by way of compensation any adequate sentiment related to the larger entity.[1]

When we come to the third criterion, the answer is a clear and decisive negative. The municipal status of the Greater London Region is non-existent, except as regards a number of *ad hoc* bodies each one of which has jurisdiction over a different regional territory. The areas of all the general local authorities, including the London County Council and the other county councils, are essentially parochial when viewed in relation to the needs and realities of the region.

This lack of municipal status in the region is a matter of the utmost importance from the point of view of civic consciousness. It explains more than any other single cause both the absence of a strong or effective regional patriotism and also the indifference of Londoners to the monstrous growth and misdevelopment of the region. "Common government," it is truly said, "is itself a powerful unifying force, setting a common stamp upon an area and its inhabitants, and operating powerfully to intensify the sentiment which attaches to the community so governed. Common government, if it is to be democratic and successful, presupposes a degree of community, but it also strengthens community."[2]

I hold it to be an indisputable proposition that if we desire to cultivate and encourage a sense of community among the citizens of the London region, it is necessary for the area to acquire a regional form of government. Only through a Greater

---

[1] H. A. Mess: "The Growth and Decay of Towns," *Political Quarterly*, July–September 1938, p. 397.　　　　　　　　　　　[2] *Ibid.*, p. 395.

London Council can a Greater London spirit arise which will attempt to envisage the problems and the welfare of the metropolis as a whole rather than the problems and the welfare of its particular parts. This, above all and beyond all, is the crying need of the time. It is by this means, and by this means only, that we may hope to achieve "the integration of all that is contained in London to express some idea or aim or purpose that alone gives value to the masses of which it is composed."[1] At present London expresses no aim or purpose of any kind whatever. It merely drifts about helplessly with the tide, like a beached whale.

Unfortunately, the trend of events continues in precisely the opposite direction. Not content with having established 28 metropolitan borough councils within the administrative boundary at the end of the last century, a step designed to create "not one London, but thirty Birminghams,"[2] the mistake has been made of incorporating numerous urban districts in Outer London in recent years. Thus, Barnes, Brentford and Chiswick, Hendon, and Heston & Isleworth became boroughs in 1932; Dartford, Finchley, Southgate, Willesden and Wood Green in 1933; Mitcham, Sutton & Cheam, and Tottenham in 1934; Malden & Coombe, Southall and Surbiton in 1936, while in 1937 borough status was conferred upon the record number of six urban districts. In 1938 Romford, Slough, Chingford, Erith and Dagenham became boroughs.

The folly of this course is obvious. The inevitable and known result of conferring borough status upon an urban district is to enhance its sense of separateness, its self-importance, its awareness of its own dignity. These, indeed, are the very motives which lead it to seek incorporation, since the difference in powers and duties is negligible. Equipped with a mayor wearing his chain of office, aldermen in their robes and a town clerk, a town hall and a borough treasurer, receiving the Charter from some dignitary such as the Lord Mayor of London, the new borough rises perceptibly in the scale of pride and self-assertiveness. It becomes an ardent protagonist of the sacred rights of boroughs against the over-weening claims of all other

[1] Frank Pick: "The Organisation of London Transport," *Journal of Royal Society of Arts*, vol. lxxxiv, p. 210.   [2] *Ante*, p. 94.

authorities save the King in Parliament assembled; and even the King in Parliament—or, more realistically, the Minister of Health—is seldom ready to override the insistent claims of the Association of Municipal Corporations, for whose membership the newcomer to the company of boroughs is now eligible. Thus, a rising standard of particularism has been established, localism encouraged, and potential opposition to regional government intensified.

No less than 34 of these incorporations were authorised by successive Governments between the years 1921 and 1938.[1] And so, as the growth and spread of the metropolis has made the need for a regional Council more marked and more obvious, so each year the number of incorporations has increased.

All that lack of imagination could do to exacerbate the situation has been done; and the problem of providing the London Region with a structure suited to its needs has been made far more difficult of solution than it was ten or twenty years ago.

[1] The charter of incorporation is issued by the Privy Council and the process is controlled by the Government.

A REGIONAL AUTHORITY

There are only a limited number of possible methods by which a regional administration could be established in London. Before considering the various alternatives, it is desirable that we should recognise which are the services requiring some form of regional organisation in the metropolis.

At the outset we should appreciate that a distinction can be drawn between the functions which need to be directly administered by a regional authority and those which require to be regionally supervised or planned but which can be carried out by local authorities within the region. There are, in fact, three categories of services: those calling for direct regional administration; those suitable for local administration under regional guidance; and those which can be left entirely to local control and administration. For the present we are concerned with the first two groups.

The services which should be regionally administered are territorial planning (i.e. town and country planning), the larger housing and slum-clearance projects, main drainage and sewage disposal, main highways and bridges, water supply, fire brigades, hospitals, specialist public health institutions and medical services, lunacy and mental deficiency, welfare of the blind, education (subject to what is said below), aerodromes, the provision of large parks and open spaces, the prevention of river pollution, the disposal of refuse, the preservation of ancient monuments, smoke abatement, river conservancy and flood prevention. A few regulatory services of a highly specialised kind might be added, such as the licensing of cinema exhibitions, the testing of gas and electricity meters, the registration and inspection of explosives, the analysis of fertilisers and feeding stuffs, the licensing of racecourses and theatres, the registration of theatrical employers, the provision of remand homes.

The grounds for assigning these functions to the regional authority will no doubt be apparent from the critique of the existing organisation contained in Part II. Broadly speaking, the need for regional administration arises from one or more of the following causes: (1) where the technical nature of the service necessitates the area being dealt with as a whole if an effective result is to be attained, as in the case of planning or smoke abatement. (2) The necessity of providing costly buildings, plant or equipment or a highly specialised and expensive staff beyond the resources of minor authorities. This applies to such functions as fire brigades, main drainage, hospitals, large-scale slum clearance and rehousing. (3) The fact that a facility situated in a particular district actually serves the entire region. This relates to main highways and bridges, aerodromes, and large open spaces. (4) A high degree of specialisation requiring a very large constituency for optimum efficiency and economy, e.g. technical education, lunatic asylums, specialist medical services, the testing of gas and electricity meters, or analysis of fertilisers. (5) The need for unified administration or central registration in order to prevent evasion of control, as in the case of the registration of theatrical employers. (6) The need for complete co-ordination of effort, as, for example, in the numbering and naming of streets. (7) The desirability of observing a uniform standard throughout the area, as in the case of the licensing of cinema exhibitions, racecourses and theatres.

Public assistance ought probably to commence as a regional service; but the aim should be to transfer it as soon as possible to the second group of functions. There is, however, much to be said both for and against regional administration of the poor law. There is no technical necessity for regional administration arising out of the nature of the service; but experience shows that the larger units of administration in the field of relief tend to be more honest, more objective, more non-political and more efficient than the smaller ones. On the other hand, detailed local knowledge is of great value in poor law work. I am inclined to think, on balance, that public assistance might in due course be transferred to the second category of functions, subject to adequate safeguards.

The question of the appropriate authority for education is not free from difficulty. It is indisputable that the London

County Council is administering all forms of education far better than they would be administered by the metropolitan borough councils. It is also clear that technical education is essentially a regional service; and the same applies to special schools for defective children. But it is by no means certain that if a Greater London Council were established it should remain permanently responsible for the direct administration of every municipal elementary and secondary school throughout the region. There is a good deal to be said for smaller units of administration in the field of school education than a council representing a population of perhaps 9 million persons. On the other hand, nothing would be more disastrous than pure localism in so vital a matter. It is possible that the proper solution may ultimately turn out to be a distribution of functions somewhat on the lines proposed by the London County Council to the Ullswater Commission.[1] There would, however, have to be a gradual process of devolution; and in the first instance the whole field of education should be entrusted to the regional organ. But eventually certain aspects of it might be transferred to the second category of functions, subject to adequate safeguards.

This second category, it will be recalled, consists of services which could be carried out by local authorities acting under the general supervision of the regional organ. Into it would go the provision of public libraries, the protection of children, the Small Dwellings Acquisition Act, the ambulance service, the operation of ferries, diseases of animals, inspection of nursing homes, surveys for overcrowding, the provision of small holdings, licensing and inspection of petroleum, enforcement of the Shops Acts, protection of wild birds, registration of births, deaths and marriages, registration of electors, adulteration of food and drugs, registration of vendors of poison, the humane slaughter of animals, provision of mortuaries, cemeteries and crematoria, parking places, the construction, maintenance and improvement of local streets, the cleansing and lighting of highways, the provision of baths and wash-houses, inspection of dairies, cowsheds and milk shops, the regulation of ice cream vendors, provision of disinfecting stations, notifica-

[1] *Ante*, pp. 292–3.

tion and prevention of infectious disease (apart from the provision of hospital accommodation), the authorisation of offensive trades, the inspection of weights and measures, removal (but not disposal) of refuse, the licensing and inspection of slaughter-houses, massage establishments and common lodging-houses, the treatment of tuberculosis in dispensaries (but not in hospitals or sanatoria), vaccination, the clearance of small unhealthy areas, the regulation of buildings, and a large number of functions relating to public health and the amenities of the district.

I shall discuss later the relationship between the regional organ and the local authorities, and the nature of the control which the former should exercise over the latter. I shall also defer further reference to the third group of functions, in regard to which the local authority would be completely autonomous. It is necessary to deal first with the fundamental question of the constitution of the regional authority.

A directly elected Greater London Council for the whole region would be at once the simplest, most straightforward and most democratic solution of the problem. It would possess the supreme advantages of being in direct contact with the citizens and of representing only regional interests; and it would thus act as a powerful stimulus to the growth of a regional consciousness which we have seen to be so necessary. By virtue of its importance it would attract able and energetic men and women to its membership. From its direct mandate it would draw the strength needed for the accomplishment of many difficult tasks.

Such a Council should not be too large. Its total membership should not exceed 150. If the constituencies were to consist of 75,000 persons each, a population of 9 millions would elect a Council of 120 members, to which might be added 30 alder-men.[1] The average number of electors for each Member of Parliament in Outer London is at present 73,000, while the

---

[1] The number of voters would of course be much smaller. In the administrative county of London the electors number 2,855,542 and the population 4,397,000—a proportion of slightly less than two-thirds. A constituency of 75,000 on this basis would contain about 50,000 electors. The members of the Greater London Council should be paid. *Post* pp. 338–9.

[2] J. C. Johnstone: "How London has Eluded Control by its own County Council," *Daily Telegraph*, 12th August, 1938.

number in Inner London is 47,000.² In Outer London the average constituency has a population of about 100,000, while in the administrative county the average population for a constituency is 72,000. So we are already operating constituencies of about this size for Parliamentary purposes in a large part of the region and much bigger ones in the rest of it. The new constitution of New York City Council provides for one councilman for every 75,000 voters, with a remainder of 50,000 votes entitling an electoral division to one additional member.¹ This is now working well, and is considered greatly superior to the old Council which had 66 elected members, each of whom represented an aldermanic district with an average population of 100,000 persons and about 30,000 voters. It is said that the very size of the enlarged constituency prevents the member from "running errands for the members of the political club in his district," which he was formerly wont to do, and compels attention to the larger public issues.

Let us now consider the alternatives to such a scheme. One of them is an indirectly elected body representing for specified purposes all or some of the existing authorities in the region. This would be a body on the lines of the old Metropolitan Board of Works, and would reproduce all the defects which brought that board to an ignominious end. We know from long experience that an indirectly elected authority is unable to arouse public interest in its proceedings; that it labours under the shadow of a secondary responsibility which is wont to develop into irresponsibility; that the members bring to the council chamber the views of the particular authorities they represent rather than the interests of the body to which they are accredited; that the policies and issues which confront the Council are never placed before the citizens for an expression of their will; and that corruption, jobbery and nepotism easily find a foothold—as in the case of the Metropolitan Board of Works. For these and cognate reasons the proposal will not

¹ The constituencies in New York are enormous. Manhattan, with a population of 1,684,543, returns 6 members to the City Council; Bronx, with 1,499,090, returns 5; Brooklyn, with 2,798,093, returns 9; Queens, with 1,346,659, returns 5; and Richmond, with 176,683, returns 1. The average is thus about 288,000 for each councillor. For further details about the New York City Council, see *post* pp. 461 *et seq.*

survive serious consideration. We must reject on similar grounds all the other ingenious devices of a federal character which are put forward from time to time. It may be regarded as axiomatic that no satisfactory regional body can be built up from the existing local authorities in the region. While their municipal independence persists, there can be no Greater London commonwealth.

Any kind of joint committee or board would be equally unsatisfactory. The financial objections to such a body representing a number of independent authorities were pointed out to the Ullswater Commission on London Government. If a joint body is able to make final decisions without reference to the constituent authorities, the finances of the latter are liable to be completely upset by the votes of representatives from other areas; while if executive and financial powers are not delegated to the joint body, every recommendation would have to be referred back to the constituent authorities and "nothing would ever happen."[1]

Another type of alternative scheme is for some kind of an *ad hoc* authority. This raises several large questions of principle and merits discussion in a separate chapter.

[1] Royal Commission on London Government, 1923: Norman: Minutes of Evidence, pp. 58–9.

CHAPTER IV

THE *AD HOC* BODY

The *ad hoc* body has been favoured as a solution of the difficulties of London government from 1839, when the Commissioners of Police for the Metropolis were first set up, to our own day, when several other statutory authorities for special purposes have been created. The Metropolitan Water Board, the Port of London Authority, the Thames Conservancy, the London Passenger Transport Board, the London and Home Counties Joint Electricity Authority, the Metropolitan Area Traffic Commissioner—not to mention the old Metropolitan Asylums Board, whose functions are now transferred to the London County Council—here surely is a sufficient number of examples to provide ample material on which to base a conclusion.

The attractiveness of the *ad hoc* idea is not difficult to understand in a situation such as that which exists in London. The ground is littered with multifarious elected authorities possessing jurisdiction over utterly inadequate areas. Each one of those authorities is a centre of potential opposition to any rational scheme of reform. On the other hand, the technical needs of a service—water, transport or whatever it may be—are easily ascertained and strongly urged by responsible administrators or independent experts who at least desire to promote the efficiency of that service. Hence the wary politician, the timid civil servant and the technical specialist readily turn to the *ad hoc* authority as the easiest way out of their difficulties. "Ministers have almost ceased to apologise," writes Mr. Herbert Morrison, "for creating Greater London authorities for purposes which, if local government were rationally organised in the area, could have been discharged under normal local government auspices. Indeed, some enthusiasts with specialist minds occasionally bob up demanding yet another special Greater London authority in respect, for example, of housing or town planning. There are people who believe that the establishment of a special authority will solve most problems for co-ordination, whereas it may have done little more than create a salary list."[1] Sometimes the more pressing technical difficulties may be assuaged for a time. But ultimately the *ad hoc* body gives rise to as many problems as it solves.

The most serious drawback of the *ad hoc* body is that there is no method of co-ordinating its work with related activities carried out by other bodies. It has one, and only one, object in view; and it is in a sense failing to discharge its duty if it attempts to take a comprehensive view of things. Yet the services of a great modern city are becoming more interrelated every day, and even their efficiency is determined to no small extent by the degree of co-ordination that is attained. Housing, planning, transport, highways—how can one separate such a group as this?[2] And housing in turn involves education, drainage, public health, gas and electricity and many other services. There are hundreds of other points of contact between the various public and social services where "the single eye" is needed to obtain the best result.

I have already mentioned, as an example of the lack of co-ordination which now exists, the complaint by the London County Council that in some parts of the county there is not a water supply adequate for fire-fighting purposes—the Metropolitan Water Board having, of course, no responsibilities in connection with protection against fire.

The point was most clearly demonstrated at a recent sitting of the Royal Commission on the Geographical Distribution of the Industrial Population, when Mr. Pick, Vice-Chairman of the London Passenger Board, was giving evidence.[3]

[1] Rt. Hon. Herbert Morrison: *How Greater London is Governed*, p. 122.

[2] A witness for the London County Council explained to the Royal Commission on London Government that the L.C.C. had sanctioned a loan by a metropolitan borough to pave a certain street with wood blocks on strong concrete foundations, because a motor omnibus route had been established on that street and motor-bus traffic destroys macadam very rapidly but has little effect on wood paving. After the sanction was given the route was changed and the necessity for repaving the street entirely disappeared. Royal Commission on London Government, 1923: Norman: Minutes of Evidence, pp. 37–8.

[3] Minutes of Evidence, 2nd March, 1938.

3370 Q. (by Mr. D'Arcy Cooper) Is there no co-operation between local authorities and the planning bodies and yourselves in regard to planning problems?

   A. (by Mr. Pick) No. There is no real effective co-operation. Certain local authorities unite together to have a common town planning scheme. That has happened in the Thames Valley, for instance. But they do not ask us about their town planning.

3371 Q. Or their transport problems?

   A. Only when they have to complain.

3445 Q. (by Sir William Whyte) Have you not a right as a statutory undertaking to represent against a scheme of a local authority before the Minister of Health?

   A. Only if we are landowners in the area.

3448 Q. And the Government Departments to whom proposals are made for housing development and town planning have never advised the Board?

   A. No.

3457 Q. (Chairman) With regard to your tube railway extension to Cockfosters and High Barnet, is there any allegation that that will cut into the Green Belt policy?

   A. No. There is no Green Belt at High Barnet. We just touch the Green Belt at Bushey Heath. We arrive in the Green Belt at Bushey Heath station.

3458 Q. And that extension has been criticised accordingly, has it not?

   A. Yes. Lord Lytton, I notice, refers to it. But what sort of criticism is it that does not want to reach the Green Belt? I say, you have to get out of London as well as get into London. Does it mean that the Green Belt is to be placed out of reach?

3459 Q. The question of reaching it is rather different from the question of affecting or destroying the Green Belt policy?

A. Yes. He really blames us for other people's faults, because it is quite right that there should be stations in the Green Belt. It is quite right also that the local authority should plan to retain the Green Belt and it is not the Board's fault if when they put a station down the place becomes urbanised and industrialised and what not.

If anyone can read these questions and answers without feeling exasperated at the waste of good intentions he is fortunate. The London County Council with its £2 millions subsidy for the Green Belt, the county councils with their much larger expenditure, the county district authority with its planning powers, the London Passenger Transport Board with its zeal for providing facilities and anticipating traffic wants: we can assume they are all well-intentioned bodies doing their respective jobs as honestly as possible with goodwill to all and malice towards none. Yet between them they are responsible for the maladroit bungling which has produced this waste of money and effort, and a further despoliation of what was intended to be preserved. This kind of disastrous muddle is inevitable in present circumstances: it arises out of the very nature of *ad hoc* bodies. Yet Mr. Pick writes that "at the moment the task of dealing with local passenger transport as a unity, as a specific function of this metropolis, is proceeding under favourable conditions."[1] As a transport executive, he is naturally satisfied with the unified administration, the statutory monopoly, the £40 millions of new capital guaranteed by the Government for expansion and improvement works. As the representative of the Transport Board, we have already seen, he desires the further growth of London to a population of 12,000,000 persons or such larger size within the 15-mile radius as the straphanger's patience will tolerate.[2] All that derives from an attitude concerned exclusively with transport considerations.

It would be quite wrong to assume that the *ad hoc* body is necessarily more efficient or economical than the omnibus or

[1] F. Pick: "The Organisation of Transport," *Journal of the Royal Society of Arts*, vol. lxxxiv, p. 211. [2] *Ante*, p. 320.

general type of authority. Indeed, the reverse is often the case. One of the main reasons for the abolition of the poor law guardians was their inefficiency and extravagance—some of it unavoidable in the circumstances. The Metropolitan Water Board is an interesting specimen from this point of view.

The Water Board consists of 66 members representing a large number of local authorities. Its administrative work is conducted through no less than 8 separate committees. Several of these, such as those dealing with finance, law and Parliamentary matters, works and stores, are similar to those found in most general local authorities. The salaried staff is about 1,150 (including about 100 temporary clerks) together with (in March 1935) more than 4,000 workmen.[1] The principal officers of the Board comprise the Clerk and Parliamentary Officer, Chief Engineer, Accountant, Solicitor, and Surveyor. The Chairman of the Board receives an allowance of £500 a year.

It is obvious that the necessity to employ separate chief officers of these categories for the sake of the water service alone, with the duplication of all the ancillary requirements, such as minor staff and premises, involves a substantial increase in overhead and establishment charges falling on the water supply. The actual result bears out this supposition. In 1920 a Departmental Committee was appointed to enquire into the working of the Metropolis Water Act, 1902. The Committee found that the transfer of the water supply from the companies to the Metropolitan Water Board had not led to economy. On the contrary, the reverse had happened. Even before the Great War there was no net saving discernible. "It might have been expected," observed the Committee, "that the amalgamation of 8 distinct and separately controlled undertakings into one single body would result in considerable savings, especially from the point of view of administration. . . . It is, therefore, at first sight, somewhat surprising that not only has there never been any saving in total cost, but that . . . the actual expenditures of the Board, even from the commencement, have been in excess of the total of the 8 undertakings whose properties were taken over and the cost of the water supplies, whether

[1] Metropolitan Water Board, Thirty-second Annual Report, p. 9.

22

measured per service or per 1,000 gallons, has risen."[1] The cost per 1,000 gallons supplied rose from 7·47d. in 1901 to 8·57d. in 1918–19; by 1933–4 it had reached 11·22d., and by 1935–6 12.03d. The cost for 1936–7 and 1938–9 remained at about a shilling a thousand gallons. For several years the ever-increasing deficit in the Board's revenue was met by a deficiency rate, but in 1921 legislation was passed to permit the charges for water to be increased.[3]

The Departmental Committee criticised the failure of the Water Board to introduce a greater measure of co-ordination and planning. They found that although 16 years had passed since the transfer, complete unification of the 8 separate water systems had not been achieved. "It is not clear," the Committee reported, "why this movement towards the unification and concentration of the undertaking has not been more rapid."[4]

The Departmental Committee did not recommend any fundamental change in the method of administration. It contented itself with advocating a reduction in the numbers of members on the Water Board and urged the appointment of a general manager to supervise the entire work of the undertaking. Neither of these proposals were carried out, although soon afterwards complaint was again made to the Ullswater Commission on London Government concerning the excessive size of the Board.[5] It is difficult to see what alternative system of organisation the Departmental Committee could have recommended, since the situation had not changed in any material respect since 1902, when an *ad hoc* Water Board was set up owing to the lack of any general municipal authority having jurisdiction over a sufficiently large area to take over the water companies' undertakings. The London County Council had been rejected on account of the small size of its territory. It was obviously beyond the terms of reference

---

[1] Departmental Committee to enquire into the effect of the Metropolis Water Act, 1902, [Cmd. 845] 1920, p. 6.

[2] *London Statistics*, vol. 40 (1935–7), p. 340.

[3] For details of the water charges, *ibid.*, p. 429.

[4] Report, [Cmd. 845] 1920, p. 7.

[5] Royal Commission on London Government, 1923: Minutes of Evidence, E. B. Barnard (Herts C.C.), pp. 200–1; Musgrave (Essex C.C.), p. 298; Middlesex, p. 401.

of the Committee to reform the whole structure of London government.

To accept an existing institution *faute de mieux* is, however, a very different thing from regarding it as based on a desirable principle worthy of extension. I have heard highly placed persons connected with the Metropolitan Water Board dilate on the excellence of the *ad hoc* system for a technical service such as the water supply. It is alleged that the method of indirect election works well; that the members of the Board are of great public experience and long service on the Board;[1] that they take a great interest in their work; that appointment to the Board is much prized and eagerly sought after; and that the administration is cheaper and better than it would be under a municipal council.

These contentions cannot be accepted as valid, nor the conclusion which they seek to support be regarded as justified. The 66 members of the Metropolitan Water Board constitute an exceedingly slow and cumbersome piece of machinery. From the point of view of administrative efficiency, a paid commissioner would probably do the work better; while from the standpoint of cost, a substantial saving would certainly occur if the water board was merged with a regional authority, or even if its functions were transferred to the London County Council. The democratic element in the Board's constitution is very small: it is no more than a name to the mass of Londoners, and its monthly meetings do not normally arouse the faintest ripple of public interest. The average attendance of members at these meetings was 75 per cent of the total membership in 1933-4, 80 per cent in 1934-5[2] and 83 per cent in 1938-9. These high percentages show that the members are assiduous in the performance of their duties. Their zeal in this respect may be compared with that shown by members of the London County Council, for which the figures are as follows:[3]

[1] The length of membership is on the average a period of nine years. Sir William Prescott: "The Present and Future Policy of the Metropolitan Water Board," in *London's Water Supply*, Supplement to *Morning Post*, 9th December, 1935.

[2] Thirty-second Annual Report of the Metropolitan Water Board, p. 8.

[3] The figures are calculated from the London County Council Return of Attendances of Members, 1934-6. Publication No. 3242/1937.

LONDON COUNTY COUNCIL. ATTENDANCE OF MEMBERS.
1934–6

	Council	Committees	Sub-Committees	Aggregate
(a) Possible member-attendances at meetings	12,019	17,312	17,504	46,835
(b) Actual member-attendances at meetings	10,885	13,862	12,545	37,292
(c) (b) as percentage of (a)	90·56	80·07	71·67	79·62

(a) represents possible "person-meetings."
(b) represents actual person-attendances."
(c) is therefore the average percentage attendance at a meeting. This is not necessarily the same as an average percentage attendance of a person at the meetings to which he is summoned, but is in general a more significant figure.

The need for co-ordination in Greater London is now so great that some of the responsible heads of the Metropolitan Water Board are beginning to have ambitions to take over functions in allied fields. As they see it, the Metropolitan Water Board should deal not only with the supply of clean water but also with dirty water—i.e. main drainage. A co-ordinating authority is badly needed for the collection, storage and distribution of clean water, and also for the collection and disposal of dirty water. The Board would promise to build a huge sewer north of the Thames and another one on the south side if it were given the chance to acquire drainage and sewage disposal powers.

This urge towards expansion into neighbouring spheres of activity reveals an implied criticism of the *ad hoc* principle; and it shows how that principle contains the germs of its own decay.

Yet despite its many disadvantages and limitations, the *ad hoc* body continues to be advocated as a short cut through the difficulties of London government. Mr. Pick, for example, takes the view that several other functions such as highways, town planning, main drainage, parks and open spaces, gas and electricity, demand treatment on lines similar to transport. "The London Passenger Transport Board," he writes, "is a model for the performance and control of commercial or quasi-commercial pursuits. For services which are not self-supporting some variant organisation may prove more desirable, but some organisation is wanted which will remove from political influence matters of business and applied knowledge whether engineering, technical or the like."[1] In his opinion, the right way to proceed is to continue to devolve functions in the London Region upon separate *ad hoc* boards. These boards would consist of technically competent executives to be appointed and not elected. The several county councils in Greater London would each nominate 3 or 4 of their members to form an appointing college, analogous to the "appointing trustees" who nominate the members of the London Passenger Transport Board. This college would appoint the members of the *ad hoc* board, but would have no further power over its administration, which would thenceforth be autonomous.

Mr. Pick recognises the ultimate need for a comprehensive body to supervise and weave the activities of all these separate organs into a single, coherent pattern; but he is willing to relegate its establishment to some far-off time in the remote future. "An aggregation of people in a single social, economic, geographical unit," he remarks, "requires a single unit for the realisation or discharge of each of the many objects or functions which it embraces. That higher organisation which co-ordinates all these functions into a whole is left at this stage as being almost too debatable for treatment."[2]

Thus in the end Mr. Pick is willing to concede the necessity for a general elected Council for Greater London, with supreme control over all these previously independent bodies. But after what a nightmare of confusion, overlapping and conflict! The 8 or 9 million inhabitants of the metropolis are apparently to be sacrificed for an indefinite period to a reign of uncontrolled *ad hoc* bodies, manned by experts pursuing their own separate specialised paths unrestrained by any consideration of the wider interests of the whole, until such time

[1] F. Pick: "The Organisation of Transport," *Journal of the Royal Society of Arts*, vol. lxxxiv, p. 218. [2] *Ibid.*, p. 209.

as the disintegration produces intolerable results. It would be a mistake of incalculable magnitude to contemplate such a course. Every addition to the number of *ad hoc* bodies would increase the confusion of policies and lack of co-ordination to a disproportionate extent. With the inception of each such organ there would be created a new and powerful obstacle to the establishment of the Regional Council which is London's greatest need. The day of its arrival would be postponed indefinitely rather than hastened.

In the realm of finance alone there are sufficient objections to rule out such a policy as unwise. Even as it is there are too many public authorities in London spending money without regard to each other's plans. The Port of London Authority, the Metropolitan Water Board, the London County Council, the Metropolitan Borough Councils, the London Passenger Transport Board, are each accustomed to embark on large capital expenditure without reference to a common programme related to the general economic situation. The disadvantages of numerous spending and borrowing authorities in a single area is in itself a very strong objection to *ad hoc* authorities.[1]

We must, then, for the various reasons that have been mentioned above, reject the *ad hoc* body as a solution of the problems of London government. The *ad hoc* bodies which are already in operation owe their existence chiefly to the fact that by reason of the obsolete structure of London government there was no municipal authority of adequate size to handle the particular functions concerned; or (and this is secondary result arising from the same cause) because there was so much petty jealousy and internecine struggle between the existing authorities that a new specialised body seemed the only practical possibility. One or other of these causes was certainly operative in the case of the water and transport boards, and the metropolitan police.

If a Greater London Council were set up, the *raison d'être* for the existence of several of these *ad hoc* bodies would immediately cease. The Metropolitan Water Board should be abolished forthwith, and so should the Thames and Lee

[1] Royal Commission on London Government, 1923: Sidney Webb, Minutes of Evidence, p. 841.

Conservancy Boards and the Roding Catchment Board. The other *ad hoc* authorities would be dealt with on their merits.

The Metropolitan Police Commission is at present defended on two grounds. First, that to grant local control of the police force in the Metropolitan Police District on the same basis as elsewhere would involve splitting up the force among 9 county and county borough councils, an obviously retrogressive step which no one would seriously contemplate.[1] But this reason would immediately disappear if a Greater London Council were created.[2]

The second ground is the alleged need for the protection of the King and the Royal palaces, the Houses of Parliament and the central government, to be under the direction of the Home Secretary, who must also be responsible for much of the work of the Criminal Investigation Department at Scotland Yard. Hence, it is argued, a separate force would be required to perform these imperial and national duties if the Metropolitan Police were transferred to municipal hands; or, alternatively, recourse would have to be had to military protection.[3]

These arguments are not convincing. When the King or Cabinet Ministers visit Birmingham, Manchester, Liverpool or Glasgow, they rely on the municipal police forces controlled by the councils of those cities: they do not take with them a special force of stalwarts under the immediate control of the Home Secretary. Nor when such personages visit the City is any anxiety aroused by the fact that the City police force is administered by the City Corporation. In the City, indeed, we find the most purely local of all the police forces in Great Britain. Yet the Sovereign and His Majesty's Ministers are not conspicuously reluctant to trust themselves to the protection afforded by the City police force.

If it be insisted that the safety of the Houses of Parliament, Buckingham Palace, and the Government Offices in Whitehall is a national or imperial responsibility which no self-respecting Cabinet could permit to pass out of its own hands, the answer is that there is always a considerable body of soldiers maintained in London, and it would be a trifling addition to their duties

---

[1] J. F. Moylan: *Scotland Yard and the Metropolitan Police*, p. 66.
[2] *Ibid.*   [3] *Ibid.*, p. 67.

to place them on guard over these institutions. Indeed, the absurd farce of the Horse Guards in Whitehall, who at present guard nothing whatever except an archway leading into St. James's Park, could be terminated by exchanging this picturesque futility for a more significant task.

There would, then, be little to be said for retaining the Metropolitan Police Commission as a separate *ad hoc* body under Home Office control if a Greater London Council were created. Much could be said, on the other hand, for conferring upon such a Council the dignity and prestige of maintaining its own municipal force. I should, therefore, be in favour of transferring control of the police. The maintenance of our police forces under a régime which is democratic both in the letter and in spirit is of the utmost possible importance; and a municipal police force is more likely to be democratic in outlook than a centrally controlled force.

The Port of London Authority should be left untouched. Its work is too closely connected with a particular branch of industry—the berthing, loading and unloading of ships—to make transfer to a municipal body a feasible or profitable proposition, although its activities impinge and even conflict with the functions of the authority responsible for bridges and the prevention of river pollution (at present the London County Council and in future the regional council).

Similar considerations apply to the London Passenger Transport Board. The Board should be left intact as a separate entity, at any rate for the present. The appointing trustees should, however, be abolished and the members appointed by the Minister of Transport after consultation with the Leader of the Greater London Council. Moreover, at least one of the members of the Board should be appointed by the Greater London Council from among their own number. A Statutory Committee should be set up consisting of three members of the Council and three members of the Board to meet at regular intervals for the purpose of co-operating on matters of common interest and exchanging information and opinions.

CHAPTER V

THE QUESTION OF DEMOCRACY

A question of great importance which arises in connection with London government is whether the prevailing methods of administration are moving towards or away from the democratic pattern which was first laid down tentatively by the Municipal Corporations Act, 1835, and subsequently developed in the legislation establishing the county councils in 1888, the district and parish councils in 1894 and the metropolitan borough councils in 1899.

The essence of this pattern is extremely simple. It consists, first, of conferring the right to vote on every man or woman who has attained the age of 21 and has occupied, as owner or tenant, any land or premises in the local government electoral area during a short qualifying period. A man or woman is also entitled to a vote if he or she is the husband or wife of a person qualified to be registered in respect of premises in which they both reside. In the second place, it consists of permitting anyone to be a candidate for election if he or she is a local government elector in the area, or owns freehold or leasehold property therein, or has resided there for a period of 12 months preceding the election. Thirdly, it involves the exercise of the constitutional rights of freedom of speech and writing, the liberty to criticise and oppose, and immunity from arbitrary arrest and imprisonment.

From these simple elements there has grown up an elaborate system of democratic government of great interest and significance. All the powers of the local authority are concentrated in the council, which is divided into a series of committees responsible for various branches of the work. The committees are brought into direct contact with the actualities of administration, and every member of the council is thereby enabled to participate in the process of decision.[1] Service on a local

---

[1] *Cf.* E. D. Simon: *A City Council from Within, passim;* H. J. Laski: "The Committee System in Local Government" in *A Century of Municipal Progress.* (London: George Allen & Unwin Ltd.)

authority in this country provides an unrivalled opportunity for comprehending the democratic process and participating actively in its responsibilities: an opportunity far greater than that afforded to the private Member of Parliament, since in the House of Commons executive power is in the hands of a small group of Ministers, while the openings in debate for the ordinary back bench member are comparatively few. The democratic local authority thus offers a unique experience in civic life which is not the least of its virtues.

As a piece of democratic machinery, the London County Council is an outstanding success which compares favourably with any other local authority in the world. Its administration in most fields—there are two or three exceptions—is highly efficient, and its integrity and competence in financial matters unrivalled. Its standing orders are excellent; it devotes a minimum of time and money to ceremony; and in general its proceedings are dignified, orderly and businesslike.[1] The amount of unpaid public service which it evokes from its members is astonishing. The institution of the Leader of the Council and the Leader of the Opposition is a unique and valuable device among local authorities in Great Britain.

The London County Council is probably by far the best organised local authority in England from the point of view of concentration of power, consistency, responsibility and leadership.

The majority party appoints the chairman and vice-chairman of all committees, which are in fact almost semi-ministerial positions. Normally the chairman answers for his committee in the Council. The General Purposes Committee of the Council consists of the chairmen of committees, the whips of the majority party, and prominent members of the minority party, corresponding to their numerical strength on the Council. It decides all questions of difficulty or dispute between committees. The Leader of the Council is not chairman of any committee nor does he aim at making speeches all the time in the Council. His real work within the Council is to organise, to co-ordinate, to give unity and stimulus to the majority

---

[1] *Cf.* A. Emil Davies, L.C.C.: *The London County Council*, 1889–1937 (Fabian Society), pp. 30–3.

party. He has to keep a watchful outlook for matters of policy which are likely to give trouble or put the Council or the dominant party in a difficult corner; and above all he must keep his hands on the big questions of principle with which the Council has to deal, or ought to deal. The Chairman of the Council occupies a non-political position.

The organisation of the Labour Party, which has been in power on the Council since 1934, is extremely closely knit in two distinct ways.[1] First, there is the organisation of the Labour councillors throughout the various committees. The policy committee of the Labour Party on the Council consists of the chairman and vice-chairman of the main committees and the Party whips. It meets every week and is in fact the central control of the Party, but there is sometimes disagreement at this meeting and the decisions of the policy committee are subject to the approval or disapproval of the Party meeting which is held every week prior to the weekly meeting of the Council. Nevertheless, the policy committee is immensely powerful and offers full opportunity for energetic leadership. No one may bring up a question in the policy committee of the Party without having given one week's notice to the finance committee. But anyone may raise a question at the weekly Party meeting.

In addition to this formal structure the chairmen of committees are in frequent informal contact with the Leader. Again, the Labour members of the hospital, education and public assistance committees have group meetings before each meeting of the full committee.

Running parallel to this Party tissue is an organisation of the officials which has the effect of correlating and correcting the work of the Council. The officers at County Hall are of course neither the officers of a political party, nor of a committee, nor of a chairman of committee, nor of the Leader. They are the officers of the Council as a whole. Each committee has a thoroughly capable committee clerk, receiving a substantial salary, who acts as the eyes and ears of the Clerk to the Council. He reports direct to the Clerk, and the Clerk to the

---

[1] The Municipal Reform Party is also closely knit, but its organisation is less highly perfected than that of the Labour Party.

Council would immediately notify the Leader of the Council in the event of anything going wrong or any serious trouble being apprehended. Thus, the leader has a double check: on the one hand through the official organisation, on the other hand through the Party organisation. Moreover, every chief officer has direct access to the Leader of the Council.

The London County Council is, of course, inevitably handicapped by the inept organisation of areas and authorities in the metropolis; and I have already dealt at length in Part II of this book with the difficulties which arise therefrom. But subject to those limitations, the London County Council can be regarded as a successful example of responsible government which democrats may contemplate with satisfaction. I shall deal later with one aspect of party government in London which I regard as seriously defective.[1] At the moment I am concerned only to emphasise the fact that, starting in the most unpropitious circumstances in 1888, the London County Council has demonstrated through half a century that municipal democracy will work well on a large scale in London.

Let us turn now to the *ad hoc* bodies of one kind or another which are either in existence or proposed as methods of solving the difficulties of London government, and consider them from the democratic point of view.

There is the *ad hoc* executive body of an autonomous kind such as the London Passenger Transport Board which is without any sort of political responsibility. Of the five trustees who appoint the members of the Board, three consist of such politically irresponsible persons as the president of the Law Society, the president of the Institute of Chartered Accountants, and the chairman of the Committee of London Clearing Bankers, the remaining two being the chairman of the London County Council and a local authority member representing the London and Home Counties Traffic Advisory Committee.

This last-named Committee is an *ad hoc* body representing local authorities and special interests connected with traffic or transport. It reports to the Minister of Transport and has no executive power. Its political responsibility is indirect and

[1] *Post* p. 351–2.

remote; and as an instrument of London government it is a step away from local democracy.

A much larger stride away from local government in the English tradition is the appointment of an executive or advisory official appointed by and responsible to the central government, such as the Metropolitan Commissioner of Police or the officer (Sir Charles Bressey) appointed by the Minister of Transport to draw up a highway plan for Greater London. Here we are definitely approaching the Continental model of central control typified by the Prefect of the Seine or the Prefect of Police in Paris.

If we wish to preserve and strengthen democracy in this country, it is obvious that we must reject expedients of these kinds and acknowledge the necessity for a directly elected regional Council for Greater London as the only type of institution which is satisfactory from a democratic point of view.

This aspect of the matter cannot be regarded as of secondary importance at a time when our democratic faith, and the institutions in which it is embodied, have acquired a new and enhanced significance in a world of competing creeds and hostile authoritarian doctrines. Democracy on the national scale can function in a healthy manner only if it is supported and nourished by democratic local government. It will be a disaster if the reform of the essential structure of London government is delayed and evaded by the introduction of piecemeal expedients introduced without regard to their undemocratic or anti-democratic character. Irreparable injury will eventually result if no attempt is made to bring the elected authorities in the metropolis into organic relation with the realities of social, geographical and economic life. Unless a halt is called to present tendencies the time may not be far off when a fifth of the nation will be living under a system of "Government by Commission," by which I mean some form of administration consisting of centrally appointed officials or irresponsible nominated boards.

Before leaving the subject of democracy in London government, a word must be said concerning the relations between the political parties and the electorate. I do not share the views of superior persons who look down from Olympian heights with displeasure and contempt at the activities of political parties. Since the electors are divided on many urgent questions according to their opinions and interests, it is desirable that these divisions of interest and outlook should be canalised and organised in a coherent and clear-cut way. It is only by means of political parties that this can be done.

In general, the development of party government in English local government during the past 20 years has been a good thing, bringing fresh vitality, interest and energy to the work of the local authorities. There have been and still are frequent abuses of the party system by fatuous persons who, not understanding that party conflict should be confined to matters where there is some genuine difference of principle in dispute, conceive it to be their task to oppose and obstruct every proposal which emanates from their opponents as a matter of righteous and inflexible duty. We have had such instances in London, where, for example, the question of rebuilding Waterloo and Charing bridges was quite wrongly made a political issue although there was no inherent reason why it should have been so treated. But on the whole, the work at County Hall has been stimulated and vitalised by party organisation; and the same is tiue in those metropolitan boroughs and out-county authorities where there is a substantial opposition.

Despite the existence of active political parties, the interest of the public in local government elections within the metropolis remains apathetic. The figures opposite give the percentage of the electors voting at the last six elections for the London County Council, metropolitan borough councils, and Parliament respectively within the administrative county.[1]

On the whole, prior to 1939 interest in L.C.C. elections seemed to be rising, and the proportion voting in 1937 was nearly 13 per cent greater than in 1925; the average poll for the metropolitan borough councils, on the other hand, has never

[1] For full details see *London Statistics*, vol. 40 (1935–7), p. 18–26; and the Return of Metropolitan Borough Council Elections, L.C.C. No. 3312/1937. See E. C. Rhodes: "The Exercise of the Franchise in London," *Political Quarterly*, January–March 1938, vol. ix, p. 113. See also "Voting in Municipal Elections." *Political Quarterly*, April–June 1938. vol. ix, p. 271.

approached the highwater mark of 42·5 per cent in 1925. The extremely low figure of 26·4 per cent in 1946 is difficult to explain. It was the second lowest percentage recorded during the existence of the London County Council. The lowest was in the election of 1919, held just after the first World War, when only 16·6 of the electorate voted

General Elections	Percentage Voting					
	1925	*1928*	*1931*	*1934*	*1937*	*1945–6*
London County Council	30·6	35·6	27·8	33·5	43·4	26·4
Metropolitan borough councils	42·5	32·3	31·3	34·3	35·4	35·1
	1923	*1924*	*1929*	*1931*	*1935*	*1945*
Parliament	60·0	71·0	65·9	65·8	61·5	68·2

The London figures are not inspiring[1]; but is it not possible that the multiplicity of authorities confuses the citizen and disperses his civic enthusiasm? If a Greater London Council were set up, I should expect to see an increase in the proportion of voters exercising the franchise, this reflecting an enhanced interest aroused by a regional authority representing a true emerging entity rather than the nominal communities into which the metropolis is at present divided.

I believe it to be desirable, however, that the two great political parties in London should make some effort to interest the electors in the vital problems of the metropolis as a whole, and to make them aware of their citizenship of the Greater London community. Hitherto nothing has been attempted in this direction by either party. Yet this is the kind of educative leadership which political parties are well fitted to give, though one must admit that the reform of London government ought not by rights to be a party question at all. Nevertheless, education and propaganda by the political parties might make a distinct contribution towards leading public opinion, and creating a popular demand for changes in areas and authorities which are absolutely necessary, and in helping to overcome the persistent lethargy in regard to this matter which has afflicted successive Governments, Parliaments and Ministers of Health.

[1] The comparable figures for the leading provincial cities as given on p. 172.

CHAPTER VI

THE REFORM OF THE COUNTIES

The question of the precise territory which the regional Council ought to administer is one to which it would be possible to devote endless pages. I do not propose to discuss it at length, since it is a highly technical matter which might well be examined by a boundary commission. There are, however, a few general points worth consideration.

The first is that the Metropolitan and City Police Districts —the so-called Greater London used by the Registrar-General for census and other purposes—is not a suitable area, although it has the advantage that many persons are familiar with it. It is much too small for regional needs and has been far exceeded by more recently created areas. Moreover, the Metropolitan Police District has one of the worst boundaries in detail it is possible to conceive.[1] Both the London Passenger Transport area (1,986 square miles) and the London Traffic Area (1,820 square miles) are far larger and better;[2] but neither of them is the best that could be designed. It may be emphasised, however, that the Greater London Council should have jurisdiction over an area of about the size of the Metropolitan Traffic Commission. This area comprises 2,419 square miles and is more comprehensive than the area of either the London Passenger Transport Board (1,986 miles) or the London and Home Counties Traffic Advisory Committee (1,820 miles), particularly in the north-east and north-west. It embraces the territory situated between about 25 to 30 miles distance from Charing Cross. It is $3\frac{1}{2}$ times as large as the Metropolitan Police District and a little over 20 times the size of the administrative county of London. It coincides with the boundaries of existing local government areas to a much closer extent than any of the other regional areas at present existing, but this by itself is not a specially important advantage. For it is

---

[1] Royal Commission on London Government, 1923: Minutes of Evidence (Webb), p. 835.

[2] For details and maps, see *London Statistics*, vol. 40, pp. 14–15.

clear that the establishment of an elected regional authority of the kind proposed will in any case make it necessary to reform drastically the areas of a number of neighbouring authorities: in particular those of the home counties. This may be an unfortunate necessity, but it is inevitable.

We must not shirk the fact that if a Greater London Council were to be given the functions it ought quite definitely to possess, the existing county councils within the metropolis would become not merely futile but positively obstructive. A double-deck structure in the region is indispensable. It would be impossible to administer so vast a territory and population save through the mechanism of numerous local authorities in addition to the regional council. But a triple-deck structure, composed of regional, county and district authorities, would be intolerably cumbersome, inefficient and costly. We could not afford it on the grounds either of efficiency or of expense. Moreover, when the essentially regional services are transferred to the Greater London Council, the county councils will be so denuded of powers and duties that they will be reduced to a mere shadow of their former selves. Hence, the continuance of the county councils within the metropolis is incompatible with the establishment of the new regional organ.

The London County Council and the Middlesex County Council are both situate entirely within the proposed region; and in consequence they would have to be abolished. Most of Surrey, a substantial part of Hertfordshire and Essex, and a smaller portion of Kent would be included. A small piece of Buckinghamshire would also probably fall within the region.

There is no insuperable difficulty in remodelling the administrative counties in the neighbourhood of London—or anywhere else, for that matter—so as to provide sensible local government areas, unless the view is taken that the county boundaries are eternal and immutable, and that it is sacrilegious to lay a finger on them. This attitude is apparently adopted by the County Councils Association, judging by the fact that it publicly opposes any and every modification in county areas which is proposed and has never put forward any constructive suggestion of its own. On this assumption, any sort of reform is of course ruled out. But it is scarcely possible that such a view will be permitted to obstruct the public good indefinitely.

The county councils were established only 50 years ago; and the boundaries of the administrative counties differ largely from those of the ancient historic shires. The existing counties are in a number of cases hopelessly obsolete in relation to the needs of local government today.[1] But no revision of them has so far ever been instituted, although the districts within the counties have since 1929 been made subject to review and reform at regular intervals.[2] The only modifications which have occurred to the counties themselves are those resulting from county borough extensions and creations.

The time has come when the desire of the county councils to maintain their existing territories must yield to the proved needs of the public and the wider considerations of municipal efficiency. A point has been reached in the metropolis where substantial progress can only be made at the expense of county government as it now exists. And it is worth noticing that the counties near London are particularly crabbed in size and awkward in shape: e.g. Bucks, Berks, Herts and Bedfordshire. The sooner this major issue is decided, the better it will be for everyone concerned. It is not county government as such which is challenged, but only county government in its present archaic form. The existing counties are already beginning to be passed over or denuded of their functions in regard to a number of services requiring regional organisation: one may mention, as recent examples, the establishment of the Area Traffic Commissions in 1930, the transfer of trunk roads to the Minister of Transport in 1937, and the great electricity regions formulated by the Electricity Commission. If the county councils desire to preserve their collective status, power and prestige, they must recognise the need to adapt their areas to the social and economic changes of contemporary life. A mere resistance to progress, a refusal to face obvious facts, will ultimately prove more destructive to their position than a

---

[1] See W. A. Robson: *The Development of Local Government*, Part I, for a full discussion. (London: George Allen & Unwin Ltd.)

[2] This was introduced by the Local Government Act, 1929.

readiness to evolve in accordance with the needs of our developing society.

What is required, in short, is the abolition of county government as it now exists within the new metropolitan region; or, rather, its merging in a Greater London Council; and the drastic re-drawing of county areas outside the region. The number of county councils should be greatly reduced and their areas in most cases substantially enlarged in size and rationalised in shape.[1]

This may be a hard pill to swallow, but what other feasible alternative exists? Take the county of Middlesex, for example. What conceivable justification can it have as an area of local government within the metropolis? It is utterly meaningless from every point of view. The Middlesex County Council is admittedly an energetic body making an active bid for favour in the public eye and for gathering power and influence unto itself. It is at present seeking to acquire a large number of powers from minor local authorities within its territory and has recently promoted ambitious and costly schemes of drainage and sewerage carefully limited by its own boundaries. In pursuit of the Green Belt policy the Middlesex County Council has been purchasing land not merely within its own confines, but also in other counties: e.g. they recently purchased 309 acres of the Denham Court Estate in Buckinghamshire at a cost of £48,000; and also contributed a quarter of the cost of acquiring 350 acres of Moor Park, Rickmansworth, in Hertfordshire.[2] It was announced two years ago that the Middlesex County Council had agreed to purchase a piece of land and some blocks of old buildings facing Parliament Square, which were threatened with a grandiose building development, for a sum of £365,000. In spending these large sums of money in preserving amenities, the Middlesex County Council was certainly earning the gratitude of everyone who cares for civic beauty.

[1] It would take me too far away from my main purpose to put forward detailed proposals on this point.

[2] The contribution of the Middlesex County Council to the acquisition of Moor Park is in excess of £30,000. The Council has also agreed to contribute to the acquisition of other open spaces outside their area to form part of the Green Belt.

Yet despite all these praiseworthy actions, Middlesex remains an anachronism in the body politic of London government. And the paradox is that the more able and energetic the County Council of Middlesex becomes, the more anachronistic its area grows. The anomalous position of Middlesex is emphasised by the situation of its Guildhall, which is placed in the City of Westminster, a stone's throw from County Hall.

Considerations of a somewhat different kind apply in the cases of Hertfordshire, Essex, Kent, Surrey, Buckinghamshire and Berkshire, because these counties are partly within and partly outside the effective limits of the metropolis. But this very fact leads to a similar conclusion as to their anachronistic character. The western part of Kent, for example, consists of pure dormitory areas such as Penge, Beckenham, Bromley, Chislehurst and Sidcup, which are part of the continuous built-up mass of London. The eastern part of Kent consists of seaside towns such as Walmer, Deal, Folkestone and Dover, dependent either on local residents, coal mining, summer tourists, or cross-channel traffic; together with self-contained inland towns such as Canterbury, Maidstone and Tunbridge Wells. There is no genuine unity between the eastern and western parts of the county and little, if any, community of feeling. The proper course of action is to sever the metropolitan from the non-metropolitan parts of the county.

A movement in this direction is actually beginning to arise from the logic of events. The heavy expenditure incurred by the Kent County Council during the past ten years on new or expanded services necessitated by the enormous influx of people from London in the west and north-western portions of the county has involved a continuous increase in the rates throughout the county. The mounting resentment of the population of East Kent at being taxed to provide these benefits for the denizens of the Kentish suburbs of London, has recently led the local authorities in East Kent to demand the severance of the two parts of the county and the establishment of a separate East Kent County Council.[1] It is contended that if East Kent paid only for its own services the rates would be nearly 2s. in the £ less than the present county rates. The irrational and indefen-

[1] For details, see *The Times*, 29th November, 1938.

sible structure of the county is thus becoming liable to explode by the simple process of internal combustion.

If the preservation of the county areas be insisted upon at all costs, then a solution on the following lines might be adopted. A Greater London Council would be elected in the manner already described, but the regional area would be drawn so as to comprise the home counties in their entirety. Provision would be made for statutory county committees to administer certain services; and these statutory committees would consist of the regional councillors returned by each county sitting together for the conduct of county business. Thus, one body would function for the whole region when it was fully constituted; but its members would assemble also in separate groups according to their constituencies for the conduct of county business through the Statutory County Committees. Middlesex would have its committee consisting of the members returned for Middlesex constituencies, and so would the other counties, including the administrative county of London. It is important, however, that the members should be elected to the Greater London Council and not to the county committee, in order that the regional council should not consist of a mere assemblage of county councillors under a different name.

I do not consider this the best solution. It would have many of the disadvantages of a triple-deck municipal structure which have already been pointed out. It would be a far weaker solution of the problem of London Government than the one previously described, whereby county government would be eliminated inside the metropolis and reformed outside. But I mention it as a second and inferior alternative because there has never yet been found a stout-hearted statesman or a far-sighted Parliament able to envisage and to apply the proper remedy for London's ills. So it is necessary to have a more feeble palliative at hand, suited not to the sickness of the patient but to the weakness of the doctor.

But of one thing I am unfortunately convinced. No amount of concession is likely to reconcile the county councils to the voluntary acceptance of any form of regional government likely to prove an effective remedy to the troubles which have befallen the metropolis. The only concession they will recognise is for matters to be left as they are. So the Minister who treads warily in case he shall disturb sleeping dogs may save himself the trouble of his caution. The path to safety and sanity is more likely to be gained by boldly facing the obstacles with a determined demeanour than by trying to slink round by the side door when no one is looking and then abandoning the attempt when the dogs growl. The dogs are not asleep. It is rather Parliament, the Ministry and the public which are dormant.

CHAPTER VII

LOCAL AUTHORITIES IN THE REGION

(a) *The Metropolitan Borough Councils*

Having indicated the general lines on which a Greater London Council should be constructed, we can now proceed to the question of the proper organisation of minor authorities within the region.

We may take first the administrative county of London. The metropolitan borough councils present an extraordinary diversity of feature and size; they do not appear to have been based on any principle whatever. They vary enormously in population, from Holborn with 38,000 inhabitants, to Islington with 321,000 and Wandsworth with 353,000.[1] They vary in territorial size over an equally wide range. Finsbury contains only 587 acres and Chelsea 660 acres, while Woolwich stretches over 8,282 acres and the acreage of Lewisham is 7,015. The rateable value of Westminster (as at April 1936) is £10,441,146, giving a rateable value per head of £83·93. The corresponding figures for Bethnal Green are £526,913 and £5·27. Marylebone has a rateable value of £3,365,844 and a rateable value per head of £36·51; Poplar levies a rate on an aggregate value of £771,611 or £5·37 for each local inhabitant.[2] The character of the metropolitan boroughs is also exceedingly diverse. Some, like Bermondsey, consist almost entirely of manual wage-earners, while others (such as Hampstead, St. Marylebone and Kensington) are predominantly middle and upper class. Some are natural units of government, while others (of which Lambeth is an example) are amalgamations of a number of smaller communities with little common interest to unite them.[3]

The metropolitan borough councils are the successors and inheritors of the powers of the vestries and district boards, so

---

[1] *London Statistics*, vol. 40 (1935–7), pp. 28–9. See *ante* p. 170.
[2] *Ibid.*, pp. 426, 449.
[3] C. R. Attlee: *Metropolitan Borough Councils, their Constitution, Powers and Duties*, Fabian Tract, No. 190, p. 2.

CITY OF LONDON
AND
METROPOLITAN BOROUGHS.

London County Council
February. 1936

WOOLWICH

GREENWICH

LEWISHAM

HACKNEY

POPLAR

STEPNEY

BETHNAL GREEN

SHOREDITCH

BERMONDSEY

DEPTFORD

CAMBERWELL

STOKE NEWINGTON

ISLINGTON

FINSBURY

CITY OF LONDON

HOLBORN

SOUTHWARK

LAMBETH

St PANCRAS

WESTMINSTER

BATTERSEA

WANDSWORTH

HAMPSTEAD

MARYLEBONE

PADDINGTON

CHELSEA

KENSINGTON

FULHAM

HAMMERSMITH

Scale

3 Miles

far as their primary sanitary functions are concerned.[1] They are full highway authorities for their areas except for bridges, tunnels and ferries. They administer the maternity and child welfare service, public cleansing in all its branches, libraries, art galleries and museums, baths and washhouses, electricity supply (if authorised) and a host of regulatory services.[2]

The London Government Act, 1899, which set up the metropolitan borough councils, did not seek to make them partners in the common task of governing London. It aimed, on the contrary, at undermining the growing prestige and power of the London County Council by conferring the greatest possible amount of dignity and autonomy on the 28 metropolitan boroughs, in the expectation that they would become centres of resistance to, and conflict with, the slowly emerging sense of unity which the London County Council was beginning to evoke and to express.[3] Hence they were given ceremonial insignia denied to the larger body, such as a mayor, a town clerk, and a town hall. The statute deliberately omitted to provide any means of securing co-operation either between the councils themselves, or between the several councils and the London County Council. The obviously desirable course of requiring the minor authorities to subordinate their actions to the needs of the larger entity was not merely ignored: it was actually flouted by numerous provisions requiring the major body to consult the district councils and to wait upon their pleasure. Thus the metropolitan borough councils have to be consulted on many points by the London County Council. Their consent is needed for town planning, highway and improvement schemes. The procedure is intricate beyond description. The Minister of Health will normally refuse to approve or promote any scheme unless the metropolitan boroughs have been consulted; and great Parliamentary opposition is likely to be incurred to any scheme to which they have not given their consent.[4] Even in regard to so elementary a

---

[1] The vestries were still the local health authorities when the Public Health (London) Act, 1891, was passed.

[2] I. M. Bolton and S. W. Jeger: *London's Borough Councils* (London Labour Publications pamphlet). [3] *Ante*, pp. 94 *et seq.*

[4] The main highway extension of Cromwell Road may be cited as an example.

matter as refuse collection there is lack of co-ordination. For example, the London County Council prescribes the methods to be adopted by tenants on the blocks of flats they have erected, but the local metropolitan borough council does the actual work of collection, and the two methods may be inconsistent.

It thus comes about that the London County Council has no more to do with the paving, cleansing and lighting of the London streets than it has with those of Manchester or Tokyo; it is neither sanitary nor burial authority. It does not collect its own rates but must precept on the metropolitan boroughs. It is almost powerless in the matter of valuation and assessment,[1] for in London alone of all the counties there is no county valuation committee to prevent anomalies and correct inconsistencies in valuation between one rating area and another, though one is badly needed. The principal municipal authority in London is not permitted even to prepare or supervise the registration list of the voters who elect it.[2]

It is impossible to defend either the areas or the powers of the metropolitan boroughs as they exist at present; and many of the members and officers who serve on them are prepared to admit in private conversation that they are unsatisfactory. Any survival value which they may possess is due not to administrative excellence but rather to political influence.

The only organ which is available for mutual co-operation between them is a body called the Joint Standing Committee of Metropolitan Borough Councils. This is a feeble and insignificant body, especially in its dealings with central departments and outside authorities. Even as a defence organ it does not bear comparison with a body like the Association of Municipal Corporations. Its proceedings are vitiated by the fact that the metropolitan boroughs are divided into two camps according to the political complexion of the party in power.

---

[1] The Assessment Committees are appointed by each metropolitan borough council and the City Corporation from among their own members. The London County Council has only one representative on each committee, the numbers of which vary between 10 and 21 members. *London Statistics*, vol. 40 (1935–7), p. 2.

[2] *The London County Council: What it is and what it does*, Fabian Tract, No. 61.

Thus, the Labour Councils usually vote together and so do the Conservative Councils. At present, Labour has control of 15 boroughs while the remaining 13 have Municipal Reform majorities. There is thus only a narrow margin between the two sides. But under its constitution the Joint Standing Committee must have a two-thirds majority for every decision. In consequence the Committee is unable to take steps of importance where a controversial question is concerned.

In any case the Joint Standing Committee is in no sense a planning body, or an organ for technical and administrative co-operation. It is here that the need is greatest and therefore its shortcomings and limitations most conspicuous. As regards the out-county authorities, London is a walled town.

Within the small area of the administrative county 28 autonomous authorities are responsible for providing such costly institutions as swimming-baths, washhouses, public libraries, etc.; yet no attempt is made to see that these are placed in the most suitable situations in each district having regard to similar provision in neighbouring districts. There is no mutual consultation or common plan to determine the site of a swimming bath or a maternity and child welfare clinic, with the result that it is mere chance whether they are suitably placed or not. In one instance which recently occurred, a washhouse in a rather poor quarter was just managing to justify its existence and pay its way. A new and more up-to-date washhouse was then built a short way away by a neighbouring authority. This rival attracted many of the consumers from the older institution which must now either close down or carry an unnecessarily high loss.

Overlapping, waste and competition of this kind clearly demonstrates the need for regional planning and co-ordination in a number of fields where the actual administration can quite well be left to the local authorities to carry out. The failure to recognise this essential principle is one of the greatest weaknesses in London government. Our thinking has apparently never advanced beyond the idea of either giving a local authority complete autonomy in a particular sphere or totally excluding it from that domain.

What is wanted today in the metropolis is a number of

local authorities of sufficient size and resources to be entrusted with a larger share of administration than they at present possess in certain departments, but subject to overriding control by the Greater London Council in ways which I shall presently describe.

The general opinion of the councillors and officers with whom I have discussed the matter is that in a giant city such as London the optimum size of the district authorities is a population of about 350,000 to 450,000 persons. This, of course, is only one factor, but it is a starting point. Rateable value is of vital importance, but is much more difficult to "organise" than population. The size and shape of territory must also be considered, and so, too, must community feeling where it exists, which is by no means always the case in London. Lastly, it is desirable where possible to have areas containing a mixture of classes and a blending of either industry or commerce with residential quarters.

The administrative county could be reorganised so as to produce a dozen large local authorities by the mere process of consolidating the existing metropolitan boroughs. The following table shows a possible combination:

TABLE A

	Population (1)	Area (acres) (2)	Rateable value (3) £	Rateable value per head (4) £
1. Fulham	150,928	1,706	1,261,876	8·76
Hammersmith ..	135,523	2,287	1,223,110	9·53
Total ..	286,451	3,993	2,484,986	8·68*
2. Kensington ..	180,677	2,290	3,330,661	18·61
Chelsea	59,031	660	1,230,568	21·21
Paddington ..	144,923	1,357	1,821,352	13·25
Total ..	384,631	4,307	6,382,581	16·59*

* Figures for totals obtained by dividing totals in col. (3) (i.e. rateable value at April 1937) by totals in col. (1).

TABLE A (continued)

	Population (1)	Area (acres) (2)	Rateable value (3) £	Rateable value per head (4) £
3. Hampstead ..	88,947	2,265	1,520,919	16·68
St. Marylebone ..	97,627	1,473	3,407,205	36·51
St. Pancras ..	198,133	2,694	2,188,496	11·74
Total ..	384,707	6,432	7,116,620	14·68*
4. Bethnal Green ..	108,194	760	525,871	5·27
Stoke Newington	51,208	864	427,074	8·51
Hackney ..	215,333	3,287	1,498,395	7·06
Total ..	374,735	4,911	2,451,340	6·54*
5. Islington.. ..	321,795	3,092	2,203,354	7·23
Finsbury ..	69,888	587	1,211,673	18·68
Shoreditch ..	97,042	658	811,663	9·21
Total ..	488,725	4,337	4,226,690	8·65*
6. Poplar	155,089	2,331	772,120	5·37
Stepney	225,238	1,766	1,733,193	8·21
Total ..	380,327	4,097	2,505,313	6·59*
7. Westminster ..	129,579	2,503	10,515,220	83·93
Holborn	38,860	406	1,646,287	46·55
Total ..	168,439	2,909	12,161,507	72·20*
8. Wandsworth ..	353,110	9,107	3,243,84⁻	9·33
9. Battersea ..	159,552	2,163	1,134,04⁴	7·72
Lambeth ..	296,147	4,083	2,334,727	8·36
Total ..	455,699	6,246	3,468,771	7·61*
10. Bermondsey ..	111,542	1,503	861,920	8·55
Southwark ..	171,695	1,132	1,328,467	8·52
Deptford ..	106,891	1,564	668,026	6·77
Total ..	390,128	4,199	2,858,413	7·33*
11. Greenwich ..	100,924	3,858	972,490	10·16
Woolwich ..	146,881	8,282	1,166,849	7·80
Total ..	247,805	12,140	2,139,339	8·63*

* Figures for totals obtained by dividing totals in col. (3) (i.e. rateable value at April 1937) by totals in col. (1).

TABLE A (*continued*)

	Populati)n	Area (acres)	Rateable value	Rateable value per head
	(1)	(2)	(3)	(4)
			£	£
12. Camberwell	251,294	4,480	1,726,285	7·38
Lewisham	219,953	7,015	1,783,680	7·89
Total	471,247	11,495	3,509,965	7·45*

Col. (3). Figures for individual boroughs from *London Statistics*, vol. 40 (1935–7), p. 449, col. 24, relate to 6th April, 1937. Figures for totals = sum of individuals.

Col. (4). Figures for individual boroughs from *op. cit.*, p. 426; obtained by dividing rateable value at 6th April, 1936, by population at mid 1935.

* Figures for totals obtained by dividing totals in col. (3) (i.e. rateable value at April 1937) by totals in col. (1).

It is not claimed for a moment that these are ideal areas. They are merely put forward to show that even without cutting up a single one of the existing boundaries it would be quite possible to improve immensely the present structure. In consolidating the areas regard has been had so far as possible to the question of geographical compactness, and the desirability of obtaining diversity of character and classes. Thus, the poverty, slums and industry of St. Pancras would be offset by the well-to-do dormitory of Hampstead and the refined gentility of St. Marylebone. The vast mass of lower middle class inhabitants of Islington would be helped out by the higher rateable value of Finsbury, which is due mainly to commercial premises.

Some problems are insoluble on the basis taken for this revision. It is impossible, for instance, to remedy the low rateable value of Poplar and Stepney without including them in an area comprising all or part of the square mile of the City.[1]

A solution which would overcome this difficulty is available by taking the division of the county into ten areas made by the London County Council for administering public assistance. These groups are as follows:

[1] The question of the City Corporation is omitted entirely from this part of the discussion and will be dealt with separately.

TABLE B

	Population (1)	Area (acres) (2)	Rateable value (3)	Rateable value per head (4)
			£	£
1. City	10,999	640	8,614,494	836·00
Poplar	155,089	2,331	772,120	5·37
Stepney	225,238	1,766	1,733,193	8·21
Total	391,326	4,737	11,119,807	28·42
2. Hackney	215,333	3,287	1,498,395	7·06
Stoke Newington	51,208	864	427,074	8·51
Bethnal Green	108,194	760	525,871	5·27
Total	374,735	4,911	2,451,340	6·54
3. Islington	321,795	3,092	2,203,354	7·23
Finsbury	69,888	587	1,211,673	18·68
Shoreditch	97,042	658	811,663	9·21
Total	488,725	4,337	4,226,690	8·65
4. Hampstead	88,947	2,265	1,520,919	16·68
St. Marylebone	97,627	1,473	3,407,205	36·51
St. Pancras	198,133	2,694	2,188,496	11·74
Holborn	38,860	406	1,646,287	46·55
Total	423,567	6,838	8,762,907	16·74
5. Westminster	129,579	2,503	10,515,220	83·93
Chelsea	59,031	660	1,230,568	21·21
Fulham	150,928	1,706	1,261,876	8·76
Total	339,538	4,869	13,007,664	38·31
6. Hammersmith	135,523	2,287	1,223,110	9·53
Kensington	180,677	2,290	3,330,661	18·61
Paddington	144,923	1,357	1,821,352	13·25
Total	461,123	5,934	6,375,123	13·83
7. Battersea	159,552	2,163	1,134,044	7·72
Lambeth	296,147	4,083	2,334,727	8·36
Wandsworth	353,110	9,107	3,243,847	9·33
Total	808,809	15,353	6,712,618	8·30

PUBLIC ASSISTANCE
ADMINISTRATIVE AREAS

In the Administrative County of London,
as from 1.iv.30, by Scheme under Local
Government Act, 1929.

The Administrative Areas indicated by
roman numerals and heavy border lines,
the Metropolitan Boroughs by names
and light border lines.

London County Council.
February, 1936.

MAP SHOWING THE DIVISION OF THE ADMINISTRATIVE COUNTY INTO 10 AREAS AS SET OUT IN TABLE B. (pp. 367-9).

TABLE B (*continued*)

		Population (1)	Area (acres) (2)	Rateable value (3) £	Rateable value per head (4) £
8.	Southwark	171,695	1,132	1,328,467	8·52
	Camberwell	251,294	4,480	1,726,285	7·38
	Total	422,989	5,612	3,054,752	7·22
9.	Bermondsey	111,542	1,503	861,920	8·55
	Deptford	106,891	1,564	668,026	6·77
	Lewisham	219,953	7,015	1,783,680	7·89
	Total	438,386	10,082	3,313,626	7·56
10.	Greenwich	100,924	3,858	972,490	10·16
	Woolwich	146,881	8,282	1,166,849	7·80
	Total	247,805	12,140	2,139,339	8·63

N.B.—See footnotes to tabulation of 12 groups on p. 366.

It would clearly be possible to divide up the county in a much more satisfactory way if the existing boundaries, which are often highly irrational, were ignored and the territory treated as a clean slate. Here I have been concerned only to demonstrate the ease with which ten or twelve new and better areas could be obtained by the simple process of consolidation.

The councils of the metropolitan boroughs should continue to be directly elected, as at present, but the number of councillors should be reduced. The size of the councils at present ranges from 30 to 60, with an addition of from 5 to 10 aldermen.[1] The size of a council is purely haphazard. The 29,063 electors of Finsbury return 56 councillors whereas in Lewisham only 53 are elected by 121,725 electors. In Camberwell, 120,738 electors return 60 councillors, while in Kensington a similar number of councillors is returned by 80,232 electors. The Hackney Borough Council has reduced its numbers from 70 members to 48 councillors with about half a dozen aldermen, but most of the metropolitan boroughs have never given the matter any thought.

[1] For full details, see the Return of Metropolitan Borough Council Elections, 1937, L.C.C., No. 3312/1937.

24

It is obvious that the size of the council should not be too large if it is to do its work efficiently, and that the constituencies should be approximately the same size. I should be inclined to suggest a proportion of 10,000 persons to each elected member, which would give a total varying between 16 and nearly 50 for the dozen authorities proposed above, exclusive of aldermen.

The aldermanic element could be retained. It is, however, desirable that the metropolitan boroughs should be shorn of some of their ceremonial trappings. A mayor and the accompanying paraphernalia of a mace and so forth, is entirely out of place in a metropolitan district. Such devices make for particularism and were introduced with that express intention.

Provision should be made for the inclusion on each local authority of the members of the Greater London Council elected for that area. The object of this arrangement would be to institute machinery for the fullest co-operation between the regional and local authority. It would not necessarily involve the attendance of the regional members at every meeting of the local council. But it would enable them to be present whenever a local matter which concerned the region was under discussion, so as to explain the views or policy of the Greater London Council. Conversely, they would be better informed of the local outlook on matters coming before the Regional authority. There would be no objection to the same person being elected to both Councils. Some of the ablest and energetic members of the London County Council have been and are members of a metropolitan borough council.

(b) *The Outer London Councils*

The out-county authorities should be dealt with on more or less similar lines. The problem with them is, however, complicated by the fact that they are not homogeneous but belong to several categories of constitutional rank.

As I have not stipulated the precise boundary of the Greater London Region but merely stated that it ought to cover an area of approximately the size of the Metropolitan Traffic Commissioner's area, it is not feasible to dispose in detail of the local authorities within the region (apart from the county councils), since they cannot be ascertained until the frontier is finally determined. I shall, therefore, take the Metropolitan Police District (the Greater London of the Registrar-General) as an illustration, and suggest the method of organising local authorities therein. As the metropolitan boroughs have already been dealt with we shall now be concerned only with Outer London—that is, the area lying between the boundaries of the administrative county and the Metropolitan Police District.

There are, in this modest Outer London, 3 county boroughs, 35 municipal boroughs, 30 urban districts, 4 rural districts and 6 parish councils. If the territory of the London Passenger Transport area or the Metropolitan Traffic area be taken there are probably another 40 or 45 county district councils[1] together with a considerable number of parish councils, to be added to the total.

The heterogeneity of constitutional status which now prevails in Outer London is largely the outcome of historic accident. In most cases it no longer corresponds with true differences of character and is therefore unscientific. It should not be permitted to continue.

This is not the only reason for compelling a measure of simplification. County borough status, for example, is based on the concept of a self-sufficient and autonomous unit of local government independent of all control other than that exercised by Parliament and the central departments. It is, therefore, quite incompatible with the idea of regional administration, which postulates in certain important matters the subordination of the parts to the interests of the whole. The county boroughs of East Ham, West Ham and Croydon would accordingly have to submit to a change of status, though it might be possible to confer upon them some special title, such as "metropolitan county borough" or "city borough," as a distinguishing sign of their former constitutional rank. But the powers of these three areas would become substantially the same as those of the other boroughs in the region in their new form. As their respective populations and rateable value are smaller than will be those of most of the enlarged metro-

[1] i.e. municipal boroughs, urban district councils or rural district councils.

politan boroughs, there is no reason why they should feel mortified in sharing a similar status.[1]

The next step is to eliminate the rural district councils. The rural district council and the parish council were introduced in 1894 in the hope of reviving village life and awaking the community spirit in the English countryside. They are quite out of place in a vast conurbation such as the metropolitan region, which is virtually urbanised or sub-urbanised throughout its length and breadth. Even within the 2,000 square miles of London Passenger Transport area there is scarcely a rural district which would not be better off with urban powers. Barnet is rapidly becoming industrialised and several large new factories are in course of erection there. The electrification of the Southern Railway is making such places as Godstone, Dorking and Horley, dormitory areas to an ever-increasing extent. To the west, the rural district of Eton will obviously not survive the urbanising influence of its neighbour, Slough, which has lately become a borough.

In any case there is no point, whatever, in these days of rapid transportation, in preserving the dichotomy between urban and rural administration—at all events in the metropolis. Technical advances are removing the age-long distinction between town life and country life, between the simple and the sophisticated ways of life. Those who dwell in the rural environs of a great city enjoy such amenities as the telephone, telegraph, and broadcasting; regular motor-bus services and frequent deliveries from the shops; excellent postal services and a morning newspaper; gas, electricity and a piped water supply; quite often main drainage and a local cinema.[2] And a car or a motor-cycle is usually to be found in the garage.

The distinction between urban and rural organs of local government has, therefore, become meaningless in the shadow of a great town. There is no point in separating the municipal borough of St. Albans from the rural district of St. Albans;

[1] The population of East Ham is 142,394, West Ham 294,278 and Croydon 233,108. Their areas contain 3,324, 4,689 and 12,672 acres respectively. The rateable value of East Ham is £753,190, of West Ham £1,509,025 and of Croydon £2,284,119.

[2] *Cf.* two articles on "Territorial Planning," by W. A. Robson, *New Statesman and Nation*, 28th December, 1935, and 4th January, 1936.

nor in dividing Dorking Urban District from the Dorking and Horley Rural District, nor in separating the rural district of Dartford from the borough of the same name. Paradoxical though it may sound, the preservation of rural amenities would possibly be more jealously safeguarded by urban authorities whose members are beginning to appreciate these things, than by groups of farmers whose eyes are dazzled by the prospect of building sites soaring up in value—witness the example of the London County Council in sponsoring the Green Belt Scheme.

Middlesex County Council has already eliminated all the rural districts in its county; and common sense points to the conclusion that the remaining authorities of distinctively rural type in the metropolis must be abolished. They are much too dependent on their respective county councils to be suitable for a regime in which the latter will disappear and be replaced by a regional organ responsible only for regional services. A stronger and sturdier authority is required in place of the rural district council.

The parish council is of such minor importance in the scheme of things in the metropolis that its future existence is almost a matter of indifference. In the ordinary course of events the parish council would disappear with the rural district council. I should be inclined to suggest, however, that if the members of any of the less developed village communities still remaining in the region desire to have a parish council, they should be permitted to set one up with the permission of the Greater London Council. But the present principle of establishing a parish council wherever the number of parishioners exceeds 300 should be revised.

We are left now with urban districts and municipal boroughs, which constitute the great bulk of the local authorities in Outer London. A large measure of simplification has already been attained. It remains to complete the final stage in the process by abolishing the distinction between these two kinds of county district. The present differences between them are mainly those of prestige and dignity. Thus, the borough has a mayor and aldermen, while the urban district has to be content with a chairman. The borough has a town clerk and a town hall; the urban district only a clerk to the council and council offices. In the past, a municipal corporation was automatically entitled to provide elementary education if it had a population in excess of 10,000 at the census of 1901; while an urban district was required to have a population of 20,000 to acquire this power. But since 1931, this no longer applies unless special provision is made by statute for new education powers to be conferred.[1] The only other powers which a municipal borough possesses over and above those of an urban district are the provision of bridges (other than over main roads), the maintenance and lighting of highways, the prevention of animal disease, and a few regulatory functions relating to advertisements, shops, explosives, food and drugs and weights and measures.

It is not suggested that urban districts shall become boroughs, nor municipal corporations revert to urban district status. What is proposed is that out of the existing boroughs and urban districts a single new type of local authority shall be created throughout the metropolis. The powers, responsibilities and duties of this emerging authority will be uniform throughout the region; and so, too, will be its relations with the Greater London Council. The physical extent of the territory over which it will have jurisdiction will necessarily vary according to its geographical situation, and also its social and economic character. Some of the local areas will be purely residential, others mainly industrial; some densely populated, others more sparsely peopled; some completely built-up and congested, others containing large tracts of open country; some rich and prosperous, others poor and struggling; some rising in importance, others declining.

The Outer London circle with which I am now dealing contained in 1931, 3,806,939 persons; and was estimated in 1935 to have a population of 4,434,600. It covers an area of approximately 574 square miles. Roughly, therefore, it has a population 7 per cent larger than that within the area of the London County Council (estimated in 1935 to be 4,141,000), spread out in a territory nearly five times as great. Its rateable value (in 1935-6) stood at £37 millions, or rather more than half that of the administrative county.[2]

---

[1] Education (Local Authorities) Act, 1931.   [2] See *post* p. 454.

This area could without much difficulty be divided so as to make about 17 suitable and substantial areas of local administration. Here again one can demonstrate the possibilities by merely amalgamating existing units. The borough of Beddington & Wallington, and Coulsdon & Purley Urban District, should obviously go with Croydon; Carshalton Urban District with the municipal boroughs of Sutton & Cheam, Epsom & Ewell, and Banstead Urban District. The boroughs of Wimbledon, Mitcham, Malden & Coombe, and the urban district of Merton & Morden would constitute a closely knit homogeneous area. Richmond, Barnes, Kingston-on-Thames, Esher and Surbiton would make a convenient unit on the south bank of the Thames; while on the north bank Brentford & Chiswick, Heston & Isleworth, Twickenham and perhaps also Sunbury-on-Thames, Feltham and Staines, would blend easily into a single riverside authority. Acton, Willesden and Wembley form a group obviously inviting amalgamation. So do Harrow, Ruislip-Northwood, Uxbridge and Bushey Urban District; and, again, Southall, Hayes and Harlington, Yiewsley & West Drayton, to which must be added Ealing in order to provide a substantial authority.

In the north, Hendon, Finchley, Friern Barnet, East Barnet, Barnet Urban District, Barnet Rural District, Potters Bar and part of Watford Rural District are a definite conglomeration calling for administrative unification. The same is true of Wood Green, Hornsey and Tottenham. Cheshunt and Enfield must join hands with Southgate and Edmonton, and also with part at least of Hatfield.

Walthamstow, which a few years ago applied in vain to extend its boundaries to comprise Chingford,[1] should be joined not only with that area but also with Leyton, and Wanstead & Woodford. Waltham Holy Cross and Chigwell would be merged, and East Ham with Barking and Dagenham. Ilford could go with Waltham and Chigwell, although it might combine better with Romford and Hornchurch lying beyond the boundary.

[1] Proceedings of an enquiry into the application of the Walthamstow Corporation for an extension of the Borough under the Local Government Act, 1929, on 7th November, 1932. Published by the Walthamstow Press.

To the south-east, Erith, Bexley and Crayford would amalgamate conveniently with Woolwich and Greenwich[1] (at present inside the administrative county), while Chislehurst & Sidcup, Bromley, Beckenham and Penge make another conglomerate in which Orpington should be included.

These proposals (with one or two exceptions) do not cover areas which fall partly inside and partly outside the Metropolitan Police District, as these would obviously have to be considered in relation to other outlying authorities.

I would again emphasise that the proposed areas are in many cases much less satisfactory than the demarcation which could be made on a *tabula rasa*. The existing units have been left intact in order that the community spirit, where it exists, shall not be destroyed but merely transferred to a larger and more appropriate entity. Thus, cement has been used rather than a knife. If any local area prefers death to what it may regard as dishonour, annihilation could easily be arranged and would no doubt be desirable for all concerned. But my aim wherever possible has been to build and not destroy.

In only one instance is there an amalgamation proposed between an out-county cluster of local authorities and one of the enlarged metropolitan boroughs (i.e. the Woolwich-Erith group, No. 16). But there are doubtless other instances where it would be advantageous. Moreover, considerable improvements could be made by agglomerating some of the local authorities outside Greater London with those inside. For example, the urban districts of Leatherhead and Dorking might be added to Group 2, and Rickmansworth, Chorley Wood and the borough of Watford be included in Group 7.

Table C on pages 377–9 shows the population, area, rateable value and rateable value per head of the proposed new areas.

It is evident that there is considerable disparity of size and resources between some of the proposed new units. This, however, is unavoidable, since it would be quite impracticable to attempt a mathematical equality in respect of any of the factors and impossible to attain it except in regard to any one of them. It is far more realistic to consolidate "areas of mutual

[1] *Ante,* p. 369.

TABLE C

	Population	Area (acres)	Rateable value	Rateable value per head
	(1)	(2)	(3)	(4)
			£	£
1. Beddington & Wallington M.B.	26,328	3,045	348,465	
Coulsdon & Purley M.B.	37,909	9,722	620,877	
Croydon C.B.	233,108	12,672	2,284,119	
Total	297,345	25,439	3,253,461	10·94
2. Carshalton U.D.	28,586	3,346	417,203	
Sutton & Cheam M.B. ..	48,363	4,338	806,689	
Epsom & Ewell M.B. ..	35,228	8,123	513,458	
Banstead U.D. (part) ..	13,089	6,546	218,023	
Total	125,266	22,353	1,955,373	15·61
3. Malden & Coombe M.B.	23,350	3,164	370,494	
Merton & Morden U.D.	41,227	3,237	479,389	
Mitcham M.B.	56,872	2,932	445,815	
Wimbledon M.B.	89,515	3,212	770,463	
Total	210,964	12,545	2,066,161	9·79
4. Richmond M.B.	39,276	4,109	564,930	
Barnes M.B.	42,440	2,519	400,685	
Kingston M.B.	39,825	1,408	473,704	
Surbiton M.B.	30,178	4,709	472,495	
Esher U.D. (part) ..	22,197	5,383	348,918	
Total	173,916	18,128	2,260,732	13·00
5. Brentford & Chiswick M.B.	63,217	2,333	680,331	
Heston & Isleworth M.B.	76,254	7,219	763,287	
Twickenham M.B. ..	79,299	7,013	459,640	
Sunbury-on-Thames U.D.	13,451	5,608	160,983	
Feltham U.D.	16,064	4,925	192,877	
Staines U.D.	21,336	8,273	275,995	
Total	269,621	35,371	2,533,113	9·40
6. Acton	70,008	2,318	774,716	
Willesden	185,025	4,635	1,562,522	
Wembley	65,799	6,290	1,111,913	
Total	320,832	13,243	3,449,151	10·75

TABLE C (*continued*)

	Population (1)	Area (acres) (2)	Rateable value (3)	Rateable value per head (4)
			£	£
7. Harrow U.D.	96,656	12,559	1,654,372	
Ruislip-Northwood U.D.	16,035	6,583	317,127	
Uxbridge M.B.	31,887	10,240	294,734	
Bushey U.D.	11,635	3,866	111,805	
Total	156,213	33,248	2,378,038	15·22
8. Ealing M.B.	116,771	8,783	1,499,363	
Southall M.B.	38,839	2,606	370,838	
Hayes & Harlington U.D.	22,969	5,160	223,945	
Yiewsley & W. Drayton U.D.	13,066	5,277	87,001	
Total	191,645	21,826	2,181,147	11·38
9. Hendon M.B.	115,640	10,373	1,799,744	
E. Barnet U.D.	18,549	2,644	276,459	
Finchley B.C.	59,113	3,475	806,129	
Friern Barnet U.D. ..	22,715	1,340	241,625	
Barnet U.D.	15,064	4,290	202,659	
Barnet R.D.	5,946	8,339	76,936	
Potters Bar U.D.	5,720	6,129	111,578	
Total	242,747	36,590	3,515,130	14·48
10. Wood Green M.B. ..	54,308	1,607	516,228	
Hornsey M.B.	95,416	2,872	1,059,572	
Tottenham M.B.	157,667	3,013	960,995	
Total	307,391	7,492	2,536,795	8·25
11. Southgate M.B.	56,063	3,763	789,167	
Edmonton M.B.	77,658	3,896	624,723	
Cheshunt U.D.	14,656	8,479	95,568	
Enfield U.D.	67,752	12,401	666,612	
Total	216,129	28,539	2,176,070	10·07
12. Walthamstow M.B. ..	132,972	4,342	825,068	
Chingford M.B.	22,076	2,868	254,503	
Wanstead&Woodford M.B.	43,129	3,842	498,987	
Leyton M.B.	128,313	2,594	772,390	
Total	326,490	13,646	2,350,948	7·20

TABLE C (*continued*)

	Population (1)	Area (acres) (2)	Rateable value (3) £	Rateable value per head (4) £
13. West Ham C.B.	294,278	4,689	1,509,025	5·13
14. Waltham Holy Cross U.D.	7,092	10,958	45,807	
Chigwell U.D.	16,338	8,971	175,284	
Ilford	131,061	8,425	1,344,053	
Total	154,491	28,354	1,565,144	10·13
15. East Ham C.B.	142,394	3,324	753,190	
Barking M.B.	51,270	3,877	654,458	
Dagenham M.B.	89,362	6,554	505,155	
Total	283,026	13,755	1,912,803	6·76
16. Erith M.B.	32,779	3,860	256,829	
Bexley M.B	32,626	4,861	466,934	
Crayford U.D. (part) ..	16,229	2,523	132,420	
Woolwich M.B.C... ..	146,881	8,282	1,141,542	
Greenwich M.B.C. ..	100,924	3,858	981,393	
Total	329,439	23,384	2,979,118	9·04
17. Chislehurst & Sidcup U.D.	27,182	8,967	431,505	
Bromley M.B.	47,698	6,513	650,977	
Beckenham M.B. ..	50,429	5,937	761,541	
Penge U.D.	27,771	770	201,411	
Orpington U.D. (part) ..	18,271	9,838	270,757	
Total	171,351	32,025	2,316,191	13·52

attraction": that is, those in which a process of assimilation is going on in regard to social or economic affairs.

The merging of existing areas into larger and more efficient units of local government would not necessarily involve the complete disappearance of the component parts. Where a community has a strong local consciousness it would be quite feasible to set up a district committee consisting of the members of the new local authority returned by that part of the area, together with a proportion of co-opted local residents; and to devolve upon this territorial committee a number of functions of special concern to the local community. In this way it would be possible to preserve local interest in the smaller area without detriment to the efficiency of local government in the larger unit. But in no circumstances should a separately elected council with autonomous powers be permitted to exist in any of the areas proposed to be amalgamated, for this would mean a triple-deck system and a recrudescence of the particularism which we are striving to avoid at all costs. The future of London government must lie in the direction of integration if any progress whatever is to be made.

The methods used for reforming the structure of authorities in the Metropolitan Police District would be equally applicable for revising the organisation of the outermost ring of local authorities lying between the boundary of the Police District and the limits of the Greater London Region, whether we take for the latter the Metropolitan Traffic area, the London Passenger Transport area, or a new frontier to be determined by a boundary commission. The principles have already been explained and illustrated in sufficient detail to render it unnecessary to pursue the matter further. Simplification, uniformity of status, amalgamation and enlargement of areas: these are the watchwords by which the adjustment of local authorities to modern needs must be guided in the metropolis.

The proposals so far advanced would give a total of 29 local authorities in place of 111 which now exist. Of these, 12 would lie within the present area of the L.C.C. and 17 in Outer London delineated by the Metropolitan Police District boundary. If the Traffic or Transport areas were taken, or some other territory of similar size, another dozen or so councils would be required, bringing the total number of local authorities in the whole region to about 40. This I believe to be approximately the correct number.

The one area for which no provision has so far been made is the square mile of the City. To this we may now turn our attention.

(c) *The City Corporation*

The ancient City presents a spectacle today which is both tragic and comic. There is something heart-rending in the thought of all the lost opportunities which resulted from the refusal of the City Corporation to share its inheritance with the wider community of the metropolis, and the memory of the pusillanimous failure of successive Parliaments and Ministries to insist on its participation in the full task of London government. There is something ludicrous in the sight of the square mile of the City, with its vast wealth and long traditions, entirely divorced from the rest of London, a heart cut off from the living body of the metropolis, like the heart removed from a dissected animal beating in a scientist's laboratory.

The City's failure to extend the area under its jurisdiction with the progressive expansion of the metropolis has made it today "a small island of obstinate medieval structure in the midst of a sea of modern local authorities."[1] London as a city within its ancient boundaries and retaining its privileges is a mere survival; and, as Sir Laurence Gomme pointed out, a survival not from the best periods of its history, from the ideal of Roman Lundinium, from the efforts of Anglo-Saxon London, from the Charter of William the Conqueror, but from the "dishonoured charters restored by the graceless necessities of James II. . . . It clings to old customs and old ceremonial as the medieval city did, in order to keep out new ideas and conceptions which it holds to be inimical to its interests."[2] It sins, Gomme truly declared, against all principles of local government. It disregards the doctrine of general utility upon which all government must be founded, and the doctrine of the greatest good of the greatest number upon which alone government by power is justified.[3]

In terms of administrative efficiency, the objections which can be urged against the City Corporation are not of great weight. The main functions which could with advantage be transferred to a larger authority are the provision and maintenance of cross-river bridges and the management of wholesale food markets. For generally speaking, the City Corporation performs only strictly local functions of a minor character;

[1] W. E. Hart and W. O. Hart: *An Introduction to the Law of Local Government and Administration*, p. 224.
[2] Sir George Laurence Gomme: *The Governance of London*, pp. 393–4.
[3] *loc cit.*

and these it carries out with considerable efficiency. The medical officer's department of the City has been a distinguished one since the days of Simon; and it discharges its unusual responsibility as port sanitary authority in a satisfactory manner. The City police force, the only one permitted to be maintained by a municipal body in the metropolis, is an exceptionally fine body of men and compares favourably with the Metropolitan Police force in regard to physique, demeanour and efficiency in controlling traffic.

The objections to the City Corporation today rest on two other grounds. One of these is financial; the other relates to prestige and pageantry.

There are two distinct but related aspects of local government, corresponding to what Bagehot called the efficient and dignified parts of the British Constitution.[1] Man does not live by bread alone, nor a local community only by its drainage system and sewage farms. The ceremony, the pageantry, and the traditions of civic institutions have a value which cannot be measured by any calculus known to the social scientist. It is no small thing that the Lord Mayoralty of London, with its ancient privileges and fine Mansion House, have been monopolised by a tiny unrepresentative fraction of London's teeming millions, divorced from the living realities of contemporary life. It is not a negligible loss that the age-long traditions of freedom and civic pride which attach to the City Corporation and are historically associated with its magnificent Guildhall, are frittered away in the tomfoolery of the Lord Mayor's procession and the dull conventionalities of the Lord Mayor's annual banquet, instead of forming the inspiring apex of a splendid municipal pyramid, broad-based on the whole of mighty London. Even the name of citizen is denied to those who are not freemen of the City; the denizens of the administrative county are in law mere "inhabitants." The long line of Royal Charters, extending over nearly a thousand years, the splendid plate, the historic primacy of the City in its relations with the sovereign, its close ties with the higher judiciary— all this is reserved for the enjoyment and edification of the

---

[1] W. Bagehot: *The British Constitution*, Chapter I, p. 8 (World's Classics edition).

thirty or forty thousand persons who live in the City, or occupy business premises there, or are freemen of the City. The right to stand for election as a Common Councilman is confined to a still narrower basis.[1]

Why the City of London should have been permitted to maintain an archaic attitude while every other ancient corporation was compelled to adjust itself to the facts of growth and change, has not and cannot be explained on rational grounds. The answer can only be given in terms of political power and influence.

The second objection is on the ground of finance. Here no doubt we find the key to the question of why the City fathers were so anxious to preserve intact the powers and privileges of the Corporation against the many assaults that were made against them during the 19th century. The property of the Corporation is a huge vested estate.

The rateable value of the City of London in April 1937 stood at £8,182,091,[2] compared with a total of £52,549,372 for the rest of the administrative county. With the exception of Westminster, this is a far larger figure than any other local government area in the metropolis; and the rateable value per head is enormous: more than £836—a figure so large, it would seem, that it is often not stated in official publications.[3] The produce of a penny rate yields nearly £34,000.

Thus, although the City Corporation raises about £1,000,000 out of the rates towards its own expenditure of about £2½ millions, and contributes a sum approaching £3 millions a year from the same source to the revenues of the London County Council,[4] it yet remains almost the lowest rated area in the administrative county. For 1936–7, it struck a total rate of 10s. in the £, which was lower than any area except Westminster (9s. 10d.) and St. Marylebone (9s. 11d.).

[1] A Common Councilman must be (i) a freeman of the City and a £10 occupier registered in the Parliamentary register at the time of the election, or (ii) a freeman householder paying scot and bearing lot.

[2] *London Statistics*, vol. 40 (1935–7), p. 449.   [3] *Ibid.*, p. 426.

[4] In 1935–6 the total expenditure of the City Corporation was £2,436,671, of which £979,808 fell on the rates. This is approximately one-seventh of the total sum raised by the L.C.C. by precept on the City and the boroughs. The precept of the L.C.C. on the City for the year ending March 1937 was £2,872,476.

The swollen rateable value of the square mile is not, however, the chief source of the City Corporation's great wealth. Its financial power is derived mainly from a vast estate, the amount of which is not disclosed, but whose value undoubtedly runs into many millions of pounds. Its magnitude can be gleaned from the fact that the balance sheet covering the total operations of the Corporation deals with £18·6 millions (for the year ending 31st March, 1938) while the balance sheet for the City's Cash, which has nothing to do with rates, shows a total of £6,514,585. The "City's Cash" is derived from the ancient estates of the Corporation acquired by gift, grant, purchase or bequest. They consist of land and premises situated not only in London but in Surrey, Essex, Oxfordshire and Middlesex. There is also the Market Fund, which is the net surplus made by the Corporation out of its market undertakings; this produced more than £25,000 in 1937–8. The City Estate produced a net income, after deducting costs of management, of £209,805.[1] The whole income at the disposal of the Corporation from the City's Cash and other "private" sources was estimated to exceed £850,000 in 1937–8.

This large income and the property from which it is derived, lies in the absolute disposal of the Corporation to deal with as it pleases. There is no inspection, audit or control of any kind by the Ministry of Health, the Treasury or any other central department. An example of the complete freedom enjoyed by the Corporation over what is called "the City's private purse" is the fact that for 20 years no allowance was made in the accounts for interest accruing from large balances in the hands of the Chamberlain. As a result, a sum of no less than £42,000 was permitted to accumulate without any entry being made in the relevant accounts until it was suddenly decided to transfer the sum to the City's Cash for ordinary current expenditure.

The detailed accounts show that the City's standard of expenditure is on a scale of lavishness, if not extravagance, which is unparalleled in any other town in the country, or, for that matter, in the world. In the year ended March 1937 the *annual* redecorations for the new Mayoralty cost £1,228;

[1] See the accounts of the Corporation, vol. 305.

the robes for the Lord Mayor—another annual sum, presumably —£171; and the erection of a pavilion on Lord Mayor's Day, £232. The Swordbearer to the Lord Mayor receives a salary of £746 plus a uniform allowance of £40;[1] the Common Cryer and Serjeant-at-Arms is paid £547 with a similar addition for his official costume; while the Ale Conners receive £40. The utterly absurd custom which is still continued of presenting bundles of cloth to Ministers of State absorbs £106.

The City Corporation is famed for its hospitality, and this is paid for out of the City's Cash. A reception and luncheon in honour of some event costing between £1,500 and £2,000 is quite usual in the Guildhall, and £9,047 was allocated to receptions in 1937–8. The recent Coronation was the occasion for spending more than £20,000.

The Corporation is accustomed to set aside substantial sums each year from the City's Cash for philanthropic and charitable purposes. In the year ending March 1937, for instance, £12,381 was spent in "Sundry Donations" on a long list of objects or institutions, including the Royal Empire Society, the extension of the Bodleian Library, the Salvation Army, the School of Oriental Studies, the People's Palace and the History of Parliament. A further sum of £1,500 went to "Charitable Donations" such as the Children's Country Holidays Fund, and the North Islington Infant Welfare Centre; while £1,466 was absorbed by "Donations for Public Purposes." This does not exhaust the story of the Corporation's generosity, for £10,000 was given in the year in question (the fourth of ten instalments) for building the new head-quarters of London University. Thus, more than £25,000 was given to deserving objects of a public nature. The chairman of the Coal and Corn and Finance Committee, in presenting the Estimates for the year 1937–8, remarked that "City's Cash is a fund generously dispensing benefits to the citizens and ratepayers, and towards those benefits the ratepayer has not been asked to contribute a single farthing. It is a fund which has been, and is being, administered for broad and generous public purposes."[2]

[1] The headmistress of the City of London School for Girls, maintained by the Corporation, receives a salary of £684 8s. 6d.
[2] *The City Press*, 25th June, 1937, p. 9.

25

It is not necessary to dissent from this statement, nor to contend that the Corporation is not benevolent and generous in its donations, in order to disagree with the system whereby these large funds are administered. The tiny electorate, the narrow basis of qualification for election, the refusal to accept an enlargement of boundaries, means in effect that the huge City Estate is in the unfettered hands of a small oligarchy. The property which was acquired or bequeathed for the benefit of all the citizens of London has passed into the disposal of a handful of their successors. There is much to praise and much to criticise in the disposal of the City's Cash; but neither praise nor criticism is really to the point. The issue is essentially similar to the one which faced the Royal Commission on Municipal Corporations in 1835,[1] namely, whether the corporators are at liberty "to do as they please with their own," or whether the public has an over-riding interest which no prior right can defeat.

If anyone should think that this insistence on the need for sharing with the whole metropolis that which was intended for the benefit of all London in former times, is a mere tilting at windmills, let him leave the prosperous commercial and financial quarters of the square mile and travel eastwards or south-east to the unrelieved desolation of Stepney, the slum-laden wastes of Poplar, the dreary squalor of Bermondsey, with their dense populations and low rateable values, their poor amenities and lack of open spaces, and ask whether the powers and property of the City could not be better used if they were attached to a wider sphere of municipal duty.

What is to be done about the City? There are several possibilities. One is to democratise the constitution by conferring the right to vote at Common Council elections, and to stand for election as a Common Councilman, on every man or woman who has worked continuously for a year in the City. This would enlarge the electorate by at least half a million persons.[2] In the second place, the objectionable system of

---

[1] First Report of Royal Commission on Municipal Corporations, 1835, *passim*.

[2] The day population of the City, according to the Census of 1921, was 436,721.

electing aldermen for life by separate wards would be abolished in favour of a 6-year term of office and appointment by the Court of Common Council. Third, the functions of Common Hall in relation to the City Corporation should be transferred to the Common Council. Fourth, the annual income from all sources which contribute to the Corporation's so-called private purse, should be deemed to be rateable value for the purposes of levying the Greater London rate. The effect of this would be to require the City to make a substantial contribution to the funds of the larger body from the City estate, while leaving a considerable portion of the income to be retained by the Corporation.

So far as the ceremonial side of the question is concerned, there is probably nothing worth doing at this late date. New attitudes and new traditions have grown up, and at this stage it would probably be a mistake to seek to associate the ancient customs and historical emblems of the old City with the modern institutions we are contemplating for the metropolis of to-day. The time for that has passed; and it would be an anachronism to constitute the decayed City as the nucleus of a new Greater London authority, even if it would consent. The attempt would in any case almost certainly fail, for there is no reason to think that the attitude of the City has changed in the slightest degree from the unwavering opposition to any kind of reform which characterised its outlook throughout the 19th century. There is, therefore, no future for the City Corporation in the Greater London we are contemplating, only its past.

The City Corporation cannot, however, any longer be permitted to have a monopoly of the insignia and the nomenclature of civic life. The denizens of Greater London must be designated citizens and not mere inhabitants. The Greater London region must have its Lord Mayor; and the Lord Mayor of Greater London must have the precedence to which he is entitled by the substance of present things rather than by the shadow of the past.

CHAPTER VIII

A NEW RELATIONSHIP

The mutual relations between the major and minor authorities in the London Region should be conceived in terms differing fundamentally from anything which now exists either in the metropolis or elsewhere. Over a wide range of functions the local authorities must be subject to over-all control by the Greater London Council. Their scope will be enlarged and their independence reduced.

For the convenience of the reader, the functions to be administered by the local authorities under the general supervision of the regional organ is again stated. The list is broadly as follows, though no attempt has been made to make it exhaustive. It comprises: public libraries, the protection of children, the Small Dwellings Acquisition Act, the ambulance service, the operation of ferries diseases of animals, inspection of nursing homes, surveys for overcrowding, the provision of small holdings, licensing and inspection of petroleum, enforcement of the Shops Acts, protection of wild birds, registration of births, deaths and marriages, registration of electors, adulteration of food and drugs, registration of vendors of poison, the humane slaughter of animals, provision of mortuaries, cemeteries and crematoria, parking places, the construction, maintenance and improvement of local streets, the cleansing and lighting of highways, the provision of baths and washhouses, inspection of dairies, cowsheds and milkshops, the regulation of ice cream vendors, provision of disinfecting stations, notification and prevention of infectious disease (apart from the provision of hospital accommodation), the authorisation of offensive trades, the inspection of weights and measures, removal (but not disposal) of refuse, the licensing and inspection of slaughterhouses, massage establishments and common lodging-houses, the treatment of tuberculosis in dispensaries (but not in hospitals or sanitoria), vaccination, the clearance of small unhealthy areas, the regulation of buildings, and a large number of functions relating to public health and the amenities of the district.

The powers of the Greater London Council would be of various kinds, which we may consider in turn.

First, the power to lay down norms or minimum standards in regard to a particular service, to be observed either throughout the region or in a specific part or parts of it. Thus, the Greater London Council could prescribe minimum standards of street cleaning and refuse collection to be observed throughout the metropolis.

Second, the power to require uniform or consistent technical methods or equipment. For example, it is desirable that the methods (as well as the standards) of street lighting[1], and the equipment used for collecting house refuse, should be uniform in character throughout the region. But subject to this overriding control, there is no reason why refuse collection or street lighting should not remain a local function.

Third, the planning of certain services should be in the control of the Greater London Council. This applies particularly to the siting of such institutions as schools, baths and washhouses, municipal cemeteries and crematoria, disinfecting stations and so forth.

Fourth, the Greater London Council should have the right to require a local authority to provide a particular service, such as a ferry, or a parking place; and to act in default if it fails to comply and to recover the cost in the usual way.

Fifth, schemes relating to services exercised by local authorities concurrently with the major organ should be submitted to the latter for approval or revision. This would apply to such matters as the construction and improvement of local streets, the notification and prevention of infectious disease, the clearance of small unhealthy areas, and any others which interlock with services to be directly administered by the regional authority.

Sixth, the Greater London Council should have power to lay down the general policy to be observed throughout the region in regard to functions falling within this category. It is essential that public assistance (if it were transferred) should

---

[1] At any rate as regards roads of the same class.

be administered on the basis of a single policy applicable to the entire metropolis. The same thing applies to the registration of births, deaths and marriages, the inspection of nursing homes, and many other services.

Seventh, the regional organ should possess the right to delegate any of its functions, or any activity relating thereto, to all or any of the local authorities; and the latter should thereupon have the duty to carry out the task devolved upon it in the required manner. Once it is admitted that the principal body has over-riding power, it will be possible to introduce a considerable amount of delegation within the region, even in regard to those functions for which the Regional Council is to be directly responsible.

A number of objections were urged against the principle of delegation before the Ullswater Commission on London Government. The Kent County Council's witness, for instance, criticised the idea on the grounds that the minor authorities to which duties were delegated would have no sense of responsibility; that their members would not be directly accountable to the ratepayers whose money they were spending; that in consequence the larger council would have to supervise the expenditure of the local authorities very carefully; that as a result the latter's work would become largely a matter of routine and detail which would not attract a suitable type of candidate for office.[1]

These objections consist mostly of mere debating points. They are unsupported by any body of facts and are in any case of dubious validity. There is nothing inherently detrimental to the sense of responsibility in the principle of delegation. The delegate is answerable to the delegating authority and can be deprived of powers if the work is not performed in a satisfactory manner. A particular task is not made more of a routine nature nor less interesting merely because it is delegated by a larger regional authority instead of being conferred directly by the law. If the present body of local councillors in the metropolitan area are only willing to serve the community on condition that the present "anarchy of autonomy" is preserved,

[1] Royal Commission on London Government, 1923: Minutes of Evidence (Cornwallis), p. 340.

it is time they were replaced by others more willing to adapt themselves to the needs of the larger community; but I do not believe that to be the case. Finally there is the argument that local expenditure would have to be scrutinised by the larger organ; but as this is in any case highly desirable on general grounds, we should welcome any step in that direction.

From every point of view the principle of contingent financial control by the major authority over the local authorities is indispensable to the proper ordering of the relationship between the two. If the local bodies possess financial autonomy, they will snap their fingers at the Greater London Council. Conversely, without potential control over expenditure, the latter will in the last resort be impotent. Furthermore, the whole basis of municipal finance in London is wrong at present.

As things now are, it is impossible to equalise the provision of quite elementary and essential local services, such as refuse collection, street lighting and cleaning, public libraries, baths and washhouses, for the simple reason that the poorer districts, with a rateable value of between £5 and £8 per head, are unable to afford the minimum standards which ought to prevail throughout the metropolis. It would, therefore, be quite impracticable to give the Greater London Council power to lay down common standards and norms applicable in every part of the area without at the same time making provision for financial aid to be given to Poplar and Deptford, Bethnal Green and East Ham, and similar areas of low taxable capacity.

The several objects in view could best be served by requiring local authorities to submit their budgets to the Greater London Council each year. The rate-borne expenditure in respect of at least some of the functions enumerated above (and others falling into the same category) would, when approved by the Council, be met by the proceeds of a rate levied over the whole metropolitan area. This arrangement would not only secure the equalisation of rates appropriated to a wide range of essential services, but it would also introduce in effect a system of grants-in-aid from the regional organ to the local authorities. For the allocation of the proceeds of the special metropolitan rate would be dependent upon the Greater London Council being satisfied, after inspection, of the efficiency with which the relative services were being conducted.

The Greater London Council would be authorised, subject to a right of appeal to the Minister of Health, to disallow any item in the budget of a local authority which, having regard to economy and good administration, it considered should not be charged on a regional fund.[1] In such a case the local council would be entirely free, if it so desired, to raise the money by means of a local rate. Conversely, the Greater London Council would be authorised to insert in a budget any item for which it considered provision ought to be made.

The Greater London Council would thus have a source of revenue for its own needs and sufficient budgetary control over the local authorities to attain effective over-all supervision.[2] Those who are shocked by this proposal may be reminded that a precedent exists in the arrangements made for the repair of main roads after 1888. The newly established county councils took over the main highways but urban authorities, if they so desired, were permitted to retain the right of repair.[3] Where this was claimed the county council was required to make an annual payment towards the cost of maintenance, repair and improvement. The normal procedure under these provisions is that the district council submit estimates each year to the county council for the work proposed to be done. The county council then approves or amends the expenditure and pays the sum agreed upon to the district council. In this way a form of budgetary control in respect of this particular function was introduced. The county council also had a statutory duty to see that the work was properly carried out.

A case still more clearly in point is the system in force in London in regard to the treatment of tuberculosis. The London County Council is the sole authority for the residential treatment of phthisis in hospitals and sanitoria of various kinds. The metropolitan borough councils, on the other hand, are responsible for the dispensary treatment of tuberculosis, which

---

[1] Cf. *London To-day and To-morrow* (London Reform Union pamphlet), 1908, p. 7.

[2] For a discussion of the essential elements in over-all control over administration, cf. the Report of the President's Committee on Administrative Management. Washington D.C., U.S.A. (1937).

[3] Local Government Act, 1888, Section II.

must be carried out in accordance with the provisions of the scheme drawn up by the London County Council, and is effected either through municipal clinics or special departments of voluntary hospitals. Each dispensary is linked up with a hospital to which special cases are referred for consultation or special out-patient treatment. The London County Council makes a grant towards the expenses of the metropolitan borough councils calculated as nearly as possible to represent 25 per cent of their estimated net average expenditure (over three-year periods) approved by the London County Council, less certain deductions.[1] Here again there is definite budgetary control with respect to this particular service; and also, of course, overhead planning and control.

It has already been pointed out that local authorities would be free to expend whatever money they desired from the rates on residual services, without let or hindrance from the regional council. Thus, the differences in wealth between various parts of the metropolis would continue to exist; but their effects would be mitigated by the financing of a large part of the local authorities' expenditure out of a common metropolitan rate.

It follows from this that the business of assessment and valuation for rating purposes could not be left almost entirely in local hands as it is at present. In the administrative county there are now 29 assessment committees appointed by the metropolitan borough councils and the City Corporation respectively, the size of which varies between 10 and 21 members. The London County Council has only one representative on these committees. In the region of the future the assessment committee for each area ought to consist of an equal number of representatives of the Greater London Council and of the local authority, with a sprinkling of co-opted persons drawn from outside. There should in addition be a regional valuation committee, analogous to the county valuation committees which exist everywhere except in London, charged with the task of promoting consistency in valuation throughout the region. This would consist of one representative nominated by each assessment committee together with whatever number of regional members the Greater London Council should think fit.

[1] *Ante* pp. 219–20. *London Statistics* (1935–7), vol. 40, p. 88.

In regard to capital expenditure, the system of loan control laid down by the Metropolis Management Act, 1855, should be retained and extended to cover the entire field. The Greater London Council would become the sanctioning authority with a right of appeal to the Minister of Health in case of refusal, failure to approve within six months, or the attachment of conditions to the sanction.[1] The alternative system of direct application to the central government should be abolished.[2] The Greater London Council should be able to follow the practice of the London County Council of advancing money on capital account to the local authorities.

So far as the borrowing powers of the Greater London Council itself are concerned, the magnitude of the operations involved would make a yearly review desirable. But the matter ought to be considered by the central government in the first instance rather than by Parliament and the requirements of all other public authorities in London should be assembled simultaneously. For this purpose, the Chancellor of the Exchequer should set up a standing Treasury Committee presided over by himself, and containing representatives of other interested departments such as the Ministry of Health, the Board of Education and the Ministry of Transport. This committee would receive the proposals in regard to capital expenditure and the raising of loans during the next 12 months of the Greater London Council (and the local authorities within the region), the London Passenger Transport Board, the Commissioner of Police for the metropolis, the Port of London Authority, and any other *ad hoc* bodies which may exist. The Committee would then make the appropriate recommendation to Parliament, which could be embodied in the annual Money Bill. In this way the "comprehensive annual survey"[3] of large expenditure in London which has been sought in vain for more than sixty years might be achieved.

We are now in a position to enquire into the relations which should subsist between the proposed organs of municipal administration in London and the central departments concerned with local government functions.

Here again it is evident that innovation will be needed. For

---

[1] *Ante*, p. 265.   [2] *Ante*, p. 266.   [3] *Ante*, pp. 263-4.

if the Greater London Council is conceived on the right lines, it will itself be a kind of "central authority" in the metropolis in relation to the local authorities within its territory, at any rate over a considerable range of services. It would therefore be highly wasteful and inefficient to duplicate the processes of inspection, direction and supervision which have been outlined in this section. Yet obviously the central government departments have large responsibilities, not only in regard to the general conduct of various services in the region, but also in regard to the distribution of grants-in-aid voted by Parliament.

This problem would be met by authorising the central departments to delegate to the Greater London Council some or all of their functions in regard to the local authorities within the region. Thus, the granting of almost all the sanctions and approvals required from central Government departments by the minor authorities in the metropolis could be devolved upon the Greater London Council, and would, indeed, in any event take place if the arrangements previously outlined were accepted.[1]

The departments in Whitehall would remain immediately responsible for the inspection and supervision of services directly administered by the Greater London Council itself (i.e. those coming within the first of the three categories into which the functions of London government have been divided).[2] The quality of the central departments' work in this sphere would doubtless gain in excellence and thoroughness by their release from the mass of minor detail which occupies so much of their time at present.

No one with any knowledge of the existing state of affairs will imagine for a moment that the delegation to the Greater London Council of duties relating to the local authorities which has been here suggested will denude the central departments of their rightful work and relegate them to a condition of genteel and ineffectual idleness. Indeed, quite the contrary is to be expected. The departments have for years been so inundated and overwhelmed with comparatively small matters of day-to-day routine arising from the hundred or so local authorities in the metropolitan region that they have never had

[1] *Ante*, pp. 389–90.   [2] *Ante*, p. 327 *et seq.*

either the time or the freedom of mind to give attention to the fundamental questions of London government. This is obviously true of the Ministry of Health, whose responsibilities in the field of local government are the greatest, and whose preoccupations are most distracting.[1]

The central departments are at present in a situation which must be extremely exasperating to the more constructive, imaginative and energetic type of men in Whitehall. They have a thousand separate and unrelated activities to perform in reference to the local authorities in the metropolis; but at no point is a central department able or authorised to deal with any of the large and burning problems of London government. The reorganisation of areas, the distribution of powers, the planning of the metropolis, the co-ordination of services, the improvement of personnel, the equalisation of finance: these are matters in regard to which the Ministry of Health or the Board of Education, the Ministry of Transport or the Treasury, are devoid of effective authority. With one or two rare exceptions, these vital questions are matters with which the Whitehall departments do not concern themselves.

It is just these larger questions requiring a wider outlook, careful enquiry into what is being done at home and abroad, advice on administrative and technical planning, the provision of information and criticism, that should form the substance of a central government department's work. It is just these very matters that they most persistently ignore. The amount of intelligence work done by the departments is almost negligible.[2] What, for example, have they done to keep public opinion informed—or even to keep themselves informed—on the striking changes in the government of New York City, or the planning of Stockholm or Moscow, or even (to come nearer home) the remarkable achievement of Wythenshawe? Yet these things are all relevant to the immediate task confronting London.

---

[1] It may be recalled that one of the advantages claimed by the Minister of Health for the introduction of the block grant in 1929 was that it would relieve the Ministry of the need for much detailed supervision of local authorities in the realm of public health.

[2] I would except from this statement the medical reports and studies of the Ministry of Health; and the reports of the Advisory Committee to the Board of Education.

The one really valid answer to a charge of this kind is that the Government departments are overloaded with an intolerable burden of day-to-day administrative paraphernalia, which has so choked up their blood stream that they are prevented from taking part in the higher activities of central direction. The re-ordering of relationships within the metropolis which has been outlined here would provide an effective remedy for this malady. It would leave the central departments free to devote most of their time and energies, so far as London is concerned, to co-operating with and supervising the work of the Greater London Council in regard to the major functions for which that body would be directly and primarily responsible. It would not be necessary for them to abandon all control of the local authorities within the region. It would still be desirable for Whitehall to institute occasional inspections of the metropolitan councils, to call for information, to make comparisons with other regions, to hear appeals, to conduct public enquiries. But broadly speaking, the regular supervision and routine inspection of the local authorities within the metropolis would be delegated to the Greater London Council, with the proviso that if the latter body failed to discharge any of its duties satisfactorily, the delegated power would be withdrawn and would revert to the central department concerned.

In general, this would mean that the Greater London Council would be normally responsible for the conduct of negotiations with the central government. This alone would introduce a degree of simplification and a reduction of work which can scarcely be grasped by those who are not familiar with the voluminous correspondence and lengthy conferences involved by the tedious and cumbersome arrangements now prevailing, whereby the Ministry of Health has to deal with more than a hundred separate authorities in regard to every detail of maternity and child welfare work or food inspection. I have heard it suggested that the recent decision of the Minister of Health to confer powers relating to midwives on the London County Council rather than on the metropolitan borough councils was based chiefly on the desire of the Ministry to deal with one local authority rather than twenty-eight. It is scarcely surprising that the Ministry should display this understandable preference.

We may turn now to the question of personnel. The method of electing members of the major and minor authorities has already been dealt with, but there is one question which has not so far been mentioned: namely, the payment of members.

The position in this respect is already difficult in connection with the London County Council. A member of that body must devote the equivalent of at least two whole days a week to the work of the Council. There is not only the weekly meeting of the Council on Tuesday afternoons and the meetings of the standing committees to which every member belongs; there are sub-committees and sub-sub-committees to attend;[1] institutions to inspect; committees of management of schools, hospitals and asylums on which members must serve; public functions to be attended; and the regular and special meetings of the party organisation, which we have already noted to be an indispensable part of the effective functioning of the Council. In addition to all this there is a heavy mass of reading to be got through in the way of minutes, reports and memoranda, and also the needs of the constituency to be considered.

The demands made on an alderman are normally even heavier than those falling on an ordinary member. The chairmen and vice-chairmen of committees, and of course the leader of the Council and the leader of the opposition, are the most hard-worked of all; while the chairman of the Council has an immense burden of ceremonial duties to perform. Some chairmen give the greater part of their time to the work of the Council.

The fact that all this devoted service should be given without material reward of any kind is a remarkable indication of the public spirit with which the members of the Council are imbued. And so, too, is the remarkably high attendance of members at meetings of the Council and its committees and sub-committees.[2] But it also has an unsatisfactory side. It is obvious that the circle of persons who can afford to give up so much of their time to the public business without recompense

---

[1] The number of committees, sub-committees and sub-sub-committees of the London County Council is 1,500, and there are 14,000 places on them. This was stated by Lord Snell in a lecture at University College of the South West at Exeter on 9th September, 1938.

[2] Detailed information is given on p. 340.

is comparatively small. There is no doubt that the high proportion on the Council of well-to-do persons of independent means
on the one hand, and of married women and trade union
officials on the other, is partly explained by this fact.[1] However
much we may admire the gratuitous public service which is
rendered at County Hall, it would seem desirable on balance
that in future, if a Greater London Council is established, the
members of it should be paid. In no other way can the circle
of possible candidates be widened to the extent required by
the democratic ideas of today. Moreover, the demands which
a Greater London Council would make on the time and
energies of its members would be still greater than those made
on the London County Councillor, for the distances to be
traversed would be far longer than in the administrative county,
and the members would have the additional responsibility of
belonging *ex officio* to the local authority in which their
constituency was situated.

I should suggest, therefore, an annual remuneration of £250
a year for councillors and aldermen, with £500 a year for the
leader of the Council and a similar sum for the chairman of
the Council. Such a sum would not be large enough to attract
men and women to the council chamber unless they were
genuinely interested in the work, but it would be sufficient to
save a certain amount of hardship which now occurs and to
make it possible for many to enter the council chamber who
are now debarred.

In regard to the officers of the Council, the proper development would be in the directions recommended by the
admirable report of the Hadow Committee on Local Government Officers.[2]

The personnel of both the London County Council and the
metropolitan boroughs is curiously isolated from the rest of
the country. Mobility between local authorities elsewhere in
Great Britain is already very common, especially among the
higher grades, and has been greatly facilitated by the recent

[1] See the detailed Survey by Eleanor Ernst: "The Personnel of the
London County Council" in the *Political Quarterly*, July–September 1935,
p. 417.
[2] Report of the Departmental Committee on the Qualifications, Recruitment, Training and Promotion of Local Government Officers, 1934.

legislation making superannuation ubiquitous and compulsory.[1] But with a few notable exceptions London remains a separate enclave. The work of the London County Council is on so large a scale that it tends to be more highly specialised than elsewhere. Hence, provincial authorities feel disinclined to take men and women from County Hall, whose experience they believe is likely to be on narrower lines than that which results from employment by a smaller authority. This in turn inevitably reacts on the staff policy of the London County Council, since it narrows the openings for the Council's officers elsewhere and thus strengthens their claims to promotion at County Hall. A few of the principal officers at County Hall have been appointed from outside local authorities or taken from the Civil Service, but this is a rare occurrence and gives rise to a quite undue amount of resentment.

So far as the metropolitan borough councils are concerned, there is practically no interchange with the staff of the London County Council, mainly owing to the jealousy and friction which has always prevailed between the respective authorities. At the same time, it is often difficult for the metropolitan boroughs, with their small areas, restricted populations, and circumscribed powers, to attract able and energetic men from provincial towns. This is particularly noticeable in the field of public health. The work of an assistant medical officer in a large provincial city, or of the medical officer of even a small town, is more varied and interesting than the job of the medical officer in a metropolitan borough.

The work of the municipal civil service in London must necessarily differ from that in the rest of the country, quantitatively and perhaps also qualitatively. The small amount of movement between the metropolis and provincial authorities must, therefore, be recognised as probably inevitable, at any rate for some time to come. But this makes it all the more desirable to increase the integration of the local government staffs within the metropolitan region.

The local government officers within the metropolitan region should be regarded—and should so regard themselves— as members of a great London municipal service: employed, it is

[1] Local Government Superannuation Act, 1937.

true, by separate local authorities but forming part of a unified service informed throughout by an essentially co-operative and consistent *esprit de corps*.

The first step in this direction would be the establishment of a personnel commission for the metropolitan region, containing a few representatives of the major and minor authorities, together with a sprinkling of outside persons experienced in matters of education and staffing. The task of the commission would be to hold open competitive examinations, and to conduct tests of other kinds, for the purposes of recruitment. The principal examinations would be at the three ages corresponding to the three clearly marked stages of our national system of education.[1] The regional and local authorities would notify their vacancies to the commission. The latter would hold examinations and publish the lists of the successful candidates. The local authorities and the regional council would then arrange interviews and be free to appoint whoever they desired from the list.

The personnel commission would co-operate closely on the one hand with the Central Advisory Committee set up by the Minister of Health, and on the other hand with the establishment committees of the municipal councils concerned. In conjunction with the latter, it should be possible to work out methods of grading, salary scales, arrangements for holiday and sick pay, and even an agreed basis of promotion applicable throughout the region. Transfer between the Greater London Council and the metropolitan boroughs, and between the boroughs themselves, should be encouraged to the utmost possible extent. The officer who has been for 30 years in one place would become a rarity. Interchange with the Civil Service would also be regarded as highly desirable, especially in the case of departments concerned with local government, such as the Ministries of Health, Education and Transport.

The arrangements which have been here envisaged are designed to bring the different parts of the system of London

---

[1] *Cf.* Report of the Departmental Committee on the Qualifications, Recruitment, Training and Promotion of Local Government Officers, 1934, pp. 19–20.

government into organic relationship with one another and with the central government. In place of the chaos and confusion, the overlapping and inefficiency, the conflict and friction, which results from the senseless independence and irrational separatism which now exists, we might expect to see a smooth-running municipal organisation, working harmoniously and consistently for a common end.

The large-scale services would be in the hands of the regional organ, the local services administered by the metropolitan borough councils subject to over-riding direction or control, on questions of policy. The vital questions of planning, finance and personnel would be safeguarded by over-all control on major points. All else would be left to the free working of locally elected authorities, larger in size, stronger in resources, more adequately staffed, better equipped in every way to deal effectively with the difficult problems of local government and to render worthy service to the citizens.

In this section I have been concerned almost entirely with those services which the local authorities would carry out under the general supervision of the regional organ, or which they might administer under delegated powers conferred by the latter. It should, however, be clearly understood that this would by no means exhaust the list of their activities. There is a host of functions which the metropolitan borough councils could carry out in their unfettered discretion without interference or compulsion of any kind from the Greater London Council. I do not propose to enumerate these in detail. They include the provision and maintenance of playgrounds and small parks, the planting of trees, the regulation of outdoor advertisements, the provision of crèches, the inspection of bakehouses, the supervision of disused burial grounds, the licensing of seamen's lodgings, the maintenance of public conveniences, the prevention of nuisances, the destruction of rats and mice, the control of street markets, the regulation of noxious trades and many other matters dealing with the health, comfort and amenities of the locality.

CHAPTER IX

THE ESSENTIALS OF PLANNING[1]

The chaos of administrative areas and authorities in London is reflected in a corresponding chaos in the physical development of the metropolis. The matter is of such profound importance, and most people are so little informed on the subject, that I propose to deal with it at some length.

Let us start by trying to look at the matter in very broad terms. Here is the capital of the British Empire, one of the oldest living cities in the world, with a continuous history of self-government stretching back for nearly a thousand years. Its existence has been undisturbed by flood, famine, sack, pillage, or invasion. The only major disasters that it suffered were the Plague in 1665 and the Great Fire in the following year.

At no point in the history of the City has any attempt been made to direct the growth or evolution of the town by the light of a general plan. A magnificent opportunity occurred after the Great Fire, when Sir Christopher Wren produced a splendid plan for the rebuilding of the City. If Wren's plan had been put into practice, it would have served as a towering beacon to guide and light the footsteps of all the generations that followed him down to our own day. "If Wren's London had been built the tale of the English cities and towns would have been very different. Inspired by this central example they would have taken on the urbane grace that we now associate only with continental towns."[2] Unfortunately, Wren's fine scheme was consigned to the dust by the interests of shopkeepers. The vested rights of property were allowed to insist that the City should be reconstructed on the same narrow lines as before, and the mounting vision of the 17th century was lost amid the narrow alleyways marked out in medieval times and now rebuilt.

A new hope arose in the late 18th century, when the squares of Bloomsbury, Mayfair and Belgravia were laid out by a few

[1] I have left this chapter as it was written in 1939, except for a few factual changes on pages 412-3. The recent developments in regard to planning in London are discussed at length in the Epilogue.

[2] Thomas Sharp: *Town and Countryside*, p. 137.

imaginative landowners; and in the early years of the 19th century the sun began to shine on the elegant and gracious plans for the construction of Regent Street, Portland Place, Park Crescent, Carlton House Terrace and the terraces in Regent's Park which Nash conceived. These were the first comprehensive plans for the creation of beauty, dignity and spaciousness on a large scale in the heart of London. It is significant that they mainly affected Crown lands, although the purchase of a certain amount of private property was also involved.[1]

Nothing comparable to Nash's scheme or even remotely approaching it in excellence has ever again appeared on the horizon. Much of his work has recently been demolished and replaced by a patchwork of independently designed buildings devoid of any mutual relationship; and what remains is threatened with treatment of a similar kind in the near future. Since Nash there has been no town planning worth the name in the metropolis, apart from an isolated island remote from the mainland like the Hampstead Garden suburb.

The structure of cities, Le Corbusier observes, reveals two possibilities. One is a progressive growth subject only to chance, with the resultant characteristics of slow accumulation and a gradual rise. Once it has acquired a gravitational pull it becomes a centrifugal force of immense power, bringing the rush and the mob. Such is London. The other possibility is the construction of a city as the expression of a preconceived and predetermined plan embodying the known principles of the science.[2] One might cite Washington or the new Moscow as examples of this second type.[3]

The growth of towns in this country, stated the Departmental Committee on Garden Cities and Satellite Towns in 1935, has hitherto taken place either by sporadic and haphazard expansion round the circumference; or by casual building on detached areas of land, the gaps in which become gradually filled up.[4]

---

[1] The advantages Nash foresaw from his scheme are set out in the appendices to the First Report of the Commissioners of Woods, 1812, and the Second Report, 1816. *Cf.* the Report on Metropolitan Improvements, 1844, p. 5.

[2] Le Corbusier: *The City of To-morrow*, pp. 91–2.

[3] *Cf.* the chapter on the planning of Moscow by Sir Ernest Simon in *Moscow in the Making.*   [4] Report of the Committee (1935), p. 7.

PLATE VII. Sir Christopher Wren's Plan for re-building London after the Great Fire of 1666. It was "consigned to the dust by the interests of shopkeepers." (*Reproduced by permission of the Warden and Fellows of All Souls College, Oxford*)

This is particularly noticeable in the case of London. "The general sporadic extension of building is not only costly as regards services but has also the effect of destroying the amenities of the surrounding country, the area deteriorated being often far greater than the area actually built upon. With the closing up of development which follows this haphazard intrusion into the country, open spaces which should have been jealously guarded in order to provide adequate room for playing fields and parks for the ever growing urban population are lost for ever. The result is that the working population, virtually imprisoned in the central districts, are unable without serious expenditure of time and money to enjoy adequate healthy recreation in their leisure hours."[1]

The Committee complained that during the last decade we have seen the anomaly of large dormitory housing estates being developed without provision being made for local employment, while at the same time industrial areas have been constructed without provision being made for housing the workpeople in their vicinity. Slum clearance schemes are being undertaken with insufficient attention being paid to broad replanning principles, while ribbon development proceeds on the main roads without regard to its detrimental effect on their capacity to bear traffic or the excessive cost to which local authorities are put in providing the necessary public services.[2]

It is worth while recalling for a moment the rapid rate at which modern cities are rebuilt or extended at the present time. In the square mile of the City, property forming no less than 42 per cent of the rateable value of the whole area has been rebuilt since 1905;[3] yet in these 30 years or so the City Corporation has made no attempt to plan its area, although town planning legislation has been in existence since 1909 and the City Corporation could certainly have obtained whatever powers it needed. The amount normally spent in rebuilding London is stated to have been about £50 millions a year during

[1] Report of the Committee (1935), p. 7.
[2] Ibid., p. 6. This complaint is as valid today as when it was written in 1935, despite the passing of the Ribbon Development Act, 1935. This statute has had a very small effect in the London Region.
[3] Paper read by E. E. Finch, Engineer to the City Corporation, to the Town Planning Institute, The Times, 2nd October, 1937.

the last decade, from which it would appear that possibly between a quarter and a third of the administrative county has been reconstructed in that time.[1] From this it will be seen that the assumption which many people make that the built-up areas of old towns must be regarded as there, once and for all, and in consequence that nothing can be done to improve the planning of them, is quite unfounded.

At present, in addition to the ordinary annual expenditure mentioned above, a sum of £35 millions is being spent by the London County Council on a vast programme of slum clearance and improvement, calculated to take 10 years to complete and involving the displacement of 250,000 persons now living in slums. The London County Council declare that there is a reasonable prospect of the slum problem in London being solved well within the period of 10 years originally contemplated, although it is possible that additional areas not at present scheduled for treatment will be found to require attention during that period.[2]

This vast effort will constitute, when complete, an achievement of great magnitude and importance. Unfortunately, however, it is not being undertaken in conformity with any large-scale town planning scheme for the metropolis, nor evei. brought into relation with any general traffic plan relating to highways and transport. The fault is due to the fact that in the past no attempt whatever was made to formulate a town planning scheme for London.[3]

[1] Mervyn O'Gorman, M.P., in *The Nineteenth Century and After*, May 1937.

[2] *London Housing*, L.C.C. publication, No. 3272/1937, pp. 12, 24. The programme was adopted by the L.C.C. in 1933.

[3] London is not alone in its heedless ways. In Paris, too, short-sighted action of a similar kind is common, with even less reason, it might be thought, having regard to the magnificent town planning in the past which has made Paris so fine a city. A few years ago Le Corbusier was complaining that "in a number of highly strategic points in Paris immense blocks of decayed and out-of-date buildings are being demolished, and on their sites new buildings and blocks of offices are being erected. The street is left exactly as it was: occasionally the building line is taken 6 or 12 feet further back. Nothing else is done." These profitable demolitions and erections are establishing in the centre of Paris certain fixed points which will form the basis of the new 20th-century city; but they have in nowise been dictated by the actual problem of the layout of the town. Le Corbusier: *The City of To-morrow*, pp. 204–5.

The misuse of land has occurred on a large scale in the metropolis as the result of the lack of territorial planning. The scattering of building patches is enormously extravagant in the use of land and gives the impression that far more land is occupied by houses than is actually the case, and that there is no room for the population to spread outwards. It was pointed out by the Departmental Committee on Garden Cities and Satellite Towns that the entire population of the administrative county could be housed in cottages at ten to the acre in 377 square miles of the 1,729 square miles available in Outer London. This would be equivalent to a belt only 6½ miles wide added to the county (which is itself about 6 miles in radius), leaving a belt of unoccupied open country more than 11½ miles wide before the boundary of Greater London was reached.[1] No one suggests that this should be done, but it shows the waste of space in the disposition of the metropolis, which at present spreads like a blight over half a dozen counties. The problem is thus not one of space, but of the proper utilisation of land.

Another aspect of the misuse of land concerns the purpose of the utilisation. An example of this is the construction of the housing estate at Becontree and Dagenham on unusually good market garden land, the supply of which is rapidly diminishing in the environs of London, although it would have been quite easy to have chosen a site on less valuable land a little farther north. The trading estate at Slough is another example of the same kind. The Government acquired a piece of good corn land during the Great War as a repair depôt for motor lorries. After the War the site was privately developed as a trading estate of about 580 acres in which about 200 firms now employ 20,000 workmen. The area has recently become a borough. It represents a colossal blunder from a land utilisation point of view.[2] A more recent mistake of the same kind was the acquisition by the Metropolitan Water Board of 500 acres of exceptionally fertile market garden land at Walton for the construction of a new reservoir.

---

[1] *Report*, p. 7. For a good account of the misuse of land in the metropolis, see E. C. Willatts: *Middlesex and the London Region*; The Report of the Land Utilisation Survey of Britain (edited by L. Dudley Stamp), Part 79.
[2] E. C. Willatts: *op. cit.*, p. 170.

Land values and compensation are the two great obstacles in the way of imposing social control on the use of land as things now are in Great Britain. Sir Raymond Unwin has shown that although wise town planning will alter the apportionment of land values, it will tend to increase them as a whole rather than lead to a reduction. "It should therefore be possible to devise a fair system under which town planning improvements could be carried out without injustice to the landowners, but without being hampered by the one-sided policy of compensating the individual in all cases where his land is deprived of value and making him a gratuitous present in all cases where the value of his land is improved, which represents very much our present system. We must recognise that this unwise course is a serious obstacle to the proper development of towns."[1]

The obstacles in the way of effective town planning, and the provision of adequate highways, playgrounds, open spaces and other necessities have become almost insuperable in London as a result of the preposterous system we have hitherto adopted in regard to compensation. In this matter England is an extraordinarily backward country. In New York, for example, the City Council long ago decided that where the underground railways were extended, part of the cost should be paid by an assessment on the increased land values due to the extension. The figures show that the whole cost of the railways could have been paid for out of the increased values on land opened up by their operation.[2] We may contrast the position in London by taking as an illustration land bought in 1923 at Southgate at £160 an acre. This now sells at £1,500–£1,600 an acre. The tenfold increase is due almost entirely to the extension of the tube railway to Southgate. Yet the whole profit goes into the pockets of landowners and building speculators. Neither the municipal authorities, nor the London Passenger Transport Board, nor the Government (which has now had to guarantee the interest on £40 millions required by the Board for other extensions) receives one penny of the unearned increment.

In Kansas City, all improvements such as parks, boulevards,

---

[1] Sir Raymond Unwin: "Some Thoughts on the Development of London" in *London of the Future* (edited by Sir Aston Webb), pp. 191–2.
[2] *Ibid.*

playgrounds and so forth have for many years been paid for by assessing the greater part of the cost of those improvements on sites which have derived direct benefit from them. The City Council determines the proportion of the cost to be borne by the City and the proportion to be recovered from the landowners whose property has been enhanced in value. The apportionment of the latter among individual owners is left to be worked out by the expert representatives of the owners themselves. In 1921, out of $11 millions spent on improvements for the City, 83 per cent was assessed on sites, leaving only 17 per cent to be paid for out of the local rates. The results show that the increase in the value of the sites benefited greatly exceeds the amount of the assessment.[1] In London, by contrast, the total sum collected by the London County Council in 1935-6 by way of betterment charges was £1,739. This was supposed to cover the improvements at Kingsway, Tottenham Court Road, Tower Bridge southern approach, Westminster and Putney![2]

A position has now been reached in London in which the municipal authorities are not permitted to acquire land in advance of requirements to hold for an increase of value while at the same time the compensation charges for public improvements are so enormous as to be almost prohibitive. Moreover, there appears to be neither sufficient power nor determination to recover betterment charges, save on a negligible scale.

We are therefore approaching an *impasse* of a serious kind. Any improvement carried out by the London County Council or a local authority is immediately exploited by acquisitive landowners whose cupidity tends to defeat the very object of the improvement; while other urgently needed improvements cannot be undertaken because of the enormous compensation costs involved. The Green Belt scheme initiated by the London County Council, for example, has greatly increased the value of land in the vicinity of the open spaces concerned; and in consequence the proximity of the Belt is mentioned in advertisements of land and houses for sale. Much speculative house building is taking place on the fringes of the Belt. Yet no

[1] Sir Raymond Unwin: "Some Thoughts on the Development of London" in *London of the Future* (edited by Sir Aston Webb), pp. 191-2.
[2] *London Statistics* (1935-7), vol. 40, pp. 293, 398.

assessment has so far been made on the landowners in question for the increase in the value of their property, though the amount which has been added would no doubt more than pay the cost of the scheme.

The principal highways of Central London consist, with few exceptions, of short narrow lanes utterly unfit to serve the needs of a modern city in the Petrol Age. Main thoroughfares of a broader width are found only at the outskirts, where the traffic is far lighter. The public is at long last beginning to be conscious of the great deficiencies in the London highway system; but very few people realise the underlying causes. The main obstacle to improvement is once again the enormous cost of compensating landowners and occupiers. If there is to be no recovery of betterment values on a corresponding scale, the financial burden of modernising the chief metropolitan arteries will prove insupportable. The Bressey Report ignored this essential question.

The capital outlay of the London County Council during the past fifty years has been dominated by expenditure on education, tramways and housing; and the same applies to the revenue account.[1] Main highway improvements have been conspicuously neglected. One reason for this is that no political party at County Hall has been prepared to ask the electors to spend millions of pounds in compensating landowners in order to make main highway improvements which will, *inter alia*, enormously increase the value of the land in which they are situated. This does not excuse or justify the neglect of London highways. It merely helps to explain it.

It should be remembered by those who feel disposed to look lightly on the problem that the London highway system is deteriorating in serviceability almost daily. This is due not only to the increasing torrent of motor vehicles which floods the streets, but to the less noticeable fact that the density of population is rising rapidly in the central area. Huge blocks of flats and great new office buildings are springing up every month, with the result that the average building heights in the metropolis are increasing considerably while the present street

---

[1] See the diagrams on pp. 197–8. For details, see Sir Harry Haward: *The London County Council from Within*, p.109.

widths are being retained or even reduced.[1] We are thus deliberately courting disaster so far as the traffic problem is concerned.

Other forms of transportation are suffering to an equal extent through our failure to lay down a proper plan for London development. In regard to railways, for example, there is no public authority considering the very important question of facilities and terminals, although the matter could presumably be referred to the London and Home Counties Traffic Advisory Committee by the Minister of Transport, if he thought fit. London has far more main line terminals than any other city in the world; and one expert has suggested that 9 of the existing main line stations could be removed or altered in their use without loss to the public convenience.[2] It has been announced that certain of the principal stations are about to be rebuilt; but there appears to have been no public discussion of the question as a whole nor consultation with the municipal authorities in London. The matter is left entirely to the discretion of the railway companies and no one has so far even suggested that the proper disposition of railway stations should form one of the major features in the planning of London.

The railways have a long history of anarchy and confusion behind them going back to the early years of the 19th century. The persistence of this tradition might be invoked to explain, if not to defend, the present position in regard to the main line stations in the metropolis. But the same cannot be said of aviation, the newest form of locomotion. Yet here again there has been a conspicuous lack of vision, if not actual mismanagement, in the provision of aerodromes.

The practical possibilities of civil aviation were first demonstrated from about the beginning of 1912. An aerodrome was established by the Grahame-White Aviation Company at Hendon, which was then in the heart of the rural environs of London, although only 6 miles distant from Marble Arch. The

---

[1] The building line is being brought *forward* in Euston Road, Marylebone Road and Queensway!

[2] H. J. Leaning: "London Railway Construction" in *London of the Future* (edited by Sir Aston Webb), p. 76.

Hendon aerodrome was an excellent one, except that it was situated on the wrong side of London for Continental traffic, which was then scarcely considered. The point is, however, that London had from 1912 onwards a constant reminder of the growing possibilities of flying; and, therefore, of the need to make provision in advance for aerodromes, by reserving land which would certainly be needed at a later stage. No such foresight was shown. After the Great War, Hendon was permitted to become a country club and was later acquired as the headquarters of the Metropolitan Police College. Croydon was chosen as the centre for civil aviation, and later Heston. These were the two principal aerodromes for commercial air transport to and from London before the second World War.

The aerodrome at Croydon is more than 12 miles from Charing Cross. It is badly served by roads and tube railways. Access to and from Central London is slow and difficult. The average time taken by the special motor service to reach Victoria Station is 45 minutes; and it takes nearly an hour to reach Park Lane. Heston is situated 10 miles from Victoria, and the travelling times from the centre of London are equally long. Hence, much of the advantage of air travel is lost through waste of time in getting from the aerodrome to the centre of London. The great new London airport at Heathrow, which will be Britain's principal air terminus, is as far from the centre as Croydon and is at present much worse served with communications. When improvements are effected, it will take 25 minutes by the quickest route to reach Victoria. The 7 other civil airfields which are proposed for London are situated at distances varying from 12 to 28 miles from Victoria, with corresponding travelling times to the centre.[1]

It is now recognised that an aerodrome serving a large city should be situated as close to the centre of the town as possible, having regard to climatic and topographical conditions. A transportation time to the centre in excess of 15 minutes is likely to prove a serious drawback. Time is, of course, more important than distance, and therefore good means of com-

---

[1] Greater London Plan: para. 182.

munication form an essential factor in determining the location of an airport.[1]

It cannot be said that either Croydon, Heston or Heathrow shows up favourably if tested by this standard. But since there has never been a territorial plan for Greater London, it is not surprising that better provision has not been made for aerodrome accommodation.

On the social side, the drawbacks of an unplanned city are of equal magnitude and their effects still more far-reaching. They are, however, of a more imponderable character and less susceptible of precise formulation. I shall mention only two of them. The position in London in regard to playing fields has become hopeless. The situation could have been saved as late as 1930; but after that the rate of building in Outer London became so rapid and the selection of sites for building so haphazard that an adequate quantity of land within a reasonable distance of the administrative county ceased to be available.[2] Yet there was nothing obscure or difficult to deal with in the problem. The Greater London Regional Planning Committee had made it abundantly clear in its first report (issued in 1929) that a crucial point had been reached in regard to the provision of playing fields and open spaces generally, and that drastic action was required immediately if the welfare of countless millions of Londoners was to be safeguarded now and in the future. The opportunity was lost and the warning passed unheeded.

Another detrimental feature of London development which could have been avoided by planning is the coalescence of suburbs lying just beyond the boundary of the administrative county. Such places as Ealing, Chiswick, Willesden, Finchley, Wood Green, Wanstead, Croydon, Beckenham, Wimbledon and many others could have retained a certain compactness and physical individuality which they formerly possessed if a little care and foresight had been exercised by a large planning authority. Instead, they were permitted to lose their coherence

---

[1] Cf. Thomas Sharp: *Town and Countryside*, p. 199. A central position will become increasingly important with the development of helicopter devices for vertical landing and taking off.

[2] *Ante*, pp. 183–4.

and the built-up area of London has become a shapeless, formless, meaningless wilderness of bricks and concrete.[1]

I have already referred in an earlier part of this work to the multitude of local authorities in the metropolitan region which are at present entrusted with planning powers.[2] Many of these organs had prior to 1937 scarcely even attempted to exercise their functions, and less than a third of the administrative county had been town planned. Today nearly the whole area is theoretically "subject to town planning restrictions."[3] But this signifies very little in terms of actual control, and still less from the point of view of basic conceptions.

Professor Adshead has recently complained that town planning in London "is rapidly descending into a position more correctly occupied by a series of super-sanitary byelaws."[4] The London Building Acts deal with the height of buildings, with building lines, in some cases with angles of light, and up to a point with the space in the rear of buildings. The Town Planning Acts also deal with the height of buildings and the space about them; and in addition the use to which they may be put, the appearance, and the materials to be employed. Professor Adshead states that the relationship between the London Building Acts and the town planning legislation has never been thought out, although the London County Council has one Committee to deal with both subjects.

Mr. Pick brings a charge of a different kind. The following passage is taken from his evidence before the Royal Commission on the Geographical Distribution of the Industrial Population :[5]

3357 Q. (by Professor Abercrombie): When replying to an earlier question you gave rather a gloomy picture of what is happening at the moment in the planning of London with these 133 authorities.

[1] *Cf.* Evidence of A. T. Pike before the Departmental Committee on Garden Cities and Satellite Towns.
[2] *Ante*, pp. 186-7. H. Berry: "Town Planning the County of London" in *The Town and County Councillor*, November 1937, p. 21.
[3] *London Statistics*, vol. 40 (1935-7), pp. 189-93.
[4] "London under Statutory Town Planning," a Chadwick lecture delivered in London, 7th May, 1937 (unpublished).
[5] 2nd March, 1937.

It rather suggested that at the present moment instead of getting anything like the wide belt that Sir Raymond Unwin envisaged and you yourself mentioned as desirable, and the concentration of London to a unit, you are getting really a general spread under town planning powers over a very much wider area. Would you agree?

A. Yes. It is the town planning which I think is leading us astray at the moment. For instance, take Ruislip and Northwood (producing map). There is the town planning scheme of those places. When they drop their pieces of industry in, it is all a mosaic—one kind of housing here, another kind of housing there, little bits of shopping, little bits of industry, little bits of open space— when they all make their plans on those lines we get a spattering of all kinds of things all around the area under the name of town planning. I have a map of Ilford that shows the same thing. That is one of the chief sources of trouble. They put industry down in little bits, they have a big block here and a piece there, all without relationship to what is going to happen in Brentford on this side and Woodford on the other side. It is that kind of town planning, town planning by sections, which amounts to no town planning at all in regard to London as a whole.

Town planning, he writes elsewhere, does not really exist at all in London at present. "What goes by the name is almost idle and useless so far, and has not yet arrived at a conception of the congeries of towns that make up a metropolis."[1]

The Greater London Regional Town Planning Advisory Committee, which has recently been reconstituted for the third time, will almost certainly be as futile as its predecessors. It possesses no technical adviser or staff of its own but is dependent on a technical committee consisting of town planning

[1] F. Pick: "The Organisation of Transport," *Journal of the Royal Society of Arts*, vol. lxxxiv, p. 210.

officers drawn from the constituent authorities. This technical committee will only consider questions that are referred to it and it is unlikely that any of the larger matters will filter through.[1]

The plans which have at last been formulated by the London County Council for the administrative county are purely negative in character. They may prevent certain mischiefs from occurring, but they will achieve little of a creative or positive kind. There is no co-ordination between housing and town planning because most of the L.C.C. housing work is being done on estates lying outside the county boundaries, where the London County Council has no town planning jurisdiction. Hence the housing committee and the town planning committee do not co-operate or consult with one another. Moreover, the housing committee of the London County Council does not consult the town planning committee of the relative out-county authority, nor the Regional Town Planning Advisory Committee. The valuation officer has hitherto been in charge of town planning at County Hall, with the result that most of the wider and more important social aspects of planning have received little attention.

It is abundantly clear that an entirely new approach to planning is necessary if anything serious is to be accomplished in London after 30 years of muddle and indifference. So far we have been merely playing with the problem.

First of all, a Planning Commission for Greater London is an absolute necessity. It should cover either the Traffic or Transport areas, or whatever other territory may be selected for the government of the metropolitan region. This Commission should preferably form part of the new government of Greater London. A majority of the members would be appointed by the regional authority from among their own number, and one member each would be nominated by the London Passenger Transport Board, the Port of London Authority, the Minister of Health, the Minister of Transport and the Minister of Agriculture. The Commission should be kept rather small: a total of about 12 members would be the best size, of which all but 5 would represent the Greater London Council. The

---

[1] *Ante*, p. 190.

chairman would be appointed by the Commission itself either from among their own members or from outside. He should be prepared to give his whole time to the job, and the Commission should be empowered to pay him a substantial salary.

The object of creating a Commission for the work rather than leaving it to a statutory committee of the Council is to emphasise the importance of the task and to lend weight and prestige to the body entrusted with it. Moreover, a Commission would enjoy a certain degree of independence, although it would be ultimately responsible to the Greater London Council. The latter body would approve and adopt the plans, and scrutinise and pass the budget of the Commission. The Minister of Health[1] should be authorised, on appeal, to adopt the plans of the Commission, in case of failure or refusal to do so by the Council. New York and Moscow, the two great cities where the most promising large-scale territorial planning is now in process, both have City Planning Commissions. If the general structure of London government is not reformed, despite the crying need for modernisation, an *ad hoc* Planning Commission should be set up for the region. But this would be a poor alternative to the method suggested above.

The Greater London Planning Commission would require a more comprehensive staff than any now existing in this country. It would need not only town planners in the technical sense, surveyors, civil engineers and so forth, but statisticians, social scientists, economists, population experts, transport specialists, research workers, landscape artists, and public relations officers.

The Planning Commission would be an executive body. We have already had far too much experience of the futility of advisory committees. There is an unquestionable need for conferring full administrative powers both in regard to the formulation of plans and their enforcement, subject only to the necessity of obtaining adoption by the Greater London Council or (in case of refusal) by the Ministry of Health.

The Greater London Planning Commission would be primarily concerned with questions of regional importance. These

---

[1] If a new national planning organ were to be set up, it would replace the Minister of Health in this and cognate matters.

would include the provision of open spaces and the restriction of building, the disposition of main roads and bridges, large-scale zoning for use, the location of aerodromes, railway stations, goods yards and motor-coach terminals, the regulation of height, density and character of buildings in certain areas, the siting of housing estates, satellite towns or garden suburbs, the position of schools, hospitals, libraries, museums, swimming baths, fire stations, markets and other public institutions, and matters of a similar kind. The Greater London Council would devolve on the Commission whatever powers it might possess under the Restriction of Ribbon Development Act, 1935, and the London Building Acts.

Each local authority within the metropolitan region would have a statutory town planning committee for the purpose of making a detailed local plan covering its own area. The local plans would be required to conform with the regional plan, and the Commission would have power to approve, reject or modify the former.

The Greater London Planning Commission would in turn be required to co-operate closely with the Ministry of Health and other central government departments concerned. In the event of a new central organ being established for the purpose of formulating a national territorial plan or of controlling the location of industry, the Commission would have to co-ordinate its own work with the activities of that body. Such a development is greatly to be desired. It has been much discussed in recent years, and several proposals have been put forward. It would, however, take me too far afield to examine them.[1]

Hitherto it has been assumed in this country that the correct procedure for territorial planning is to start at the bottom with comparatively small units. For any larger type of organisation we have relied on the hope that local authorities would co-operate through voluntary regional groupings.

This is the exact opposite of the proper procedure. Planning should start on the national scale, and work down through

[1] See, for example, the evidence given before the Royal Commission on the Geographical Distribution of the Industrial Population by the Town Planning Institute, by the Garden Cities Association, by W. A. Robson and others. See also the Report of the Departmental Committee on Garden Cities and Satellite Towns, 1935, p. 9.

regional and local schemes. The regional and local planning
authorities ought to be required to observe the broad lines laid
down by a master plan drawn up by a national body.

A discussion of the principles which should inform a plan
for the metropolis lies outside the province of this book. It
would require much technical knowledge which the author does
not possess. The suggestions made by Sir Raymond Unwin in
his path-breaking Report of the Greater London Regional
Planning Committee in 1929 are so masterly in conception and
so admirable in method that one would suppose they must
form the starting point of any civilised approach to the problem.
The new emphasis on the appropriate use of land, having regard
to its physical characteristics, which have resulted from the
work of the Land Utilisation Survey, also demands attention.
In the last decade a considerable literature on town and country
planning has been produced in England and elsewhere, much
of it highly competent and illuminating. The essential difficulty
lies not in discovering what ought to be done but in clearing
away the obstructive forces which at present impede the path
and in devising efficient administrative machinery for positive
action.

It is to these aspects of the question, therefore, that the
present study has been confined. In the following pages I shall
deal with the administrative side of the satellite town idea, and
the question of controlling the distribution of population, since
both these have a distinct bearing on the opportunities which
would lie before a London Planning Commission.

In case the reader may think that my preoccupation with
the administrative side of the question is of secondary importance or at least somewhat remote from the centre of gravity,
I will conclude this chapter by recounting a discussion which
took place in 1934 before the Departmental Committee on
Garden Cities and Satellite Towns.

The representatives of the London Regional Planning Committee (Alderman A. T. Pike and Mr. R. Hardy Syms) informed
the Departmental Committee that the existing advisory powers
were insufficient, and asked for statutory powers to enable the
Regional Committee to prepare a plan of development for the
whole area, in consultation with the constituent bodies. Such

a plan would at present have only what Mr. Chuter Ede (a member of the Departmental Committee) called "a purely artistic interest." It would not be enforceable.

This request met with opposition of the strongest kind from the Surrey County Council. Mr. Ede, who was in the chair at the meeting of the Surrey County Council at the time when it passed its resolution of disapproval, observed to the Departmental Committee that "To ask local authorities to agree to sterilisation and the loss of rateable value by some body outside is really asking more than I have been able to discover in human nature in local government."

The implication of this remark is that unless and until we get some form of regional government in London to replace the existing units, the planning of the metropolis is an impossibility. There are, on Mr. Ede's showing, three alternatives. First, to continue with the present conflict, waste and confusion. Second, to change human nature. Third, to reform the structure of London government so that control will not be imposed by "outside" bodies.

CHAPTER X

THE BRESSEY REPORT

An attempt has recently been made to cut through the Gordian knot of our complex difficulties in London by segregating the specially pressing problem of highways and subjecting it to separate treatment. The Highway Development Survey for Greater London, undertaken by Sir Charles Bressey and Sir Edwin Lutyens at the instigation of the Minister of Transport, is an event of such an unusual kind that it merits careful consideration in connection with the planning of the metropolis.

The story of the neglectful treatment accorded to the London highways during the past 50 years has already been told at some length; and attention has been drawn to the waste, inadequacy, and stunted growth occurring in a highway system left to the general mismanagement of a multitude of autonomous district authorities devoid of any common purpose.

It will, therefore, be unnecessary even to question the fundamental postulate on which the Bressey Report is founded: namely, the indisputable need for the unified handling of the highway situation in Greater London on bold and comprehensive lines. If further proof were needed, it is to be found in the Report itself. "The want of uniformity and consistency that marks our road system," it states, "is, of course, mainly due to the parochial control exercised in the past by so many different authorities." An example of this is the 20 changes of width which occur in the two miles of main road passing through Bushey and Watford.[1]

The aggregate length of the arterial roads in the London Traffic Area, we learn from the Report, is 525 miles. These constitute the principal radial outlets from London. Nearly a third (32·1 per cent of the length) consist of a single carriage-way of 2 lanes not exceeding 26 feet wide. Almost half the length (47·5 per cent) consists of a single carriage-way of 3 lanes varying between 27 feet and 35 feet in width. Only 10 per cent of the length comprises a double way of 4 lanes,

[1] *Highway Development Survey (Greater London,)* (1937), H.M.S.O. p. 28.

from 36 feet to 44 feet in width.[1] These figures constitute a serious indictment of the system of London government. The total economic loss which is entailed by such a failure to cope with modern needs can scarcely be computed; but one or two illustrations will serve to show the magnitude of its dimensions. In 1929 the London General Omnibus Company estimated that if the average speed of traffic could be raised from 8 to 10 miles an hour, the saving to the Company alone would amount to £1,000 a day or more than £300,000 a year. The delays from congestion in 1927 cost the Company £1,000,000 in out-of-pocket expenses, because the services could not be run according to schedule even at 8 miles an hour. Messrs. J. Lyons & Co., the caterers, find that congestion doubles the cost of delivery, making the cost for each call 6s. 8d. as compared with 3s. 4d. a call during the night, when there is no congestion. The Ministry of Transport estimates the cost of delay at the Iron Bridge, Canning Town, to be £1,000 a day. The President of the Commercial Motor Users Association estimates the cost of congestion within a 3-mile radius of Charing Cross to be about £35,000 a day.[2] At the same time Colonel O'Gorman, M.P., an expert on the road question, remarks that looking down from an aeroplane, the surprising fact is the unexpected emptiness of miles of London streets. Many are blocked, but the congestion is concentrated at crossings, road junctions and other nuclear points.[3]

In such circumstances as these it was not surprising that great public satisfaction should be expressed at the publication of the Bressey Report in the early summer of 1938. After the nightmare of sloth, incompetence and failure to grapple with the metropolitan highway problem which Londoners of all classes have endured and are finding more and more intolerable, a huge sigh of relief seemed to fill the air at the appearance, for the first time, of a bold and ambitious scheme of improvement. Newspapers and weekly reviews of all shades of opinion combined to extol the merits of the Report, and to praise its

[1] *Highway Development Survey (Greater London,)* (1937), H.M.S.O. p. 27.
[2] Thomas Sharp: *Town and Countryside*, p. 184.
[3] Mervyn O'Gorman, C.B., M.P., in *The Nineteenth Century and After*, May 1937.

authors and the Ministry of Transport. Politicians of all parties gave it their blessing in and out of Parliament. Chambers of Commerce, motoring associations and business organisations of various kinds sang its praises. Every motor-bus passenger and pedal cyclist trudging home from work, no less than wealthy motorists in Rolls Royce limousines, thought with gratitude of the Bressey Report. His Majesty's Stationery Office registered a best-seller among Government publications.

The Bressey Report is no doubt a substantial step towards the improvement of the metropolitan highway system, and I do not wish to detract from the praise due to its authors. But it is impossible for a student of London government to accept it in the uncritical spirit adopted by the Press and the general public. A number of questions demand an answer.

In the first place, what is the status of the Report? Who is the responsible body to carry it out or even to approve it? There are about 100 highway authorities within the London traffic area, and all of them are autonomous within their respective districts. The Report is in form a report to the Minister of Transport, and the Minister will no doubt base his policy of grants-in-aid on its recommendations. But it is impossible to believe that this alone would suffice to implement the plan within a reasonable time.

Second, what relation, if any, does the Bressey Report bear to the town planning of London? This question is pertinent whether we have in mind the feeble piecemeal planning which is now in process, or the large-scale regional planning on bold and imaginative lines one would wish to see instituted in the future.

The only reference to this vital point in the Report is a somewhat evasive statement to the effect that several of the proposed routes entail expensive demolitions across densely built-up areas served by devious streets, lanes and alleys. These areas, the Report observes, require drastic replanning in order to bring them into proper relation with the new thoroughfares. Otherwise the route will be flanked with unprofitable and untidy sites of irregular shape, defying architectural treatment and unsuited to modern buildings. Sir Charles Bressey adds: "it is outside the province of this Report to put forward detailed plans for the redevelopment of areas adjacent to new routes, but I have assumed that the responsible local authorities will take the task in hand."[1]

Can such an assumption safely be made? The answer is an emphatic negative, judging by previous experience. And this, indeed, is borne out by a passage which occurs earlier in the Bressey Report itself. Speaking of the arterial road programme carried out during the last 20 years or so in the Greater London area, the authors remark: "The development of land for residential, commercial and industrial purposes has been fostered and accelerated to an almost embarrassing extent by the new arterial roads, and it must be regretfully admitted that town planners were often outpaced by ill-directed private enterprise. Ribbon development proceeded unchecked, and in some areas the local authorities seem to have regarded the new road frontages as a welcome source of immediate rateable value, derived from continuous rows of houses, each of which is usually flanked with a garage entrance. Land fronting the new routes has always been in keen demand, and little of it now remains unsold. As a site for industry and manufacture, arterial road frontages offer the great advantage—apart from traffic facilities—of affording a permanent advertisement of the highest value—as witness the magnificent modern factories set amid attractive gardens along the course of the Western Avenue. Could these changes have been foreseen, it is probable that more use would have been made of the powers conferred upon the Road Board under the Development and Road Improvement Funds Act, 1909, and extended by subsequent legislation to other highway authorities—enabling a belt of land to be acquired along the course of new roads. In many areas the adjoining territory has, unfortunately, been cut up into the serried criss-cross of residential streets which discharge into the main routes at intervals of 300 or 400 feet, impeding the flow of traffic."[2]

In face of this admitted record of incompetence and lost opportunities, it is difficult to understand how Sir Charles Bressey and Sir Edwin Lutyens can "assume" that a similar inept handling of the situation will not occur in the future,

[1] *Op. cit.*, p. 34. [2] *Ibid.*, p. 25.

when further openings for the exploitation of main road improvements are once again presented to local authorities and private profit-making interests. More than optimism is required to support such an assumption.

There is, however, a larger sense in which the complete separation of the „Highway Development Survey from town planning problems weakens the whole fabric of the Bressey Report. Since highways are no more than means of communication, their proper disposition should obviously be related to the directions in which the streams of traffic need to flow. These traffic streams are determined by the location of industry and commerce, the houses of the people, their recreation grounds, places of amusement, and so forth. Under town planning, highways are co-ordinated with these other factors so that the traffic streams flow quickly and smoothly without waste of time, energy or money. Without town planning, industry, commerce, housing, etc., must either accommodate themselves to the roads; or more or less ignore them and place a burden of inconvenience, delay, expense and irritation on the shoulders of those who use them. This, indeed, is one of the most obvious differences between planned and unplanned development.

The Bressey Report sets forth an ambitious and elaborate highway plan, to cover the next 30 years and calling for an expenditure of possibly £250 millions, without anyone— including the authors—having the slightest idea of what is to be the future development of the Greater London Region: where its industry is to be located, where residential areas are to be developed, where (or whether) satellite towns are to be constructed, where schools, hospitals, fire stations, railway stations, and other public institutions are to be located, what land is to be sterilised for building or acquired for open spaces. How can one plan a great highway system on satisfactory lines without knowledge of these essential factors?

Mr. F. J. Osborn has observed that "You cannot plan London by starting at the traffic end, or by an arithmetical projection of existing tendencies into the future. The Bressey proposals are one of the periodical bursts of artery-driving zeal which alternate with the periodical inertias caused by their fabulous cost and their very temporary value in postponing congestion. It is safe to prophesy that a few of the cheaper parts of the proposals will be carried out, mostly in the outer parts of the region, and that they will foster further suburban growth and fill up instantly any slight road improvements made towards the centre. It is tragic that so much work, expense, and imagination should be vitiated by the absence of a regional planning machinery or outlook."[1]

[1] F. J. Osborn: "The Planning of Greater London" in *Town and Country Planning*, July-September 1938, p. 98.

CHAPTER XI

SATELLITE TOWNS

One of the most important matters which would have to be considered in connection with a regional plan for the metropolis is the extent to which the decentralisation of population is desirable and possible.

The method which is most frequently discussed in regard to London is the building of satellite towns on the lines of Welwyn or Letchworth. The idea is gaining ground slowly but steadily that this is the most hopeful remedy for the over-growth and congestion of London, although it is by no means the only key to a complex problem.

The Departmental Committee on Garden Cities and Satellite Towns appointed by the Minister of Health issued a report in 1935 favourable to the movement. "A town may become overgrown as a single unit," the Committee declared, "just as may happen to any other organism. The community links in such case become weakened or lost. The remedy may well be found in organising growth beyond such limits by means of satellite units having some independent local life but depending on the parent town for those conveniences and amenities which only a large population can support. Such organisation to be effective must be provided for and expressed in the physical form of the town. The units of development should themselves be organised on more or less self-contained lines according to their distance from the parent town."[1] Mr. Herbert Morrison, in giving evidence recently to a Royal Commission on behalf of the London County Council, has subscribed to the view that "the urban development of Greater London already exceeds the aggregate which would have been desirable on general principles of town planning and in the interests of the well-being of the population of London."[2]

There are many possible variations in the size, character and

[1] Report, p. 9.
[2] Royal Commission on the Geographical Distribution of the Industrial Population: Minutes of Evidence, 2nd March, 1938 (Morrison).

position of satellite towns. Some experts consider the maximum size to be a population of 50,000 to 60,000,[1] while others would fix the optimum figure either above or below this limit. Some people believe that the best results can be obtained only by designing and constructing a new town on unbroken ground chosen specially for the purpose. Others prefer the idea of developing an existing village or small country town with a community life of its own into a satellite on carefully controlled lines, surrounded by a sterilised belt to prevent excessive growth. The original pioneers of the garden city movement have all been in favour of the "open development" style which is found at Letchworth and Welwyn; but a vigorous opponent of this style has recently expressed strong views on the practice of applying to urban living "a small-holdings idea which has no relation to town life."[2]

It is not necessary to insist dogmatically on the "rightness" or "wrongness" of any of these opinions. Those who hold them are agreed on the essential principles of the satellite town idea, and the substantial advantages to be derived from its application on a large scale in the metropolitan region. It is difficult, indeed, to see what other suggestion can be compared with this proposal as a remedy for the excessive size and concentration of built-up London. In carrying it out, however, there is room for an immense amount of diversity in the size, character and situation of the satellites. There will be ample scope for the expression of the best ideas in the several schools of thought.

The satellite town idea offers, to hundreds of thousands of Londoners who are now penned up in the most squalid and sordid quarters of inner London, the possibility of a healthy and spacious existence amid pleasant surroundings, with easy access to the countryside. It would involve the application of the slum clearance idea to factories that have fallen below the reasonable standard of industrial accommodation which a civilised community should allow its workers to be offered. It

[1] A. T. Pike: Minutes of Evidence before the Departmental Committee on Garden Cities and Satellite Towns, 24th January, 1934 (unpublished).

[2] Thomas Sharp: *Town and Countryside: Some Aspects of Urban and Rural Development*, p. 149. This is a most interesting book. See also *English Panorama*, by the same author.

would introduce an ordered proximity between workplace and home which would contrast favourably with the long, dreary, fatiguing and expensive journeys which must be undertaken on every working day by millions of Londoners. It provides a method of spending very large sums of public money on projects which are likely to offer reasonable prospects of becoming ultimately self-supporting—a consideration of some importance when the need for large-scale public expenditure to relieve unemployment again arises. Experience in the past decade showed that it is by no means easy to find objects of extensive public expenditure which are likely to yield even a partial return. If, however, a programme of this kind is contemplated, it cannot be formulated at short notice but must be worked out well in advance.

It must not be assumed that satellite towns are all that is required to reconstruct London on healthier and more up-to-date lines. Much else is needed on a large scale in the built-up area of the administrative county. Whole districts in the central core of the metropolis, and especially in the south, east and north, including large portions of such places as Shoreditch, Stepney, Bermondsey, Kilburn and Islington, require to be completely cleared and rebuilt.

The satellite town method has been referred to at some length not merely because it is a promising idea but because the creation of such towns by local authorities in the London region is today impracticable. If we desire to promote satellite towns by municipal enterprise it will be necessary to introduce substantial reforms in the structure of local government.

It is quite unreasonable to expect satellite towns to be wholly or mainly built or financed by local authorities in places lying outside their own areas. If the London County Council were to sponsor a satellite town, it would have to spend millions of pounds, and to embark upon a huge administrative task, for the purpose of constructing a town in which it would exercise no municipal functions and over which it would possess no powers of taxation. It is unfair to expect the London County Council to devote its resources and energies to the establishment of garden cities or even garden suburbs in places where the benefit will go entirely or mainly to the inhabitants and authorities of outlying areas.

The London County Council has already developed a number of large housing estates at various places outside the county.[1] I shall now show shortly the highly unsatisfactory results which not only have occurred, but must inevitably occur from such efforts, so long as the existing organisation continues.

[1] *Ante*, p. 169.

CHAPTER XII

BECONTREE AND DAGENHAM:
AN OBJECT-LESSON

We have already noted that owing to the inadequate size of the administrative county, and the fact that there is no building land available within the area, the London County Council has been driven to establish most of its post-war housing estates outside the boundary.[1] Well over half of the dwellings provided by the Council are situated in the areas of outlying authorities.[2]

The most extensive of these estates is at Becontree and Dagenham in the County of Essex, about 10 miles from Charing Cross.[3] This is the largest municipal housing estate in the world. It covers 2,770 acres, or more than 4 square miles of land. On it some 25,000 dwellings have been constructed to accommodate approximately 115,000 persons. The estate contains 27 churches, 30 schools, 400 shops, 9 public houses, 14 doctor's houses, cinemas, clinics, libraries, a swimming pool and various other institutions. The cost up to the present is £13,455,170.[4]

When the estate is complete it is expected to provide accommodation for 125,000 to 130,000 persons. This compares with about 250,000 persons housed by the London County Council on its other estates.[5] The estate at Becontree had a larger population in 1931 than Blackburn, Gateshead, Huddersfield, Norwich, Preston, Southend, South Shields or Stockport. Its area is larger than 21 of the metropolitan boroughs.

---

[1] *Ante*, pp. 169.   [2] For details, see *ante*, p. 215.
[3] Strictly speaking, one should refer to Becontree and Dagenham, since the estate comprises both places; but for the sake of brevity and convenience I shall speak of Becontree only.
[4] *London Housing* (L.C.C. publication, No. 3272/1937), p. 157. 8,726 of the dwellings are 3-room houses; 6,739 are 4-room non-parlour houses; 5,085 are 4-room parlour houses; 3,369 are 5-room houses; 138 are 6-room houses. The remainder are flats of various sizes.
[5] E. C. Willatts: *The Land of Britain. Middlesex and the London Region*, p. 166 (Report of the Land Utilisation Survey of Britain).

The estate clearly needed to be under the control of one local authority and was planned with this in mind. But the obvious requirement was not met and in consequence the planned development of the area has been in several ways ineffective and unsatisfactory. The estate lies within three different local government areas: Ilford, Barking and Dagenham, each of which has a separate borough council acting as a self-contained unit. The building operations were commenced in three separate parts of the estate, with the consequence that "from the start there were two, if not three, centres of population and foci of nascent local sentiment. Such beginnings of social life as tenants' associations and sports clubs grew up in duplicate. . . . The tenants in different parts of the estate have realised that their destinies are bound up in many particulars with their fellow-citizens of those municipal areas. They tend to think of themselves as living in East Barking or New Ilford. Thirdly, whilst the London County Council desired to develop a township, it was only a housing authority building outside its area; it had no direct powers to establish or foster industries, and in this respect it was in a greatly inferior position to the companies which initiated Letchworth and Welwyn."[1]

The local government authorities concerned with the Becontree Estate consist of: (1) The London County Council, which is merely the landlord exercising powers under the Housing Acts. It has no control over municipal services. (2) Ilford Municipal Borough, which has jurisdiction over the northwest portion in regard to the usual local government services. It is also the elementary education authority for that part of the estate. (3) Dagenham Municipal Borough—formerly an urban district but incorporated in 1938—administering the eastern half of the estate. It is the local authority for library purposes but not for education. (4) Barking Municipal Borough (incorporated in 1931), which is responsible for local government services in the western part of the estate. Barking is an authority for elementary education. (5) Essex County Council, which has power over the whole area for certain purposes,

[1] H. A. Mess: "The Growth and Decay of Towns," *Political Quarterly*. July-September 1938, pp. 400–1.

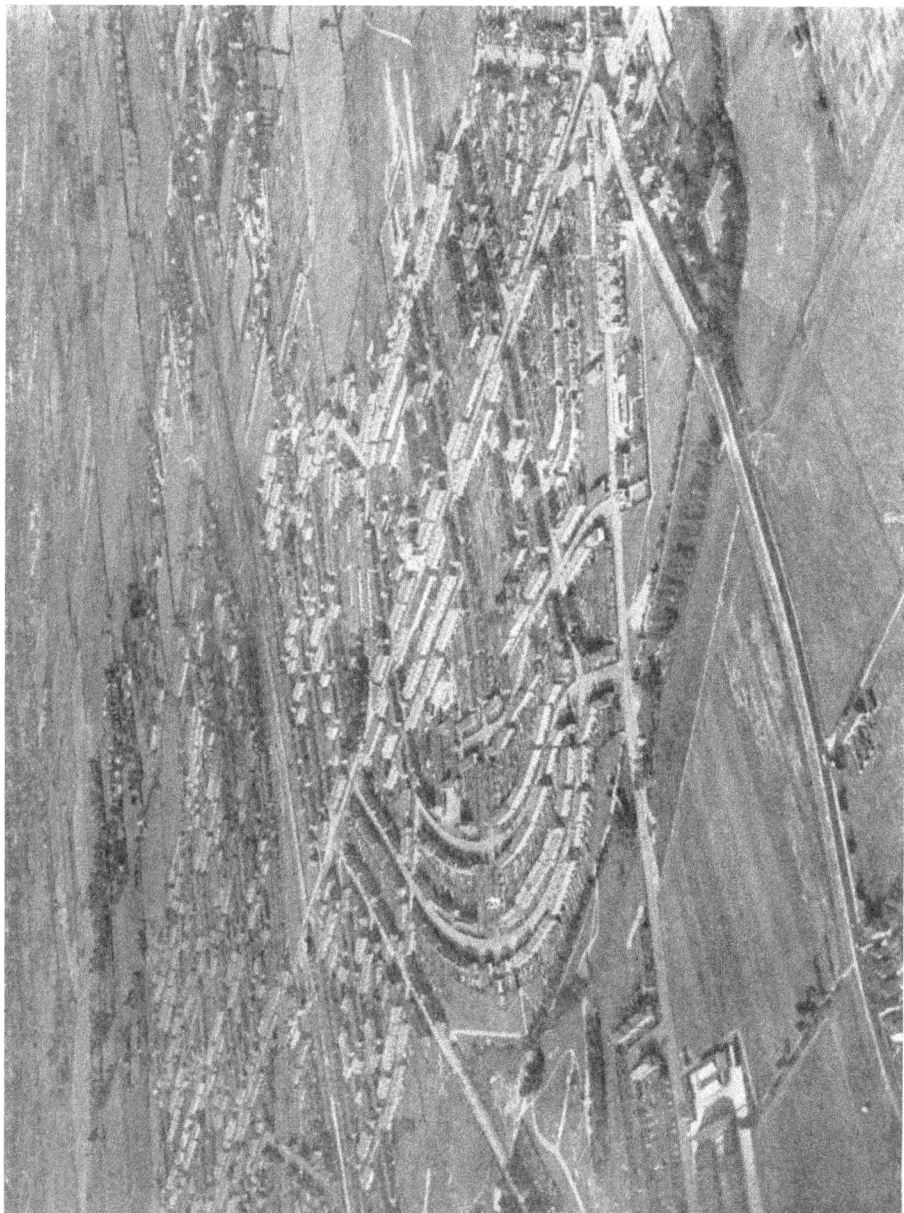

PLATE VIII. The London County Council housing estate at Becontree and Dagenham. " A wildnerness of puzzled and discontented persons " *(Aerofilms)*

such as public assistance, the supervision of tuberculosis and other medical services. It is the authority for higher education, and provides secondary and technical schools, evening institutes and classes for adults. It is also the authority for elementary education in the Dagenham area.[1]

These local authorities were confronted with a formidable problem when the London County Council decided to build this huge estate in their areas. The building of schools alone has called for a capital expenditure on their part of at least £1,000,000.[2] At the same time, the rateable value of the Dagenham part of the estate was, and is likely to remain, low, since there are few shops and no large houses to help to raise the financial level. The yield of a penny rate in Dagenham is less than the produce of a similar rate in any one of 13 local authorities with populations of the same size in Greater London.[3]

When the newcomers began to arrive in the early days of the estate they found there was no school accommodation whatever available for infants in the northern part of the area. Serious overcrowding in the existing elementary schools persisted in various parts of the settlement until 1930. Evening institutes did not appear for some years, while those who want technical instruction must still travel some distance to the institutes of boroughs much nearer London or within the administrative county itself. A careful observer remarks that the provision for central and secondary school education on the Becontree estate is on a lower standard than it is in London, although the reorganised senior school departments of elementary schools are beginning to provide a form of secondary education for all school children.[4]

It is unfair to blame the local authorities for the shortcomings of the municipal services. They were given a huge and unforeseen burden to carry without any adequate means of obtaining additional resources. Nor can the London County Council justly be criticised. They were powerless to render help in regard to local government services, although the London County Council has the largest and most comprehensive educational and public health system in England. All the

---

[1] Terence Young: *Becontree and Dagenham*, pp. 109–10.
[2] *Ibid.*, pp. 167, 275.   [3] *Ibid.*, p. 57.   [4] *Ibid.*, p. 80.

municipal organs concerned were in the grip of a malaise of mis-organisation which reduced them severally and collectively to frustration and impotence.

The London County Council did what it could. It aided the provision of public services in Becontree by setting aside 500 acres of land on the estate for use as open spaces.[1] The officials of the Council have taken the utmost pains to try to foster a community spirit on the estate as a whole and to discourage the growth of sectional sentiments related to Becontree on the one hand and Dagenham on the other. "They hoped that one of the roads facing the central open space, Parsloes, would have become the social and administrative centre for the whole Estate; they hoped that local government buildings, large shops, cinemas, churches and other buildings would have been erected there. On a small scale, this may happen but at most it will be the centre of the Barking section merely—the centre of gravity of the Dagenham Urban District Council [now a borough] is moving eastwards off the Estate."[2]

Thus, the larger aim of the London County Council of creating a "new township, complete in itself" has not been realised; and the reason for this, we are told by Mr. Terence Young, the historian of Becontree, is because of the division of the area among three local authorities.[3]

Almost all the municipal services are split up among these three authorities. Electricity is supplied by the electricity departments of the boroughs of Ilford and Barking in their respective portions of the estate and by the County of London Electric Supply Company in the Dagenham portion. Sewage disposal is split up in a similar manner.[4] The supply of water and of gas, both of them in company hands, appear to be the only services which are rendered for the estate as a whole. Private enterprise, in the shape of the Gas, Light and Coke Company and the South Essex Waterworks Company, are ahead of municipal organisation so far as the Becontree Estate is concerned.

The defects of the Becontree experiment in terms of civic

[1] Terence Young: *Becontree and Dagenham*, p. 276; *London Housing* (L.C.C. publication, No. 3272/1937), p. 157.
[2] Terence Young, *op. cit.*, p. 92.   [3] *Ibid.*, p. 280.   [4] *Ibid.*, p. 111.

spirit and the imponderable elements in community life are intrinsically more serious than its shortcomings in terms of material efficiency. It seems almost incredible that no one appears to have enquired into the difference between a human dump and a town before embarking on this huge expenditure of money and effort affecting so many thousands of lives. But the fact remains, Professor Mess remarks, that it was not realised for a considerable time that "to build 22,000 houses and to assemble 22,000 families does not of itself create a town; it creates a wilderness of puzzled and discontented persons."[1] Becontree is not a very promising nursery for the citizens of to-morrow.

The same kind of thing has occurred on a smaller scale elsewhere in the environs of London. The Watling estate, for example, established by the London County Council outside the administrative county on the north-west side of the metropolis, is sharing a similar fate. The estate comprised (at the end of 1936) more than 4,000 dwellings containing 15,000 rooms, erected at a cost of about £2,500,000.[2] It started under more favourable auspices than the Becontree estate because the Watling area was in its early stages a clearly defined urban aggregate with a rural fringe. It coincided, moreover, fairly closely with one of the wards of the borough of Hendon and hence was subject to greater unity in regard to local government services. But "the surrounding area has now been built up also, so that the boundaries of Watling have been blurred and awareness of Watling is more difficult to maintain. . . . It is doubtful whether the Watling estate is going to develop further, or even to maintain, the sense of community which at one stage seemed to be emerging."[3]

The fact that the essential failure of these efforts is due either to the defective municipal organisation of areas and authorities on the one hand, or to a lack of regional planning on the other, or to both causes, rather than to the difficulties of creating a genuine community out of an assembly of working-

[1] H. A. Mess: "The Growth and Decay of Towns" in the *Political Quarterly*, July-September 1938, p. 398.
[2] *London Statistics* (1935–7), vol. 40, p. 155.
[3] H. A. Mess: *op. cit.*, p. 402.

class families brought together without previous ties, is shown by the experience of the Hampstead Garden Suburb. The Garden Suburb represents an entirely different idea. It is a well-to-do middle-class settlement containing a rather high proportion of cultivated persons. It was started by voluntary effort based to some extent on non-commercial aims. It was commenced about 25 years ago in a favoured area adjoining Hampstead Heath and Golders Hill, and has been subject to rigid control from the standpoint of preserving amenity and architectural consistency. Here, it might be thought, are all the elements conducive to the building up of a satisfactory community. Yet Professor Mess remarks that "the consciousness and the organisation of the Hampstead Garden Suburb community are divided between two local authorities, the borough of Hendon and of Finchley. This operates adversely in such ways as that the Residents' Association has to operate through two local government committees. It may be because its political consciousness is thus weakened that the Hampstead Garden Suburb has to this day neither public library nor swimming-bath. The one public elementary school is provided by the borough of Hendon."[1]

The importance of drawing the correct conclusions from these divers examples can scarcely be exaggerated when we recall the immense scale on which public housing activity is now being pursued throughout the country, and especially in the metropolitan region.

All the foregoing difficulties are likely to be encountered, and all the mistakes of the past repeated, if the proposals of the South Essex Regional Planning Scheme—to take one important example—are allowed to go forward in their present form. This scheme calls for a series of working-class towns, with a population of about 30,000 each, to be situated on the north of the industrial belt running along the Essex bank of the Thames. It contemplates a huge expansion of industrial development in an already swollen area and looks forward to the provision of housing accommodation for at least another 500,000 persons.[2] The towns in question may possibly be

[1] H. A. Mess: *op. cit.*, p. 403.
[2] Terence Young: *Becontree and Dagenham*, p. 278.

built wholly or in part by private enterprise, and all the problems which have arisen in Becontree and the other estates will occur in an equally intractable form, unless drastic changes are made in the organisation of London government and effective methods of town and country planning introduced. If such planning were brought about, it is doubtful whether the essential features of the South Essex Scheme would survive close scrutiny.

The changes which are imperatively needed to avoid the defects of the existing type of out-county housing estate, and to a yet greater degree for the successful carrying out of the satellite town idea, should now be clear.

In the first place, it is essential that a satellite or housing estate which is promoted by a large municipal authority, should be situated within the area of that authority. Only thus can the responsible body exercise both the powers of a landlord and those of a local authority—a combination which is essential for a successful result. In no other way can the provision of municipal services on a reasonable standard be assured from the outset without an intolerable local burden being placed on the outlying local authority. Only thus can both the burden of the effort and the benefit of the achievement fall on the same community and the ends of justice and efficiency be served simultaneously. Only by this principle can friction between the initiating organ and the recipient area be avoided.

It would be, and has been, possible to construct satellite towns by other methods. One possibility is the type of company which undertook the building of Welwyn Garden City; another is a public service trust or commission. Yet a third alternative would be by the Ministry of Health or other central government organ. But so far as London is concerned, there is every reason to suppose that there would be immense advantages in looking to a Greater London Council for the essential initiative to provide the momentum. Admittedly it would be necessary for the central government to make grants or loans, and to provide assistance and supervision of other kinds. A relationship of a new type between central and local authority might be required in which responsibility for the final result would be shared through some form of joint organisation.[1] This would not, however, affect the basic conception of looking to the regional council for the essential motive force, the underlying enthusiasm and the special knowledge required for the main administrative effort. Moreover, the satellite towns would have to be in conformity with the general lines of the plan for the metropolis laid down by the Greater London Council or its Planning Commission.

The creation of a satellite town would no doubt call for the establishment of a special committee of the Greater London Council during the process of construction. And as the town began to emerge, a local authority would have to be formed for the discharge of municipal services therein (other than those of a regional character) and for the fostering of a civic spirit. Otherwise there could be no true community, but only another human dump of "puzzled and discontented persons."

The question of transferring or decentralising population as between one region and another is a special problem for which other methods would be required. A discussion of such methods lies outside the scope of this book.

[1] See my evidence before the Royal Commission on the Geographical Distribution of the Industrial Population.

CHAPTER XIII

THE BARLOW COMMISSION ON POPULATION

However wisely we reorganise the structure of London government, however sensibly we may try to plan its future development and present form, all our efforts are likely to end in
disappointment and failure unless we make up our minds to
face fairly and squarely the necessity for limiting the size of
London.

A considerable part of this book has been devoted to depicting
the immense growth of the metropolis both in terms of population and of physical extent. I have shown in detail the way in
which London has for many decades outgrown the machinery
of local government and the serious disadvantages which have
resulted therefrom. It should not be difficult to obtain general
agreement to the proposition that the metropolis has reached
an excessive size, whether regarded from a social, administrative, political, economic or military point of view. This is true
whether we consider the dimensions of London absolutely, or
relatively to the total size of Great Britain in terms of area and
population. It holds good no matter whether we have in mind
the health and welfare of the inhabitants of London or those
of the rest of the country.

The problem of London, it must be emphasised, is a problem
which concerns the whole nation. The over-concentration of
population, wealth, industry, commerce, drama, music, art and
many other elements of civilised life in the capital city, has
produced a serious lack of balance in the country as a whole.
The extraordinary aesthetic and cultural poverty of provincial
life in England is in large measure due to this phenomenon.
There are relatively too many art exhibitions, concerts, lectures,
theatres and so forth in London; and too few in the provinces.
London has sucked the rest of the country dry from a cultural
point of view, while the attraction of its economic pull has
helped to deprive the depressed areas of their prospects of
revival. The over-nourishment of the metropolis has been
partly at the expense of South Wales and Tyneside. The

embarras de richesse of the capital has its counterpart in the economic, cultural and aesthetic malnutrition of Liverpool and Newcastle, Durham and South Wales.[1]

London itself has also suffered acutely in the quality of its own life. The ever-widening spread of its vast sprawling mass means that an increasing number of its inhabitants have been driven into the remote outlying dormitories where they lead a petty suburban life far removed from the magnificent cultural, scientific and artistic institutions which are the chief glory of London and the principal justification of a giant capital city. It would be interesting to discover what proportion of the residents of Barnet or Malden, Chingford or Bromley are accustomed to attend a West End theatre, or visit the British Museum or the National Gallery or the Wallace Collection, or even to take their children to the Zoological Gardens in Regent's Park. The quality of London life inevitably becomes attenuated as one recedes from the centre to the periphery.

Even social life becomes empoverished when the distances to be traversed are so great. How can social intercourse take place or friendship be fostered between persons and families separated by an hour, or even two hours of exhausting travel in crowded tubes and motor-buses? Thus Londoners in the outer suburbs must necessarily limit their circle of friends to local people.

The better-off classes, and particularly those who own motor-cars, are impelled by the increasing urbanisation of the metropolis to acquire country houses or cottages in distant retreats where the countryside is still unspoilt; and in these places they seek refuge every week-end. The result is still further to diminish the general quality of social life in London; and, incidentally, to reduce the interest of these influential sections of the community in the fortunes and the welfare of the metropolis. They are, in a sense, escapists.

The problems of local government which are raised by the emergence of a metropolitan centre the size of London are so formidable that we have never seriously attempted to solve them. A unified type of city government which may be worked with success in a town as large as Manchester or Birmingham,

[1] See the article by Brinley Thomas on "The Drift of Industry" in *The Times* Supplement, 31st May, 1938, p. xii.

cannot be applied to a vast, incoherent mass of 7 or 8 million persons. We have, therefore, drifted along from one expedient to another without ever really grasping an intelligible principle of policy. In the past fifty years, for example, two opposing tendencies have been manifested in the structure of London government. On the one hand, a merging tendency exemplified by the absorption of the Poor Law Guardians, the Metropolitan Asylums Board, and the London School Board by the London County Council. On the other, a tendency towards separation of function reflected by the establishment of the Metropolitan Water Board, the Port of London Authority and the London Passenger Transport Board.

A comprehensive scheme of reform has been put forward in the foregoing pages which is designed to overcome, to the maximum extent, the difficulties and disadvantages of the present situation. If it were adopted, we should have a regional form of government very different from anything which has hitherto existed in Great Britain. It would be an experiment in democracy of a new and promising kind. But the larger and more amorphous London becomes, the more unwieldy and incoherent its sprawling mass is permitted to grow, the more difficult the experiment will be. On these grounds alone—and there are many others of great cogency—there is ample reason to check the further growth of London until we have solved the extremely difficult political, constitutional and administrative problems which have arisen from its immense size. The view put forward by Mr. Pick that London can remain a healthy organism only if it continues to grow appears to me to be the exact opposite of the truth.[1] On the contrary, London is not a healthy organism at present for the very reason that it has grown to so excessive a size; and the more it is permitted to grow the more unhealthy it will become.

If restriction of the further growth of London be accepted as a desirable aim, the one certain method of achieving our purpose is through control over the location of industry. Up to the present no attempt has been made to guide the location of industry, apart from recent belated efforts to attract enterprise to the derelict areas. The traditional assumption is that

[1] See *ante*, pp. 319–20.

the selection of a site is the concern only of the entrepreneur, who is supposed to base his actions on the most delicate calculations and elaborate economic analysis. This assumption is, if not unfounded, at least untested and unproved.

The annual survey of industrial development made by the Board of Trade gives the following motives which influenced industrialists in the choice of sites for their factories in two recent years.[1]

	No. of cases	
	1937	1936
(a) Accessibility of raw materials	26	27
(b) Proximity to markets	34	24
(c) Suitability of labour	67	56
(d) Cheap land, low rent or low rates	34	23
(e) Proximity to other factories in the same industry	41	38
(f) Convenience of premises	212	224
(g) Proximity to employer's residence	2	13
TOTAL	416	405

These figures relate to Great Britain as a whole, but there are also details given for various regions. Thus in Greater London, for the year 1937 there were 5 cases coming under cause (a), 13 under cause (b), 22 under cause (c), 18 under cause (d)—these are in the Outer London areas and probably represent the attraction of these areas compared with the inner London districts where land is more expensive and rents higher. Under cause (e) there are 14 cases in Greater London, half of them being in the clothing trades. Cause (f) accounts for more than half the total and in no less than 181 instances the suitability of existing buildings was the determining influence. These 181 instances cover a wide range of industries and 97 of them refer to Greater London.[2]

The figures are based on oral information which is admittedly vague and incomplete; and the Board of Trade does not claim complete reliability for them. But even allowing for a margin

[1] Where more than one reason was given the chief one is taken for purposes of classification.
[2] *Survey of Industrial Development*, 1937, Board of Trade (H.M.S.O., 1938), pp. 4–5.

of error they indicate how fallacious it is to assume that factories must be placed where they happen to be placed, if economic harmony is to prevail. In nearly half the cases the mere convenience of existing premises was the determining cause. The actual location of the site, considered as an immobile factor, refers only to those few instances, of which there were 5 in London, where accessibility to raw materials is the predominating cause. And even this can scarcely be regarded seriously in a small country such as Britain, equipped with excellent transport facilities. The same applies to cause (b): proximity to markets. As regards the next factor, the labour supply is obviously mobile, or can be made so. The question of cheap land or buildings does not arise in London as compared with other parts of the country.

We can see, then, how cautiously we should accept loose statements concerning the economic necessity for industrial development in any given area; and, conversely, with what reserve we should accept estimates of the economic damage which would result if such development were restrained or prohibited in any district or region. As regards the frequently alleged need for proximity to markets or accessibility to raw materials, Mr. Harold Macmillan, M.P., has proposed the solution of this particular difficulty by eliminating the price of transport from the cost of production. He would do this by nationalising the transport system and providing the transport of goods as a free service. Distance would thus cease to be an economic factor in any particular case.[1]

Even if economic considerations did at present determine in every case the exact location of industry, this alone would not be conclusive as to the practice which should be permitted to prevail. There are highly important social, political and administrative considerations which must also be taken into account. Moreover, the new mobility of industry is a factor of great potential significance of which little use has so far been made. Factories and mills were formerly chained at first to the waterways and later to the coalfields. Now they are free to go almost anywhere, for electrical power is ubiquitous, or can easily be made so.

[1] Harold Macmillan, M.P.: *The Middle Way*, pp. 233 *et seq.*

We have already for long been committed to the principle of controlling the location of industry within a local area, since every territorial plan contains zoning provisions to regulate the amount and character of building development which may be constructed in the various parts of the district, and the uses to which it may be put. But so far we have not attempted to control the location of industry as between different areas.

This in turn is the result of the fundamental conception which has hitherto guided, or misguided, the planning movement in Great Britain. We have left the essential initiative in the matter of territorial planning to the local authorities. And not even the larger ones were selected for the purpose, since the county councils (apart from the London County Council) were not primarily responsible for town and country planning, though they might be given powers in certain circumstances. Thus it has been assumed that the correct procedure is to start at the bottom with comparatively small units. For any larger type of organisation we have relied on the hope that local authorities would voluntarily form themselves into regional authorities. This policy has been a complete failure in the metropolis.

In any case such a policy is obviously useless to guide the location of industry. It is asking too much of local authorities to expect them as isolated units to take effective steps to preclude industrial development from their areas to any substantial extent. Practically all local authorities regard industrial development as desirable in almost any circumstances. They accept without question the belief that a new factory going up in their area is necessarily a good thing. The immediate advantage is obvious and direct; the disadvantages, where they exist, are indirect and more remote, no matter how inexorable they may become. Hence local authorities openly compete with one another for the privilege of having their area chosen by entrepreneurs on the look-out for industrial sites. This attitude is inevitable in the present state of affairs.

Moreover, local authorities, whether urban or rural, are hesitant and lacking in confidence in the exercise of their planning activities for another cause of which full account has not so far been taken. This is the immensely important fact that, for the first time in history, the city has lost its physical coherence. For hundreds of years the boundary of a town was a definite thing, a clear organisation of roads, gates or walls, with streets leading from the periphery to the centre. The advent of mechanical traction during the past 20 or 30 years, the development of the electric railway, the motor-car, the motor-bus, tramway or trolley-bus, has robbed the municipal boundary of its former significance. There no longer *is* a boundary, except in name, save for administrative purposes.

Social and economic life flows over the frontier with increasing intensity and has submerged the lines of demarcation. People live outside the limits of the town and flood into its area on every working day in thousands, or vice versa. Even the suburb, which formerly possessed recognisable characteristics and a certain compactness of its own, is growing diluted and inchoate as the townsman goes farther and farther afield in search of a home removed from the noise and hustle of the town. The effect of this on territorial planning is that a local authority covering only part of an economic and social community feels reluctant to discourage industrial expansion within its area unless the neighbouring authorities having jurisdiction over contiguous territory are willing to follow suit, which is not always the case. The difficulty is overcome only in those places where regional planning authorities have been set up.

If the territorial planning movement is to be strengthened so as to play an important part in determining the distribution of the population, two essential changes will be necessary. In the first place, planning will have to start on a national scale, and work down through regional and local schemes. In the second place, the future location of industry will in the first instance have to be determined on broad lines by a national organ.

The national plan would not attempt to specify where any new individual factory is to be situated. It would be unnecessary to do more than (a) prohibit industrial development in certain scheduled areas, (b) authorise industrial development in certain other areas. There would remain a wide latitude of choice open to the entrepreneur. It would, however, be a great mistake to limit the national direction over the location of industry to mere prohibition of further expansion in overgrown areas. From the point of view of amenity and good social planning, there is everything to be said in favour of express authorisation of industrial areas.

The local or regional planning authorities would proceed to make detailed plans for their areas in conformity with the broad lines laid down by the master plan. The latter would, of course, be compelled to deal with such matters as trunk roads, harbours, and docks, etc.

The regulation of the location of industry need not necessarily be carried out by a single uniform method. There are, in fact, several possible methods, all of which could usefully be applied in different parts of the field.

In some cases it will be essential to impose definite prohibitions over particular areas. For reasons already given the metropolitan region falls into this category.

Even in these cases it would probably not be practical to impose an absolute prohibition. There would no doubt have to be provision made for permission to be granted in exceptional cases where, for example, it could be proved that it would be economically impossible to carry on the enterprise profitably elsewhere, *and* that it would not lead to an influx of workers into the region. There would also be the question of extensions to existing factories to be dealt with.

In effect, this would mean, in legislative terms, that the construction or conversion of premises for manufacturing purposes would be forbidden in the prohibited areas except under licence. A licence would not be granted unless the exceptional conditions were satisfied, and the burden of proof would lie on the applicant.

A second method worth consideration in regulating the location of industry would be to use the public services as instruments of persuasion and dissuasion. If electrical power were offered by the British Electricity Authority at specially favourable rates to manufacturers who were willing to start factories in the derelict areas or wherever else the central planning body wanted to encourage industrial development, and at specially high rates in places where it was desired to discourage it; if municipal houses, roads, hospitals, water supply, schools and other public services were promised in some places and withheld in others in order to assist in directing the location of industry, business men would discover very cogent arguments in favour of compliance, without the use of any compulsory powers whatever. Few people realise that we have powerful instruments of control already in our hands if we will but use them.

A third method would be to provide an advisory service designed to assist manufacturers in the choice of a site. The amount of information at the disposal of the entrepreneur must in many cases be extremely inadequate even from the purely economic point of view. An advisory service could operate in two different but related ways. On the one hand it could issue general reports and recommendations concerning the location of industry from the economic and social points of view; on the other hand it could be asked to report on every proposal to establish or extend a factory. Manufacturers might be required by law to submit their proposals to such a body even if its recommendations were of a purely advisory nature. But the reports, both general and particular, should be published.

In the proposals which have been made above, the reader will see that the regulation of industrial development in London is linked up with the territorial planning of the rest of the country. This is a necessary and inevitable connection. It is impossible to conceive the restriction of London's growth as a purely local or even regional problem. It is clearly a matter of national concern calling for central administration.

The relation between the national planning organ and the regional planning authorities (in the metropolis, the Greater London Council); the administrative powers which both the former and the latter should possess; the extent to which their policy should be laid down explicitly by legislation; the need for a right of appeal to some kind of administrative tribunal; the basic principles which should inform territorial planning and the location of industry on a national scale; the desirability of establishing national parks and preserving the sea coast; the role which the central government should play in promoting decentralisation of population as between one region and another: these and cognate matters are complex problems which call for detailed treatment in a separate study.[1] To attempt to deal with them here would overload these pages and possibly obscure the main issue, which is that some drastic restriction on the growth of London is essential to ensure both the good government of the metropolis and the welfare of its inhabitants. That result, I have tried to show, can be effectively achieved by controlling the location of industry. Such control must be projected on a national scale. Those are the essential facts to be borne in mind.

The urgency of the task is shown by the fact that between 1924 and 1934 the area in the administrative county of London devoted to industry alone (not including docks) increased from 3,000 to 3,660 acres, while the area used for other business purposes has also been greatly extended.[2] In outer London there is immense activity in factory building and extension. Some of this is due to movement from the inner core of the metropolis, but most of it is either new or consists of transfers from other parts of the country. Industrial development in Greater London is disproportionately large compared with the other regions into which the Board of Trade has divided the country.[3]

The whole subject was considered by the Royal Commission on the Geographical Distribution of the Industrial Population. This Commission was appointed in the middle of 1937 under the Chairmanship of the Rt. Hon. Sir Montague Barlow, a former Minister of Labour. Its terms of reference were "to enquire into the causes which have influenced the present geographical distribution of the industrial population of Great Britain and the probable direction of any change in

[1] See my pamphlet, *Planning and Performance* (J. M. Dent, 1943); "Planning Administration and Planners," by W. A. Robson in *Homes, Towns, and Countryside*, edited by Gilbert and Elizabeth McAllister; and my chapter on "Government," in *Physical Planning*, edited by Ian R. M. McCallum.

[2] F. J. Osborn: "The Planning of Greater London" in *Town and Country Planning*, July–September 1938, 9. 98.

[3] *Survey of Industrial Development*, 1937, Board of Trade, H.M.S.O. (1938), p. 2.

that distribution in the future; to consider what social, economic or strategical disadvantages arise from the concentration of industries or of the industrial population in large towns or in particular areas of the country; and to report what remedial measures if any should be taken in the national interest."

The Barlow Commission gave special attention to the problem of the metropolis. It described the social, economic and strategical disadvantages arising from so vast a concentration of population and industry, and analysed the detrimental effects of its magnetic attraction on other parts of the country. The Commission declared that the disadvantages arising under all three heads—social, economic, and strategical—presented themselves with a special degree of urgency in the case of London and the Home Counties. Not only the obvious danger of attack from the air in case of war, but also the continuing tendency of population and industry to migrate to the metropolis, together with the sheer immensity of London's size, spread, and rate of growth, convinced the Commission that steps should be taken forthwith to check its further expansion.[1] The Commissioners therefore unanimously concluded that the continued drift of the industrial population to London and the Home Counties constitutes a social, economic and strategical problem demanding immediate attention.[2]

The Report of the Commission emphasised that a regional system of government would materially assist the solution of several of the problems with which the Commissioners were confronted. In particular, the problem of planning would be greatly simplified. "The Regional Council would become the principal planning authority for the region, certainly for major regional requirements, leaving probably to joint committees where existing, or to existing local authorities, the detailed administration of schemes. Planning would receive a great stimulus and [proceed] on more comprehensive and better organised lines than is at present possible with the multiplicity of small planning authorities; and housing could be better

[1] Report of the Royal Commission on the Distribution of the Industrial Population. Cmd., 6153/1940, paragraph 426.
[2] Ibid., paragraph 428 § 5.

related to industry. Larger financial resources would be available, and decentralisation in proper cases could be encouraged, e.g. to satellite towns."[1] Regionalism would also help to secure the balanced distribution and diversification of industry. The location of industry would be more effectively regulated if regional areas were established.

There was a considerable divergence of opinion among the Commissioners regarding the machinery for governmental action at the national level. The majority favoured a National Industrial Board composed of a chairman and three other members experienced in industry and commerce appointed by the President of the Board of Trade after consultation with the Ministers of Health, Labour and Transport, and the Secretary of State for Scotland. The minority called for a new Government Department, or one evolved from an existing Department.

There was also a substantial difference of outlook among the Commissioners on the question of the general powers of the central authority. The majority envisaged a weak body possessing mainly advisory functions while the minority urged the need for strong executive control over the location of industry throughout the country. Even the majority of the Commission were divided on this question, for three of the members who signed the majority report added a note of reservations on this and other points of principle.

But the Commission was united on the need for immediate and effective action to prevent the further growth of London. As the drift of the industrial population to London and the Home Counties (Bedfordshire, Buckinghamshire, Essex, Hertfordshire, Kent, Middlesex, and Surrey) constitutes a social, economic and strategical problem which demands immediate action, declared the majority report, the Board should be vested from the outset with powers to regulate the establishment within that area of additional industrial undertakings. The Board should have power to refuse consent to the establishment of new industrial undertakings except where the entrepreneur could show to their satisfaction that it would not be possible to conduct the proposed enterprise elsewhere on

[1] *Ibid.*, paragraph 379.

an economic basis.[1] Professor Jones, Mr. G. W. Thomson, and Sir William Whyte, while signing this report, disagreed with the proposal that the Board should be given only advisory powers outside London and the Home Counties. In their view the Board should be given general executive power to regulate industrial growth throughout the county. "It should," they said, "be an executive body for the whole country, with similar powers for dealing with all parts. . . . We regard the scheme for the London and Home Counties area, recommended in the Report, as no more than an immediate application, in one area, of a wider or national scheme to be brought into operation at the earliest possible moment."[2] Moreover, the dissentient Commissioners did not consider that the right way to prevent or check the further growth of the metropolitan region was merely by exercising restrictive powers. In their view a more fruitful approach would aim at creating more favourable conditions of life and work in other parts of the country, thereby weakening the inducement to seek work in or near London.

The minority report, signed by Sir Patrick Abercrombie, Mr. H. E. Elvin, and Mrs. Hichens, took a still wider view of the problem. In regard to London and other prohibited areas they proposed to impose more drastic tests than those suggested by the majority. Every manufacturer desiring to build or extend a factory, or to occupy an existing one not already occupied by him, or to convert other premises for manufacturing or processing purposes, would have to obtain a permit from the new Department. A permit would *prima facie* not be granted for a prohibited area. But permission might be given to an applicant who could satisfy the Department (1) that it would not be economically profitable to establish the factory elsewhere; (2) that the workpeople required can wholly or mainly be obtained from amongst persons living or normally seeking their livelihood within the area; (3) that reasonable housing and municipal services are available in the area for the workpeople, or can be provided without undue expense to the ratepayers; (4) that the establishment of the factory on the proposed site will not cause a substantial increase of traffic

[1] *Ibid.*, p. 206.   [2] *Ibid.*, p. 211.

congestion in the area; (5) that the proposed site is not objectionable on strategic grounds; (6) that the proposed site will not be destructive of amenities which ought to be preserved, such as existing open spaces in towns, or special features of natural, historic or architectural beauty.[1]

It can be seen that all the members of the Barlow Commission agreed that the further growth of London must be checked; that the best method of restriction is by control over the location of industry; and that appropriate machinery with executive powers should be set up for this purpose without delay.

The Commission thus clearly understood the correct principles which must underlie any intelligent policy for future planning. It grasped the need to arrest the meaningless accumulation of people and buildings in the metropolis as a preliminary to positive and creative development. It perceived the most effective instrument for carrying out the negative aspects of this policy.

On the question of administrative machinery and powers, the minority report was unquestionably based on surer ground than the majority report, and the creation of the Ministry of Town and Country Planning signifies acceptance of their recommendations. An advisory board would merely have spelled a decade or two of impotence and delay. When the majority really meant business, as in regard to London, they did not rely on advisory functions, but recommended executive powers.

It was not within the Commission's terms of reference to inquire into London government or to consider the relation between their proposals for restricting the entry of new industry into the metropolis and the authorities responsible for other services. But the Report gives explicit and unequivocal support to the conception of a Regional Council for planning and allied functions; and the regional idea is implied throughout both majority and minority reports. For these reasons alone—and there are many others—the Barlow Commission is rightly regarded as marking a turning point in public recognition of the basic requirements of good planning. The Reports form great State documents of enduring value.

We may now trace the attitude of successive governments

[1] *Ibid.*, p. 228.

towards the principles set out in the Reports. On 7th June, 1944, Mr. Dalton, speaking as President of the Board of Trade in the war-time Coalition, told the House of Commons that the Government did not accept the Barlow recommendation relating to London as it stood, but would consider each case on its merits. They recognised that London was not one of the areas in urgent need of factory development. Sir Stafford Cripps, who succeeded Mr. Dalton at the Board of Trade in the Labour Government, in a speech to the National Union of Manufacturers in November 1945, said that we must not repeat the mistake of over-concentrating industries in confined districts of London. Neither of these statements offered any positive assurance that the Barlow recommendations would be carried out.

The White Paper on Employment Policy devoted a chapter to the balanced distribution of industry. In this the Government made it clear that they would seek in future to guide the location of industry in order to prevent a recurrence of the depressed area problem, and to secure diversification of industry in those areas which had been unduly dependent on the more vulnerable industries. Industrialists contemplating the establishment of new factories or the transfer of existing plants, would be required to notify the Government at an early stage. This, said the White Paper, would enable the Government to exercise a substantial influence over the location of new industrial developments, as contemplated by the Barlow Report. In pursuit of this policy the Government would ask for power "to prohibit the establishment of a new factory in a district where serious disadvantage would arise from further industrial development"– a sidelong reference to London— while offering positive inducements to industrialists who are willing to establish factories in development areas.[1] In influencing the location of industry the Government would take account of strategic, industrial and social considerations.

These intentions were not realised at the legislative stage. The Distribution of Industry Act, 1945, as originally drafted, contained a clause giving powers to restrict factory building in areas where serious disadvantages would arise from further

[1] Cmd., 6527/1944, Chapter III, paragraph 26.

454 GOVERNMENT AND MISGOVERNMENT OF LONDON

development, but this was dropped during the Bill's passage. Hence the statute confers powers only to enable the Board of Trade to take positive action in certain specified areas, and omits the important principle of restriction. As Lord Balfour of Burleigh remarks, the Distribution of Industry Act seems, unfortunately, "to take control of location only as a means of avoiding unemployment in certain areas."[1] That aim is important, but it does not justify disregard of planning considerations, or neglecting decongestion and dispersal.

The White Paper envisaged that the main responsibility for formulating and administering the policy relating to the distribution of industry would rest with the several Departments concerned, namely, the Board of Trade, the Ministry of Labour and National Service, the Ministry of Town and Country Planning, and the Scottish Office. Arrangements would be made for supervising and controlling under the Cabinet, and as part of the central Government machinery, the development and execution of the policy as a whole. There would also be a regional organisation to bring together the representatives of these central Departments. The Board of Trade would be the leading Department responsible for all general questions of industrial policy. The President of the Board of Trade would be answerable to Parliament, and his Department would deal with enquiries and representations from interested sections of the public.

The present policy of the Board of Trade is based on three main principles: (1) to secure full employment throughout the country; (2) to secure diversification of industry; (3) to restrict the entry of new industry in London to applications for work of national importance which cannot be done elsewhere. The only positive method which at present exists of enforcing the restriction on industrial expansion in the metropolis is the withholding of a building licence. This is a temporary power which is available only during the present emergency situation in the building industry. It would be quite insufficient for a long term policy.

The main grounds on which the machinery calls for criticism

[1] "The Planning Act, 1944, and National Policies," in *Planning and Reconstruction Reference Book, 1946*.

is its extreme centralisation. The headquarters' organisation consists of two panels: Panel A, which deals with industrial building projects, and Panel B, which determines the allocation to industrialists of surplus government factories. The former panel reports to a ministerial sub-committee on the Distribution of Industry, which forms part of a Cabinet Committee. These panels consist entirely of officials of various central government departments. In each region outside London there is also a Regional Distribution of Industry Panel made up of the senior regional officers of the central departments; and a Regional Board for Industry composed of representatives of employers and trade unions together with the regional officers of the central departments. These organs have various advisory functions which we need not consider in detail. The Regional Distribution of Industry Panels also have executive powers which have been delegated to them by the headquarters panels, except in the London region.

It can be seen, therefore, that the entire organisation is exclusively related to the central government. There is no attempt to devolve any of the functions to a regional body representing the teeming millions of the metropolis; nor even to bring into the picture the local authorities who are responsible for the planning and local government of London. I shall revert later to the defects of these arrangements.

CHAPTER XIV

THE PROSPECTS OF REFORM

If one contemplates dispassionately the situation described in Part II of this book, it is impossible to resist the conclusion that drastic and far-reaching changes are required in the organisation of London government. But however urgent and imperative the need may be, it cannot be assumed that it will be fulfilled by some automatic process of social reform, operating under the influence of unknown forces. There is nothing inevitable in the adjustment of political and administrative institutions to contemporary needs. There are plenty of instances to be found both in the past and in the present where adjustment has not taken place. The price to be paid for such failure is, of course, another question.

It may be worth while, therefore, to spend a little time enquiring into the possibility, or the probability, of some kind of order being brought into the chaos of London government in the near future.

The first thing one can say is that regional reorganisation is not necessarily a political question in the party sense at all. It would be possible for either or both of the principal political parties in the metropolis, without any sacrifice of principle, to include in their programmes the demand for a Greater London Council and the other related reforms; and to awaken a strong desire for such changes among the electorate. There is, however, nothing in the subject which marks it down as specially suited to the philosophy of the London Labour Party, or peculiarly adapted to the outlook of the Municipal Reform Party. Both parties are in favour of administrative efficiency, as they understand it, and neither of them is given to factious opposition to improvements in the machinery of local government. It is true that a great part of the responsibility for the failure of the Ullswater Commission on London Government in 1923 to achieve any tangible result must fall on the Municipal Reform Party, which was then in power at County Hall with a huge majority.[1] For as we have seen, the fiasco of that enquiry was chiefly due to the manner in which the witnesses for the London County Council presented their case and the Council's refusal to make any concrete proposals.[2] But much water has flowed under the bridges in the 15 years which have elapsed since the Ullswater Commission; and the Municipal Reform Party, like the London Labour Party, is probably unquiet at the present situation and would like to see an improvement.

Incidentally, there is little reason to believe that either of the political parties would stand to gain or lose much by the establishment of a regional organ. So far as one can see, the political colours which dominate the various quarters of the administrative county tend in general to be continued beyond the boundary into Greater London, and persist even outside the Metropolitan Police District. Certainly the north and north-east districts in Outer London retain the working class, Labour-voting flavour of the metropolitan boroughs in north and east London, such as Stepney, Poplar, Bethnal Green, Hackney and Islington; while to the north-west and south-west are well-to-do suburbs whose interests and sympathies are doubtless on the Conservative side.

Similar tendencies exist in regard to finance. Precise information exists as to rates and rateable value in Greater London (i.e. the Metropolitan Police District); and for the present purpose we may therefore confine ourselves to this area. The "rating map"[3] shows that the highly rated areas (over 12s. in the £) extend from the metropolitan boroughs in the east, north-east and south-east to the outer ring of districts lying beyond the administrative county in the same directions. Essex, indeed, has had a burden placed on its County Council from the overflow of population from London. To the north, north-west and south, on the other hand, lie the areas where rates are seldom above 10s. in the £ and often below 9s. In the west is an intermediate zone, situated partly inside and

[1] The strength of the parties on the London County Council in 1922–5 was Municipal Reform 82, Liberal 26, Labour 16. See W. A. Robson: "Thoughts on the L.C.C. Election" in *Political Quarterly*, April–June 1934, p. 167.   [2] *Ante*, pp. 295–6.
[3] The Rating Map is included at the end of this volume.

partly outside the county, where the rates are between 10s.
and 11s. in the £.

The table on page 459 gives details over a period of years.

From this it can be seen that for long the average rate in
Outer London was above that of the administrative county.
In 1935–6 this position was reversed for the first time; but the
difference between the average rate in the administrative
county and that in Outer London was trifling until 1940–1.
The average rate for Greater London as a whole was within
a penny of the average rate for the administrative county until
1941, when the position became complicated by wartime
conditions.[1] Formerly the difference between the two in a
converse sense was much larger.

It would appear from this that, *ceteris paribus*, the effect of
taking Greater London as the area of administration and
taxation would be almost negligible, so far as the average rate
is concerned. Prior to 1935–6, the effect would have been to
increase the rates levied in the administrative county.

This, however, does not conclude the matter. The rate
levied is in certain ways less significant than the rateable value
per head of the population. The following table includes this
factor:

	Estimated mid-1946 Population	Rates made 1946–7	Rateable Value	Rateable Value per Head of Population	Total Rates levied	Rates levied per Head
		sh. d.	£	£	£	£
London	3,131,600	12 5·95	53,703,498	17·14	33,553,498	10·71
Extra-London	4,745,990	14 8·41	45,651,831	9·61	33,555,998	7·07
Greater London	7,877,590	13 6·34	90,355,329	12·61	67,205,600	8·53

These figures show that the rateable value per head in the
administrative county is much higher than it is outside (£17
per head compared with a little over £9 12s.); and that every
occupier within the London County Council's territory is paying
on the average £3 13s. a year more in rates than those who dwell
in Outer London. Thus, although the rate in the £ in Outer
London is 2s. 3d. higher than in Inner London, the inhabitants
of the administrative county are in fact paying far more in
rates than Londoners who reside beyond the boundary. This
is without doubt largely due to the heavy cost of the services

[1] For this reason, the figures for 1941–2 and 1946–7 are not included in
the table on page 459 as it is believed they are abnormal. No return was
made for the years 1942–3 to 1945–6.

AVERAGE RATES FOR THE PARISHES IN THE ADMINISTRATIVE COUNTY OF LONDON, EXTRA LONDON, AND GREATER LONDON, 1914–15 AND 1924–5 TO 1940–1, AND RATEABLE VALUES, 1946

Year	London Administrative County		Essex		Herts		Kent		Middlesex		Surrey		Extra London		Greater London	
	s.	d.	s.	d.	s.	d	s.	d.	s.	d.	s.	d.	s.	d.	s.	d.
1914–15	7	10·87	10	0·24	7	10·27	7	4·06	7	10·00	7	2·53	8	1·76	7	11·68
1924–5	11	4·78	21	4·14	12	3·46	11	11·06	12	1·27	10	7·60	13	9·44	12	0·90
1925–6	11	6·52	21	4·00	12	9·22	12	2·25	12	6·64	11	8·58	14	3·08	12	3·93
1926–7	11	7·25	21	5·62	12	10·94	12	1·83	12	9·45	11	4·39	14	3·34	12	4·05
1927–8	11	7·90	21	3·10	13	3·78	12	5·05	13	0·94	11	4·80	14	4·46	12	5·05
1928–9	10	11·59	19	4·00	13	2·64	12	5·78	12	11·36	10	11·92	13	9·55	11	9·33
1929–30	10	6·78	15	3·45	11	1·27	12	6·12	10	7·96	9	7·46	11	4·44	10	10·02
1930–1	10	9·36	15	3·29	11	1·27	10	6·69	10	7·31	9	8·23	11	3·74	10	11·57
1931–2	10	3·22	14	9·85	10	9·17	9	11·36	10	1·28	9	3·52	11	3·74	10	5·47
1932–3	9	11·55	14	8·93	9	11·33	9	3·26	9	10·53	9	1·98	10	9·91	10	2·18
1933–4	9	10·14	15	2·34	10	0·49	9	3·06	10	0·33	9	2·84	10	7·19	10	1·98
1934–5	9	7·44	14	6·54	10	3·12	9	2·84	9	9·30	9	0·49	10	9·09	9	11·30
1935–6	10	9·35	14	7·30	10	4·39	9	10·04	9	11·72	9	2·59	10	5·95	10	8·84
1936–7	11	1·95	14	9·87	10	3·57	10	4·36	10	9·07	9	10·77	10	8·01	11	2·41
1937–8	11	2·15	15	2·99	10	9·58	10	10·54	10	9·65	10	0·52	11	3·11	11	3·39
1938–9	11	10·45	15	7·53	10	8·69	11	3·63	11	4·16	10	6·18	11	5·22	11	10·57
1939–40	12	1·95	16	5·89	11	6·31	11	8·36	11	10·64	10	8·18	11	10·75	12	3·09
1940–1	13	4·35	16	9·95	13	0·01	12	4·24	12	10·07	11	2·23	12	4·69	13	3·07
Rateable value 1946	£53,703,498		£7,771,937		£1,116,780		£3,735,426		£21,982,073		£11,045,615		£45,651,831		£99,355,329	

provided by the London County Council and the metropolitan boroughs for the enormous day population which swarms into their areas.

If a Greater London Council were set up, the area of charge for certain services would be widened as well as the area of administration. The effect would be *in general* to increase the burden on the Outer London areas and to lighten it in the administrative county.

We cannot, however, expect that other things will remain equal. In the first place, the simplification and enlargement of areas, and the reduction in the number of authorities which is proposed, would almost certainly lead to a reduction in the total cost of municipal services.

In the second place, there is reason to believe that the creation of a Greater London Council would produce an increase in the value of rateable property in Outer London. The mere inclusion of a house in a London area such as the London Postal District, increases its selling value sometimes by as much as £50, and rateable values are proportionately increased. It is probable that the establishment of a regional council would lead to a substantial rise in the rateable value of Outer London, both absolutely and relatively to that of the administrative county. Hence, the increased charge on the rates in Outer London would probably be offset by an enhanced rateable value of the property on which they are levied.

This forecast is necessarily of a speculative nature. In a situation containing so many variable and uncertain factors, precision is unattainable. But so far as can be seen, the establishment of a Greater London Council would not bring about any very large changes in the rate charge as between Inner and Outer London. The principal effect would be to adjust the burden more equitably between the various districts within these major zones. In short, there would be a shifting of burden as between rich districts and poor districts no matter where situated, rather than between those in Inner London and those in Outer London.

Thus, a reasonable degree of security against sudden or acute changes in either the balance of political power or the level of rating can be given to those whose primary concern is with these matters, as well as to those ordinary citizens who would welcome a municipal reconstruction of the metropolis on general grounds provided no great disturbance in finance or political strength were involved. Neither of these matters need give rise to any real difficulty.

The strongest opposition to reform is likely to come from the local authorities of all kinds which are at present established in the metropolis. The 200 or more separately elected councils which at present cause so much waste, confusion and inefficiency, are no doubt inspired by a common desire to survive at all costs. From the great county council down to the small parish, they one and all regard the continuance of their own existence not merely as an absolute good in itself, but as something compared with which any scheme of reform intended to benefit the larger metropolitan community, is but as dust in the balance.

The appearance of vested interests in local government is a matter of great interest and importance which deserves investigation by a sociologist. It is not necessarily the result of slow growth through the ages, for it is as rampant in the United States and other comparatively new countries as in Great Britain. It is not dependent on the existence of jobbery, corruption or graft. But whatever the cause may be, municipal vested interests are defended with a passion and vigour every whit as strong as those which defend private property.

The county councils are particularly intransigeant in this respect. In the 50 years which have elapsed since their establishment in 1888, no attempt of any kind has been made to rationalise, enlarge or improve their utterly unscientific areas. The only alterations that have occurred have been through the creation and extension of county boroughs. The Local Government Act, 1929, instituted a periodical review of county districts, but in regard to the counties themselves, where change is still more badly needed, there is a tacit assumption of unalterable boundaries which is making peaceful change almost as difficult in Surrey and Kent as it is between sovereign states. Behind the individual local authorities concerned in any particular case in which revision might be brought under discussion, there loom the hostile shadows of the County

Councils Association, the Association of Municipal Corporations, and similar bodies.

These associations are mainly interested in the defence
of their members' interests against interference by Parliament,
the central government departments, or any other outside body
which may appear to threaten. They are firm believers in
collective security; and a move against one is regarded as an
attack upon all, to be resisted with all the forces at the command
of the Association.

It might well seem that the prospect of determined opposition
by the combined forces of the 200 local authorities immediately
concerned, together with those indirectly brought into the fray
through their respective Associations, is enough to make any
sort of change extremely remote, if not entirely out of the
question. It is true that any scheme for reforming London
government is politically hazardous in the sense that there are
few votes to be won by it but many votes to be endangered. It
is true that Parliament has never risen to the height of a major
proposition in regard to the good government of the metropolis.
It is true that for half a century no Government has exhibited
qualities of leadership, imagination or determination in regard
to the much-needed reorganisation of the larger local government areas and authorities. It is true that most of the metropolitan local authorities are comfortable in their autonomy.

On closer examination, however, a different possibility based
on other factors may be seen to emerge. In the first place, the
potential opposition of local authorities is not nearly as formidable as it appears on the surface. It does not necessarily
carry with it a large proportion of the voters in the areas
concerned. Governments and Parliaments are unfortunately
prone to identify the opposition of a local authority to an
administrative or legislative change affecting its area with the
voting strength of the electors in that area at the next election.
This is quite unrealistic and a serious handicap to progress.
Even assuming that every local authority in the metropolitan
region were opposed to a scheme of reform put forward by the
Government, it is doubtful whether 5 or 10 per cent of the
votes cast by Londoners at a subsequent general election would
be affected. If Parliament and Government will realise that to

attack the problem firmly is to deprive it of much of its difficulty, that alone would be a considerable step forward.

It is essential, moreover, that it should be clearly recognised that the reorganisation of London government is a national question calling for national leadership and direction. Nothing will be more fatal than to suggest that the problem is one which can safely be left to the local authorities to fight out with each other until they arrive at some arrangement mutually acceptable to themselves. Little good can come from such an approach. It was precisely this attitude which wrecked the Tyneside project. The Royal Commission which was appointed to consider the reorganisation of local authorities on the Tyne issued two Reports,[1] both of which made strong recommendations in favour of large measures of unification. The Minister of Health (Sir Kingsley Wood) then submitted the Reports to the local authorities concerned, on the ground that "it would be impossible to put forward proposals of this kind unless a general measure of assent was assured." In addressing the representatives of the local authorities in Durham and Northumberland, the Minister stated: "Under our local government system the initiative lay with the local authorities concerned, and a local Bill would be necessary to effect this proposal."[2] He hoped that the authorities would press forward with the consideration of the proposals in the Minority Report, and the Government would regard sympathetically any Bill which embodied a general measure of agreement. The Minister concluded by saying that action was necessary, for although the remedies suggested in the two Reports differed, the Royal Commission were unanimous that a measure of unification in Tyneside local government was urgently called for in the interests of progress, efficiency and economy.[3]

This negative attitude led to a complete fiasco. The local authorities failed to agree, knowing that no pressure would be exerted on them by the Central Government, and that in the absence of agreement there would merely be a continuance of the *status quo*. A similar attitude applied to London would produce a similar result. Only through the exercise of deter-

[1] Report of the Royal Commission on the Local Government of Tyneside.
[2] *The Times*, 22nd September, 1937.   [3] *Ibid.*

mined leadership and a resolute demeanour by the Minister of Health and the Cabinet can we hope to see the London problem solved. So far the Ministry of Health has given no sign that it is even aware that a problem exists.

It is historically untrue to suggest that "under our system of local government, the initiative lies with the local authorities concerned." The great reforms of the past have been carried out by the Central Government over the heads of local authorities and in face of their protests and opposition. This was the case when the Poor Law reforms of 1834 were introduced. It was the case when the Municipal Corporations Act, 1835, was passed. It was the case when the Local Government Act, 1888, established the county councils. It was the case when the Boards of Guardians were abolished in 1929.

The inexorable pressure of facts constitutes the greatest factor making for change in the metropolis. It is impossible to believe that the tendencies now operating can continue indefinitely without the chaos and confusion reaching a stage at which the alternatives to reform will be seen to be infinitely more dangerous to established expectations than a bold measure of adaptation. A policy of doing nothing is fraught with such immense disadvantages that its abandonment is only a question of time. Ultimately the structure and functions of London government will have to be harmonised with contemporary needs and accomplished facts. Further delay will merely make the situation more difficult to deal with. The path of statesmanship would be to deal with the matter before the metropolis has reached unmanageable dimensions.

Other great cities have managed to get their problems dealt with on bold and courageous lines. Birmingham was enlarged in 1909 to take in 33,000 acres from three separate counties, much of the added area consisting of rural land. In 1917 Edinburgh took in 32,000 acres—a territory four times the size of its previous area—including large stretches of purely rural country and part of the Pentland Hills.

The most conspicuous instance of constitutional reform is to be found in New York, where far-reaching improvements in the municipal structure have recently been introduced.

The population of the City of New York is estimated

(January 1937) to be 7,434,346. The whole region of Greater New York, consisting of 5,528 square miles, within a radius of 50 miles from the City Hall in Manhattan, contained in 1930 only 11,500,000 persons, so that the City comprises a much larger proportion of the Region than the administrative county of London does in relation to the Metropolitan Traffic Area, which is a much smaller region. The City area covers about 319 square miles, as compared with 116 square miles in the case of the administrative county of London.

In 1934, the State legislature provided for the creation of the New York City Charter Revision Commission. The principle of "home rule" was recognised, and the Mayor was authorised to appoint the nine members of the Commission, whose task was to study and analyse the existing governmental structure "with a view to drafting a proposed charter adapted to the present requirements of the City and designed to provide a more efficient and economical form of government."[1] After two years of public and private hearings the Commission issued a draft Charter, which was then submitted to the citizens at the municipal general election in 1936 and adopted to take effect from 1st January, 1938.

The new constitution establishes the City Council as the deliberative body, with the sole power to make and adopt local laws. It consists of a President, serving for 4 years, and councilmen elected for a 2-year term of office by a system of proportional representation. The 5 boroughs into which the City is divided are treated as separate districts; they elect one councilman for every 75,000 voters, with a right to an additional representative for a remainder of 50,000. This provides a small council of 26 members, exclusive of the President, each of whom receives a salary of $5,000 a year.[2] The Vice-Chairman receives a salary of $7,500.

The Mayor remains the chief executive officer of the City, with large administrative powers. He is a member *ex officio* of numerous boards and commissions, and chairman of the Board

[1] *New York Advancing: A Scientific Approach to Municipal Government* (edited by Rebecca B. Rankin), New York, 1936, p. 355.
[2] For details, see *ante*, pp. 330-1. The old council of New York City had 65 elected members. The present one is regarded as immensely superior.

of Estimate. He has the right to nominate the members of many of these semi-independent bodies. He can veto or approve local laws passed by the City Council, but the latter can override his veto by a two-third majority. He can appoint a Deputy Mayor, to whom he may delegate some of his duties.

Each of the 5 boroughs has a Borough President, elected for a 4-year term of office. Formerly the Borough Presidents had much greater power within their own areas, but the new Charter has transferred many of their functions to the central municipal organs, such as the Department of Housing and Buildings, and the Board of Standards and Appeals. The Presidents remain responsible within their respective boroughs for the construction and repair of highways, the filling of sunken lots, the fencing of vacant sites, the licensing of underground vaults, the removal of incumbrances, the issuing of permits to open up the streets, the construction of bridges and tunnels other than those crossing navigable streams (which are the important ones), the provision of public baths and lavatories, and the construction and management of the local sewers and drains (excepting intercepting sewers).[1]

The Borough Presidents are not members of the Council but they are linked up with the administration of the City through the Board of Estimate, which may be regarded as the General Purposes Committee. It determines policy in regard to the City's municipal affairs, local assessments, zoning and land. It fixes the remuneration, pensions and retirement of municipal officers and lays down the personnel establishment. It directs the City bureaux dealing with franchises, engineering and real estate. It is empowered to exercise "all the powers vested in the City except as otherwise provided by law."[2] It prepares the City's annual expense budgets.

The Board of Estimate is composed of 8 members, all of whom are elected for 4-year terms. The Mayor, who is chairman, the Comptroller,[3] and the President of the Council have three votes each; the Presidents of the Boroughs of Manhattan and Brooklyn have two votes each, while the 3

---

[1] New York City Charter, Chapter 4. There are no borough councils.
[2] *Ibid.*, Section 70. And see Chapter 3 generally.
[3] The Comptroller is a separately elected officer of high standing.

other Borough Presidents have one vote each. There is thus a total of 16 votes, of which 9 are in the hands of officers representing the whole City.

An outstanding feature of the new Charter is the creation of a City Planning Commission. This consists of the chief engineer to the Board of Estimate and six members to be appointed by the Mayor on staggered terms of 8 years. The Mayor designates the chairman, who becomes head of the department of City Planning at a salary of $15,000 a year. The primary duty of the Commission is to make a master plan for the City, to include not only the streets, bridges, parks, public places and transport facilities, but for "the co-ordination of such facilities in a plan which will provide for the City the most convenient means of travel between centres of residence and of work and recreation. The Commission in preparing the plan should consider not only the distribution of the population but its comfort and health and the beauty of the surroundings in which they live."[1] The development of residential areas, the location of housing projects, the disposition of parks, playgrounds and schools, the siting of fire stations, police stations, hospitals, law courts and other public institutions, and their relation to the means of transportation, will all be taken into consideration in the formulation of the master plan. An important instrument in promoting its realisation lies in the duty placed on the City Planning Commission of preparing the budget of capital expenditure for the City each year. Thus, the Commission will in effect determine the annual program of public improvements to be paid for by the issuance of capital obligations secured on the City's assets. No improvement contrary to its recommendations may be included in the general revenue budget, prepared by the Board of Estimate, except by a three-fourths majority. In effect, this means that the annual capital budget will reflect the proposals of the master plan recommended by the Commission to be carried out that year.

No small part of the credit for these promising developments in regard to planning is due to the intelligent and public-

[1] Information Bulletin, No. 34 (1st February, 1937), issued by the Regional Plan Association, New York, p. 3.

spirited activities of the Regional Plan Association and its predecessor the Regional Plan Committee. These voluntary bodies were responsible for formulating a plan for New York Region, and for presenting it to local authorities and the general public. The work was begun in 1922 and completed in 1931. It occupied the professional labours of a numerous staff of city planners, engineers, architects, economists and sociologists, among whom were men of high eminence and capacity. The cost of the work was in excess of £260,000, the money being provided by the Russell Sage Foundation.

After the Committee had published the Regional Plan in two large volumes (supported by eight extensive survey volumes) the Regional Plan Association was organised for the purpose of arousing and maintaining the interest of the public and thereby ensuring the practical realisation of the project. The Regional Plan movement has exercised a large and growing influence in New York civic life and in the physical development of the town. The City Planning Commission created by the new Charter offers reasonable hope that the aims of the movement are within measureable distance of fulfilment.

It is not suggested that the methods and machinery which are suited to New York would be appropriate to London, nor that the municipal organisation of New York is without certain defects. But in such essential matters as the population and territory coming under the jurisdiction of the major authority, the subordination of the local boroughs to the larger community, the integration of major and minor authorities, the status and powers of the City Planning Commission, and the general coherence and unity of the administration,[1] the remodelled government of New York is immensely superior to the effete arrangements which continue to exist in London. Moreover, the power of Tammany Hall, for so long synonymous with graft and corruption, racketeering and jobbery, has been shattered—one hopes for ever.

The new constitution of New York City is obviously not in accord with the accepted pattern of English local government.

---

[1] I am referring only to the omnibus local authorities. New York has numerous *ad hoc* bodies, such as the Port of New York authority, which are exceptions to this remark.

But the significance of the Charter is that it shows what can be done and has been done in the only other metropolis which can be compared in size, wealth and importance with London. It demonstrates that in the chief city of the new world there is a spirit of innovation, an ability to adapt and reconstruct municipal institutions, which is badly needed in London and has so far not been forthcoming.

Surely, one may ask, are we not as well able to introduce much-needed reforms in the government of our principal city as our friends across the Atlantic? Their achievement should be regarded as a challenge to our democracy: a challenge coming neither from a Fascist dictatorship nor from Soviet communism, but from the other great English-speaking democracy across the ocean. One would like to be able to answer this question with a clear affirmative. But unfortunately it is impossible to see any signs that the matter is engaging the attention of responsible politicians or civil servants, either national or municipal. For a few years immediately following the conclusion of the Great War of 1914–18 there was a great outburst of enquiries that seemed to betoken a ferment of intelligent dissatisfaction, and which might have led to a new creative effort. There were the Departmental Committees on the Metropolis Water Act (1920), on the Wholesale Food Markets of London (1920–1), on Unhealthy Areas (1920–1), on the Thames and Lee Conservancies (1923); there were the Royal Commissions on London Government (1923), Fire Brigades (1925) and Cross River Traffic in London (1926); and several others. But the movement was allowed to spend itself in a spate of blue books, and nothing came of it.

Now we are again faced with the possibility of another European war, in which this country might be engaged. One of the major factors in determining the outcome of such a conflict would be the vulnerability of London. The press and the bookshops are filled with discussions concerning air raids and methods of obtaining safety from them. The defence of London is a major preoccupation of the Air Ministry and the War Office. The Air Raids Precautions Department of the Home Office is anxiously devising ways and means of securing the safety of the civilian population. Local authorities have appointed Air Raids Precautions Officers. A Parliamentary Committee has considered evacuation on a large scale.

It does not so far appear to have been realised that in the event of war, one of the first necessities would be a strong unified administration for the whole London Region, able to arrive at prompt decisions and to carry them out efficiently without regard to parochial obstructions. The mass of ineffective and obsolete bodies which now litter the scene would be a national liability of the most dangerous kind. The notion that some centrally-appointed military governor or commissioner could be appointed to take charge at a minute's notice when hostilities threaten, is an idea which is both short-sighted and unintelligent. No one who has the slightest knowledge of public administration can imagine for a moment that such a step could overcome quickly and effectively the immense handicap of the disintegrated and incoherent local machinery which now exists. It is not merely centralised direction that is needed, but unified administration. The planning and organisation of efficient administrative agencies requires years of patient effort and careful thought.

The international crisis which occurred in September 1938 was highly significant in this connection. On 21st October Mr. Herbert Morrison wrote to the Home Secretary that "statements are being made to the effect that at the time of the crisis secret instructions were issued to town clerks of metropolitan boroughs intimating that in the case of war the borough councils as such would be superseded, and instructions issued direct to borough council officials as to the conduct of borough council services. These statements are causing a considerable amount of apprehension in London municipal circles, and it is, we think, very necessary that the matter should be cleared up." Mr. Morrison accordingly asked for information on the subject. He added that "while it is recognised that in the event of war local authorities would have to consider some modification in their normal procedure, any question of the supersession of the authority of public representatives would cause very deep concern."

Sir Samuel Hoare replied on 2nd November that "no communication was issued by the Government which could properly be described as intimating to town clerks that in the case of war the borough councils as such would be superseded, nor were instructions issued direct to borough council officials as to the conduct of borough civil services." But, he continued, "in time of war it would be necessary for the Government to exercise various powers which are quite outside the scope of ordinary local government, such, for example, as arranging for public warnings of air raids, enforcing lighting restrictions, requisitioning, if need should arise, certain essential things, and taking other executive action in connection with air raid precautions."

"The Government decided, in view of the suddenness of the recent emergency, that it was necessary to issue certain confidential instructions on matters of this kind to town clerks in the metropolitan area, and to officers of local authorities and other persons in other parts of the country. These instructions were not put into effect, as the end of the crisis made this unnecessary, and they have since been withdrawn."

A few days later, the report of the Kensington A.R.P. Committee was presented to the Kensington Borough Council. After describing the local arrangements made to meet the crisis, the report observed:[1] " 'The most important point to which we desire to refer is the general one as to how far it is possible to secure uniform and adequate treatment of Greater London in the matter of civilian air-raid precautions with so many different and independent bodies entrusted with substantial responsibilities in the area.' The borough councils control certain medical services and public works. They were competing against one another in the open market for important supplies."

The Committee recommended that the arrangements for Air Raid Precautions should be revised so that uniform and adequate plans can be made throughout Greater London. They emphasised the inappropriateness of charging any part of the cost of defending London on local rates, with their unequal incidence—an unusual statement from a wealthy borough such as Kensington. Responsibility for evacuation, the Committee suggested, should be in the hands of one central authority, and

[1] *The Times*, 9th November, 1938, p. 11.

a central purchasing agency should be established to obtain and distribute the principal supplies of material required for passive defence measures.

The international crisis clearly revealed the weaknesses and uneven working of the Air Raid Precautions arrangement in the metropolis. Some local authorities did reasonably well while the performance of others was deplorable.

It is quite wrong to infer, however, that because the unco-ordinated efforts of 200 or more local authorities produced unsatisfactory results as a whole, the proper remedy is to hand over the whole task to the Home Office for direct central administration. On the contrary, an intelligently planned regional system entrusted to a Greater London Council, with power to devolve a large part of the detailed administration on carefully supervised district authorities of substantial size, would be far more likely to yield good results. It would bring into potential use, in case of need, the large resources in personnel and premises, in local knowledge and goodwill, of the municipal councils. They alone could arrange for a smooth and rapid transfer of men and material from other local government functions to the Air Raid Precautions service.

If, however, we wish to build up an executive instrument which is likely to be capable of carrying out efficiently in time of war such immensely important tasks within the metropolis as the distribution of foodstuffs, the use of the fire brigade to deal with conflagrations from incendiary bombs, the organisation of first-aid clinics, the evacuation of children, invalids and old persons, and similar functions, it is essential that a start should be made forthwith, while peace prevails and time permits, with the reorganisation of London government on some such lines as those proposed in this book.

The criticisms advanced in these pages and the suggestions which have been put forward have been based entirely on grounds of social welfare. But they are equally relevant from the point of view of national safety in case of war. There is not in this matter the slightest divergence between the needs of peace and the needs of war. The measures necessary to promote the health and happiness of the people of London are precisely those which are called for to protect their lives and limbs in time of war. A great regional council enjoying the confidence of the people of the metropolis and with un-rivalled knowledge of their ways of living and habits of mind, possessing strong powers and accustomed to co-operate with and supervise a limited number of co-ordinated local authorities, would be a most valuable asset in time of national emergency. If we wish to defend ourselves against the threat of war which menaces the international situation, the reform of municipal administration in London is a safeguard not to be under-rated or despised.

Considerations of this grim kind should have the fortunate result of not merely improving the prospects of reform but of making it imperative and urgent. But once the necessity is acknowledged on the grounds of greater potential security in case of war, we can put the cause in a less immediate place in our thoughts and dwell rather on the infinite possibilities of better living which are opened up by the opportunities which would then lie in our hands.

London is not the beautiful, spacious, coherent, convenient healthy city we can make it if we desire. Let us recall the question that Le Corbusier has asked:[1] "Why do we make pilgrimages to beautiful cities if not to put gaiety into our mind and senses, if not to recognise by means of this witness in stone that man is capable of grandeur; and to feel in oneself the joy that such a certainty gives us? For all our trivialities, our comfort, money and the crease in our trousers, pale away before the rapture of such an assurance of noble feeling!"

Those are words to remember, when now at long last we resolve to make anew "the chief city of the Kingdom" and its government.

[1] Le Corbusier: *The City of To-morrow and its Planning* (translated by F. Etchells from the eighth edition of *Urbanisme*), pp. 62–3.

EPILOGUE

This book first appeared in 1939, a few months before the outbreak of World War II. For several years before then the international tension had been so acute that most thinking persons were acutely conscious of the dangers which London would face when the international situation reached its climax and the war clouds burst over Europe. The concentration of so large a proportion of the nation's population, wealth, industry, commerce and governmental institutions in the metropolitan region offered a gigantic hostage to fortune in the unknown conditions of air warfare. The lack of any coherent or efficient system of regional government greatly increased the danger of breakdown and confusion in the civil defence and other emergency services. I therefore pointed out in the concluding pages that although the suggestions for reform which I put forward were based entirely on grounds of social welfare they were equally relevant from the standpoint of national security in case of war.

Only a few months still remained before hostilities broke out and nothing was possible beyond hand to mouth expedients. What London suffered in the course of the war is known to everyone. Hitler and the Wehrmacht thought that by bombing London sufficiently hard they could force England to her knees. Fortunately they were wrong. The part played by the local authorities in evacuation, civil defence and other emergency services has not yet been told, though one hopes the story will in due course form part of the official history of the war. Such information as is already available affords no opportunity for complacency.[1] The best work was probably done by the special authorities set up to deal with purely war-time services, varying from the Regional Commissioners to the local Air Raid Wardens. The unforgettable heroes of London's ordeal by fire were the men of the London Fire Brigade, which I described here as "a body of the highest

---

[1] See E. Doreen Idle: *War Over West Ham*; R. J. Padley and Margaret Cole: *Evacuation Survey*. (1940, Routledge.)

efficiency and repute.''[1] I am glad that I recognised it to be the best fire-fighting force in Great Britain.[2]

The war has now come and gone. Almost every part of London can show its honourable scars. Some areas, including parts of the City and the East End, have been almost cleared of buildings and will have to be entirely rebuilt. Apart from the human suffering involved, that is not to be regretted.

It is not my intention to deal with the wartime experiences of the metropolis except in passing; but they form the starting point, or at least the background, for much of what follows.

One of the most striking changes in public opinion which took place during the war was the increased interest in physical planning. For the first time the planning idea was accepted as necessary and desirable by large numbers of people regardless of their political outlook and belonging to all classes of society. There were several reasons for this change of attitude. Large-scale evacuation of mothers and children, the old and the sick, the transfer of commercial offices and administrative institutions of all kinds from the great centres of population to country towns, villages and country houses in the reception areas, had undermined the belief that there is something 'inevitable' or beyond human control in the way in which the population happens to be distributed through the country; or that the unchecked growth of giant cities is part of the order of Providence. The deliberate choice of factory sites in places which were less vulnerable to air attack had the double effect of diminishing the attraction of the metropolis and of gaining acceptance for the idea of controlling the location of industry.

Above all, there was the moral and psychological need for something to which the people of London and the other afflicted cities could look forward, an ideal to sustain them through the days of privation, endurance, sacrifice and suffering.

This hope for the future was provided, so far as the shattered cities and ruined streets were concerned, by the idea of town and country planning. "Wherever the destruction has been most severe, there men and women have resolved most seriously

[1] *Ante*, p. 270.   [2] *Ibid.*

that when peace comes they will build a finer and more beautiful city. The havoc which has been wrought in some places would be almost unbearable to those who have to live among it were they not aided by the vision of a fairer habitation in days to come. It is impossible to over-estimate the value of the vision, vague and unsubstantial though it doubtless is, as an aid to the maintenance of public morale. And every vision implies a plan. Indeed, a vision is a kind of plan, or at least it is the stuff of which plans are made. Without vision there can be no planning of any sort."[1]

Nowhere did these influences make themselves felt more strongly than in the metropolis. Hence it is not surprising that the most decisive change of attitude in regard to planning should have taken place in the London region.

I have described elsewhere in this book the muddle and indifference towards the planning of London on the part of both the central government and local authorities which occurred between 1909 and 1939.[2] During the war, however, the London County Council was for the first time spurred to a belated sense of its obligations in this matter by Lord Reith when he was Minister of Works. Lord Reith requested the Council to prepare a plan to assist his Ministry (which was then the responsible Department) in considering suitable methods and machinery for effecting post-war reconstruction of town and country. The Council in consequence invited Mr. Forshaw, who at that time was its architect, to collaborate with Sir Patrick Abercrombie in the preparation of a plan for the administrative county of London. The Forshaw-Abercrombie plan was issued in 1943, commended by Lord Latham as Leader of the London County Council and subsequently adopted in principle by the Council.

A similar request was made to the Lord Mayor of London by Lord Reith in March 1941, and the Corporation thereupon set about the preparation of a plan for the City. This was completed in 1944 and submitted to the Minister of Town and Country Planning, who had superseded the Minister of Works in regard to planning matters. Mr. W. S. Morrison, the

---

[1] W. A. Robson: *The War and the Planning Outlook*, pp. 7–8. (Rebuilding Britain series.) [2] *Ante*, pp. 403–20.

Minister of Town and Country Planning, refused to approve the plan on account of its inadequacy.

So far as the rest of the metropolis was concerned, Mr. Morrison commissioned Sir Patrick Abercrombie to prepare a plan for the whole region outside the administrative County. For this purpose an area of approximately 2,600 square miles was taken extending outwards from the London County boundary to a distance of about thirty miles from Charing Cross. At some points it goes much farther, touching such places as Haslemere, High Wycombe, and Bishop's Stortford. The area includes Middlesex, Hertfordshire, Surrey, and parts of Kent, Essex, Bedfordshire, Buckingham, and Berkshire. The population of this Outer London area was $6\frac{1}{4}$ millions in 1938. By comparison the combined areas of the Administrative County and the City, comprising 118 square miles, were estimated to contain slightly over 4 million persons in 1938. The total population covered by both plans was thus over 10,000,000 persons. The Greater London Plan was published in 1945.

A rational approach to the planning process was violated at the outset by the division of the metropolis for planning purposes into three parts, like ancient Gaul. There is the City of London, the County of London, and Greater London, each with its own plan. There are thus three plans instead of one for what is essentially a single metropolitan region.

The fact that Sir Patrick Abercrombie is the author of the Greater London Plan and also joint-author of the County of London Plan does not overcome the fundamental disadvantages arising from the failure to deal with the problem of London as a whole. A planner entrusted with undivided responsibility for the whole region would envisage movements of population and shifts of industry within the metropolis on their merits from the standpoint of good planning, without regard to the interests of particular local authorities. Such a disinterested attitude cannot be taken by a body possessing jurisdiction over a smaller area. The London County Council, to be more precise, cannot be indifferent to heavy losses of population, rateable value, industry and commerce, from its area. Hence, although the London County Council may nominally have given Sir Patrick Abercrombie and Mr. Forshaw a free hand, these accomplished and eminent men must have been aware of the limitations which were inevitably imposed upon them by the limited scope and partial interests of the authority to which they had to submit the plan.

The disadvantages of attempting to plan the administrative County in isolation from the remainder of the region is clearly visible in the Forshaw-Abercrombie plan. It has deep defects which are due to the circumstances of its origin.

The County of London plan has been severely criticised by competent critics on both sides of the Atlantic. The features which have aroused the most adverse comment are its wholly inadequate proposals for de-congesting Inner London by decentralising population and industry to Outer London; the high densities which it contemplates for residential development, ranging from 100 to 200 persons an acre of housing area according to zone; the great extension of high-built flats in place of houses which these densities involve, despite the overwhelming preference for houses expressed by Londoners, and the low standard for open spaces which it accepts, amounting to only 4 acres per 1,000.

The proposals in the plan relating to decentralisation are so limited in scope as to be almost trivial.[1] The scheme aims at removing no more than about 500,000 persons from the overcrowded parts of the County—a figure which is only slightly larger than the reduction effected by the centrifugal tendency which has been going on since 1891, and which, during the years 1931–1937 led to a decrease of 348,000 persons in the central areas.[2] The plan, therefore, does little more in this vital matter than bid farewell to departing guests. After they have departed the remaining population in the central areas will be rehoused at a density of 136 to the acre, although the authors admit that they would prefer to see a density of 100 adopted.

It is, writes Mr. Lewis Mumford, with surprise and dismay that one finds that the planners of London have given little thought to the terms on which the one-family house may be

[1] Chapter II of the County of London Plan.
[2] The figure given varies between 500,000 and 600,000.

rehabilitated. "They have done just the opposite; the plans for repopulating Hackney, Shoreditch, Bethnal Green, contemplate making more than half of the new structures flats; some of them eight- and ten-storey flats. It avails not that these flats will have 'sufficient' open spaces or that their inhabitants will reach the upper floors on lifts . . . it hardly makes any less serious the outright breach with an old and healthy tradition."[1] The distribution of houses and flats in the rebuilding schemes for the East End will, in his view, prove to be a positive check to the population rate.[2] Mumford rightly regards the population aspect as the dominant consideration for wise city planning and the reform of urban life. He sees, too, the clear need for the creation of a regional authority capable of acting over a much larger area than the London County Council, and of exercising an over-all control on the entire process of building.[3]

The post-war reconstruction plan presented by the City Corporation in 1944 was a pitiful affair which followed the worst traditions of the City fathers in failing to make the Square Mile a convenient, healthy, beautiful or dignified commercial, industrial and civic centre.

The air attacks during the war inflicted immense damage on the City. Whole areas were razed to the ground; streets or districts devoted to particular trades were almost obliterated. Air-raid damage and other war-time causes reduced the rateable value by a quarter of its 1939 figure. Moreover, decentralisation and evacuation enabled many large business organisations to test the advantages, in terms of costs, efficiency, comfort and health, of settling their administrative and clerical staffs in country towns. New arrangements for warehousing and distribution induced by the emergency may prove to be permanent. "It remains to be seen," declared the Improvements and Town Planning Committee in their report to the City Corporation, "in what degree these influences, experienced during a period of abnormal control and shortage of labour and material, may exercise any restraining effect upon the redevelopment of Central London; although national policy in

[1] Lewis Mumford: "The Plan of London," in *City Development*, p. 188.
[2] *Ibid.*, p. 195.   [3] *Ibid.*, p. 197.

relation to the distribution of industry might have certain effects on London generally, we believe that it will be essential to the national interest to rebuild a central core at which the widespread activities of industrial and other enterprise can be integrated and co-ordinated, and new undertakings initiated."[1]

These high-sounding words and lofty outlook were accompanied by proposals which envisaged a day population in the City of 800,000 compared with an estimated daily influx of workers of 500,000 in 1935.

The plan showed no appreciation of the magnificent opportunity afforded by the ravages of war to rebuild the City on nobler and more rational lines; to open up its medieval alleyways on a large scale; to deal boldly and imaginatively with the crucial traffic problem produced by the seven highways converging on the Bank of England; to abolish the narrow canyons in which thousands of men and women of all classes are condemned to pass laborious days, sans light, sans sun, sans air; to provide amenities which would relieve the dismal squalor and joyless prospect which characterise most of the Square Mile. The main features of the plan comprised little more than a limited clearance of the built-up area round St. Paul's; a continuation of the Thames embankment from Blackfriars to London Bridge; and the provision of a main ring road from Holborn Circus running round the northern side of the City parallel with the boundary and then dipping south to the Tower where it would effect a junction with another new main highway connecting with an eastern extension of the Thames embankment.

The City's plan was laid before the Royal Fine Arts Commission which commented adversely on it. It was then submitted to Mr. W. S. Morrison, the Minister of Town and Country Planning, who informed the Corporation that if the plan were adopted by the Corporation "he would find great difficulty in approving it as the basis on which the reconstruction and redevelopment of the City . . . is to be carried out." The Minister advised the Corporation to make an entirely new attempt "under the guidance of some person or persons

[1] Reconstruction of the City of London, 1944. pars. 21-3.

of admitted competence and authority in town planning, to lay down the lines along which, both in the immediate post-war period and during the next 50 or 60 years the reconstruction and redevelopment of the City should be directed."[1]

This outspoken rejection of the feeble efforts of the Guildhall was the more humiliating in that it was made by a Conservative Minister of Town and Country Planning, whose attitude could therefore not be represented as due to political antagonism to the age-long Conservatism of the City Corporation. Mr. Morrison's pointed suggestion that the Corporation avail itself of the services of an eminent town planning consultant implied that he considered neither the Improvements and Town Planning Committee nor those who prepared the scheme in accordance with their instructions as competent to undertake the task unaided.

In their Report the Corporation took great pains, in a historical foreword, to exculpate the City fathers from any share in the responsibility for rejecting Wren's plan for reconstructing the City after the Great Fire of 1666. For this purpose they relied mainly on Mr. Reddaway's book,[2] seeking thereby to demonstrate, first, that "the so-called 'rejection'" is a mere legend because Wren's plan was not favoured by either the King or Parliament, and could therefore not have been rejected by the City; second, that no "lost opportunity"[3] occurred on that occasion. It is symptomatic of the City's attitude that these words were printed in inverted commas.

The Minister swept aside these pedantic observations which served only to obscure a larger consideration of the subject. "The City of London," declared Mr. W. S. Morrison, "is the heart of the greatest and most famous city in the world, the Capital of the British Commonwealth of Nations, and a centre of world trade, commerce and finance: it possesses buildings and remains of the greatest architectural, historical and anti-quarian interest, and is at the same time the work place of considerably more than half a million persons. A great

[1] Letter from the Secretary of the Ministry to the Town Clerk, dated 2nd July, 1945.
[2] T. F. Reddaway: *The Rebuilding of London after the Great Fire.*
[3] *Reconstruction in the City of London :* "Historical Foreword, p. 1.

opportunity was lost, in 1666, of laying it out afresh in such a manner as to enable it to fulfil these various functions in the most ample manner, and to develop along lines which would enable it to continue to fulfil them. An even greater opportunity has now occurred, and the Minister feels very strongly that the fullest advantage of it must be taken."[1]

In order to take full advantage of the present opportunity, the Minister continued, the makers of the City Plan must ask themselves, "what sort of place should the centre of a world capital, with an ancient and glorious history and a thriving and growing modern life, be?" The plan submitted to him appeared to have been prepared only with a view to palliating some of the major inconveniences of the pre-war City while disturbing to the least possible extent the existing layout and proprietory interests. Accompanying the Minister's letter was a note containing detailed observations on the City of London plan. Little in the plan survived the devastating criticism contained in the note.

After much fuss and fuming the City Corporation appointed Professor Holford to prepare a fresh plan in conjunction with Mr. Holden. These two eminent men have produced a plan which is far more worthy of the time, the place and the occasion. But the reputation and the prestige of the City Corporation suffered a shattering blow from the events described above. The hollowness of its pretensions to be the great guardian of the City's interests was exposed by the revelation that, left to their own devices, the City fathers would have repeated on a larger scale the follies which occurred during the rebuilding of London after the Great Fire.

It was no doubt partly owing to the weakness of the position in which the City Corporation had placed themselves by their own shortcomings that Mr. Silkin (who succeeded Mr. W. S. Morrison as Minister of Town and Country Planning in 1945) decided to abolish the City Corporation as a planning authority when he framed the new system of planning administration embodied in the Town and Country Planning Act, 1947. Henceforth, the London County Council will be the local planning authority for the City as well as for the rest of

[1] Letter of 2nd July, 1945.

the County of London. The City Corporation will, however, remain responsible for executing the plans for the square mile and it remains to be seen how far they will adhere to the principles laid down in the new plan.

The Greater London Plan prepared under the direction of Sir Patrick Abercrombie is essentially a regional plan, except that it does not deal with the dense core of Inner London, which is covered by the County of London plan. Its planning features and its administrative aspects call for separate consideration.

The Plan, it can be said without hesitation, is a great achievement. Sir Patrick Abercrombie is the most creative planner this country has so far produced, and his faculties have had full play in designing a broad pattern of human life for the vast metropolis.

The fundamental aims of the Greater London Plan are: first, to stop the haphazard growth of London; second, to effect a substantial measure of decentralisation within the metropolis; third, to introduce controlled development of housing, industry and communications. The Plan divides the Region outside the London County Council boundary into four main belts: (1) The Inner Urban Ring; (2) The Suburban Ring; (3) The Green Belt Ring; (4) The Outer Country Ring.

We have already noted that the Abercrombie-Forshaw Plan provided for the decentralisation of 500,000–600,000 persons from the County of London. The Greater London Plan adds a further 415,000 persons to this figure, making a total of about a million inhabitants to be moved. About three-quarters of them will be decentralised in and near the region by means of additions to existing towns. new sites and quasi-satellites. The remaining quarter will be dispersed outside the Region.

The additions to existing towns will absorb some 261,000 persons. The towns selected are outside the continuously built-up area. They lie within the boundaries covered by the Plan and are mostly situated in the Outer Country Ring. They include such places as Amersham, Beaconsfield, High Wycombe. Romford, Pitsea, Loughton, Cobham, Dorking, Leatherhead, and Ashtead.

The most striking proposal involves the creation of eight

new satellite towns outside the green belt ring. These towns are expected to absorb nearly 400,000 persons. The population of each new town will not exceed a maximum of 60,000, and in some instances a lower upper limit has been fixed. Ten possible sites were suggested for these new towns, but the Minister of Town and Country Planning has accepted only two of these: namely, Stevenage and Harlow. The Minister has selected Crawley-Three Bridges and Hemel Hempstead as the sites of two other towns; and additional sites for this purpose are under consideration.[1]

The new towns will presumably be of the garden city type; and although they will be satellites from one point of view they will be largely self-contained communities providing employment for their inhabitants in local industry, commerce and other occupations. The construction of these towns was authorised by the New Towns Act, 1946, which empowers the Minister to create a series of public corporations for the purpose. Several have already been started. This is undoubtedly one of the most exciting and imaginative ventures in the history of planning.

The quasi-satellites will consist of developments in the built-up area in Inner London. Professor Abercrombie frankly admits that, "On the face of it, these quasi-satellites offend against all notions of planned decentralisation; they are, in the first instance, residential, and they are too close in. Nevertheless, they are necessary features of the short-term policy of immediate post-war housing requirements; the maximum figure of 125,000 has been allotted for the purpose."

As one of the main objects of the Plan is to control development, the regulation of housing forms an integral part of the decentralisation policy. Sir Patrick Abercrombie recognises the need to correlate the housing programme with the population structure of the community, and he emphasises the need for a true census of families in order to allot accommodation on rational lines. The density standards recommended are an immense improvement over those contained in the County Plan. The proposed standards are 30 persons per acre for new

---

[1] *Greater London Plan*, Memorandum by the Minister on the Report of the Advisory Committee for London Regional Planning, paragraphs 22-8.

cities, 50 for the suburban ring, 75 and 100 for the inner urban ring. Houses will be provided exclusively for all densities up to and including 75 per acre. At a density of 100 to the acre, 80 per cent of families can be accommodated in houses, leaving only 20 per cent for flats. This compares with 1,368,000 persons to be rehoused in flats under the County Plan at densities of from 100 to 200 an acre, while only 1,932,000 persons will be accommodated in houses.

The chapter on industry is a masterly contribution, and in some respects the most important part of the Plan. Sir Patrick Abercrombie assumed that there would be a national policy based on the Barlow Commission's Report. Hence he approached the problem as essentially one of decentralisation, redistribution and orderly development. To effect this, the Plan provides for the removal of an appreciable amount of industry and commerce from central London. It bars further industrialisation north of the Thames within the inner urban and suburban rings, except in the Barking-Dagenham area on the east side of outer London. A similar prohibition would apply on the south side of the river except in a few specified places, such as Mitcham and Croydon, Erith, and the Cray Valley. New industrial development must be guided to the new towns and the existing ones which are scheduled for expansion. In no other way can they become prosperous centres of work and residence capable of absorbing the population to be decentralised. New factories would not be permitted in other market, residential and dormitory towns unless they could make out a special case on economic and technical grounds. Only industry associated with agriculture or with small-scale crafts would be allowed to settle in rural villages, and no industrial development would be permitted in the open country-side.

Improved communications are dealt with at length in a major chapter. The important proposal is made that, as in New York, all railway transport within the region should be electrified. Roads, civil airfields, inland waterways, railways and markets are all treated in a comprehensive manner.

The subject of open spaces is dealt with most adequately, and it is clear that this is a matter about which the author of the Plan cares deeply. The Plan envisages a wide green belt in which building will not normally be permitted. Beyond this there will be a brown agricultural belt. Similar girdles on a smaller scale are contemplated for the separate towns and communities, whether old or new, within the Region. The Plan proposes that the present Green Belt, which consists merely of a series of discontinuous wedges, shall be extended and made into a connected system by means of footpaths parkways, riverside walks, bridle paths and green lanes.

There are numerous detailed recommendations which deal with the preservation of the general countryside; the protection of areas of special scenic beauty; old parks, both public and private; amenities connected with the Thames and lesser rivers and waterways; the construction of parkways and footpaths, bridle and bicycle tracks; rest gardens, children's playgrounds, recreation and sports centres, town squares and town parks. Playing fields are a most important item in this comprehensive scheme. The total open space recommended is ten acres per thousand population. Of this, three acres is allocated to schools, one acre to parks, four acres to public playing fields, and two acres to private playing fields. In the more densely populated parts of the region all the open space would be publicly owned. It is only in the outer areas that private playing fields would be permitted.

It is impossible to do more here than to describe briefly the leading features of the Plan. A brief summary cannot do them justice or even mention the many other items which the scheme includes. The reader will find a study of the Plan both profitable and pleasurable. It is a great State Paper of a new kind, embodying many important sociological, aesthetic and economic concepts. As a contribution to the reconstruction of Britain it has great significance, for it applies for the first time on a large scale the general ideas and principles which have been germinating in the planning movement during the past 25 years.

The great question which the Plan poses is: How should it be carried out? Sir Patrick Abercrombie, casting his eyes around, saw that there was no body in the metropolis remotely capable of implementing the Plan. He was therefore driven to improvise machinery of an *ad hoc* character which would override or replace the 143 separate local authorities which were empowered to prepare planning schemes in the Greater London area alone. The administrative framework proposed by the Plan is, however, ill-considered and defective. It is described in a perfunctory manner and occupies only $1\frac{1}{2}$ pages in a volume of 217 pages.

The scheme recommends a statutory planning board appointed by, and responsible to, the Minister of Town and Country Planning, consisting of a small number of eminent men of affairs, several of whom would give their full time to the work. The board would have its own technical and administrative staff.

A master plan would be prepared by the Minister on the basis of the three plans for the City, County and Outer London. The board would then have the task of approving, on behalf of the Minister, all in-filling planning schemes in the area. The preparation of these local schemes and the day-to-day administration of local planning, in compliance with the requirements of the master plan, would fall on joint executive committees constituted from local planning authorities. The regional planning board would, however, not only have power to prevent the wrong use of land, but also have constructive duties to ensure the positive realisation of approved physical changes in the region. For this purpose it would be given power to buy and sell land; to utilise such land for agricultural and recreational purposes of all kinds; to undertake housing estates and trading estates; and to construct roads, buildings and public works.

The planning board would also act in a number of other executive capacities. It would, for example, serve as a regional open spaces authority; as a housing corporation; as industrial controller to deal with the location of industry and the development of trading estates; as regional public cleansing department; as population adviser; and as regional transport authority.

It can be seen that a planning board formed on this model would in respect of all its functions be indirectly an organ of the central government, since it would be appointed by the

Minister of Town and Country Planning, and be answerable to him. It would have no connection with the machinery of local government in London. It would have no democratic roots. It would be politically irresponsible. It would have no claim to represent the people of London, nor would the great community of the metropolis have any constitutional means of influencing the composition of the board, dismissing its members, discussing its proposals, or rejecting its decisions.

London government is in dire need of reform; but this is not reform. This is merely pursuing a policy of avoiding reform by establishing still more *ad hoc* authorities to deal with regional services. It postpones dealing with the municipal problem by adopting yet another highly undesirable type of organ, thereby adding to the multiplicity of unco-ordinated and competing agencies which litter the ground. To super-impose an authoritarian board of this kind on the existing chaos of areas and authorities would be a serious mistake; for only if a powerful public opinion is harnessed to the essential principles of the plan will it be possible to evoke the drive, enthusiasm and interest which are necessary to carry them out. An appointed board would certainly not succeed in this fundamental respect.

A regional planning board of the kind proposed by Sir Patrick Abercrombie would doubtless be better than nothing at all, since it would at least secure a unified outlook. But it would be immeasurably inferior to a planning organ associated with a directly elected Greater London Regional Council, and drawing its members, or a majority of them, from the latter.

On 5th March, 1946, the Minister of Town and Country Planning made a statement in Parliament about the planning of London. The Government's policy, he said, is in accord with the fourth and fifth recommendations, unanimously reached, of the Barlow Commission.[1] The plans for the County of London and Greater London between them contain a number of co-ordinated proposals aimed at achieving these objects. The Minister declared that he and his colleagues had reached the following decisions on the Greater London Plan:[2]

Firstly, the overall growth of London's population and

[1] *Ante* p. 449–50.   [2] Hansard, 5th March, 1946, cols. 189–92.

industry should be restrained. This is one aspect of the general policy for achieving throughout the country a better balance of the distribution of industry, and in particular for assisting the industrial recovery of the Development Areas.

Secondly, a planned programme of decentralisation to the outer areas of Greater London should replace the uncontrolled sprawl of the inter-war period. War damage in the congested inner areas and war-time evacuation have provided a unique opportunity for effecting this redistribution. The intention is to make provision for about a million persons and concurrently a related quota of industrial firms to be accommodated farther out—mainly in a few new towns and in selected existing towns within 20 to 50 miles of London's centre. The planned developments will be given priority according to their urgency.

Thirdly, it is proposed that the general lines of the decentralisation and resettlement should broadly conform to the proposals made by Sir Patrick Abercrombie for dividing the area surrounding the County of London into four Rings. From the County of London and the Inner Urban Ring round it, which form the congested areas, most of the decentralisation should take place. The next Ring, the Suburban Ring, should be regarded in general as static. Surrounding this built-up area a Green Belt Ring is to be carefully safeguarded, and this Ring, except in permitted cases, should act as a barrier to further suburban growth. The fourth or Outer Country Ring should serve as the main reception area for persons and industry moving out from overcrowded London into compact settlements surrounded by open country.

The Minister concluded by saying that the carrying out of these proposals would rest partly on the new town planning legislation the Government intended to introduce. While the Government endorsed the main principles of the Greater London Plan, they had not so far adopted a number of the individual development projects which it recommended, such as the number and location of the new towns and the highway proposals. These matters were being examined by the Government and by local planning authorities.

The central machinery for this purpose consists of an inter-departmental Committee composed of senior officials of the departments concerned, such as the Ministries of Health, Transport, Labour and National Service, Agriculture and Fisheries, Education, Fuel and Power, Works, the Board of Trade, etc., under the chairmanship of a representative of the Ministry of Town and Country Planning. This is a strong official body which is expected to remain in existence indefinitely.

The 143 local authorities possessing planning powers in Greater London were grouped into 23 Joint Planning Committees. An advisory Committee was set up by the Minister consisting of representatives of these bodies, the County Councils in the area of the Greater London Plan, the London County Council and the City Corporation, presided over by Mr. Clement Davies, K.C., M.P., the Leader of the Liberal Party. A technical sub-committee of officials from these authorities was appointed to examine the Abercrombie Plan in detail, and the report of the technical officers was unanimously accepted by the advisory Committee.[1]

The report of the advisory committee departed in several important respects from the Greater London Plan. The most conspicuous was the proposal to accommodate nearly 160,000 more people in the Green Belt Ring than the Plan recommended. Fortunately, the Minister refused to accept this deterioration of standards.[2] In the Outer Country Ring the Committee proposed to accommodate an ultimate population of 1,167,970 as compared with 1,013,500 scheduled in the Plan. The proposals for new towns were cut down so as to accommodate only 200,000 persons instead of 415,000. The views of the advisory committee were manifestly inferior to those of Sir Patrick Abercrombie on all these matters.

They even considered that "the whole of the population

[1] Advisory Committee for London Regional Planning Report to the Minister of Town and Country Planning. H.M.S.O., 1946.
[2] Hansard. Commons, 19th November, 1946, cols. 692–3. See also Memorandum by the Minister of Town and Country Planning on the Report of the Advisory Committee for London Regional Planning.

to be decentralised could be accommodated by the expansion of existing communities within 50 miles of London, and indeed mainly within the area covered by the Greater London Plan, without recourse to the development of new towns," which they were willing to tolerate merely as "valuable examples for guidance in planning extensions of existing towns."[1] On these issues the Minister also stood firm. Unfortunately, he felt obliged to agree to larger expansions to existing towns than Sir Patrick Abercrombie had envisaged.[2]

Mr. Silkin also appointed a committee to advise him on "the appropriate machinery for securing concerted action in the implementation of a regional plan for London as a whole." Mr. Clement Davies was also chairman of this committee, and its members included Mr. Neal, late deputy secretary of the Ministry of Town and Country Planning, Councillor Mrs. Bolton, chairman of the Town Planning Committee of the London County Council, several elected representatives of other planning authorities, and three independent persons, of whom the author was one.

The committee sat for more than two years without reaching agreement. At an early stage in the proceedings it became clear that the committee was divided between those members who desired above all to secure the creation of effective machinery for planning the metropolitan region as a whole, and those who were determined at all costs to preserve the powers of local planning authorities. The former conception involved the establishment of a strong executive organ which would be responsible alike for regional planning, planning control and large-scale development operations. The latter view meant relying in the main on the existing planning authorities, with the addition of a weak co-ordinating body with advisory functions. This would result in regional control and development ultimately resting with the central government.

A significant feature of the proceedings was the evidence tendered by central government Departments. Officials of

[1] Greater London Plan. Memorandum by the Minister of Town and Country Planning on the Report of the Advisory Committee for London Regional Planning, paragraph 18.
[2] *Ibid.*, paragraph 26.

the Board of Trade and the Ministries of Health, Fuel and Power, Transport, and other central Departments, all testified to their complete faith in their ability to administer without extraneous aid the planning features calling for regional action. Again and again we were asked to remember the perfect harmony and efficiency achieved by the Interdepartmental Committee of officials mentioned above. The fact that, with the exception of the Ministry of Town and Country Planning, no Department possesses a regional organisation even remotely corresponding with the area of the Greater London Plan[1] is not a matter which disturbs the complacency of the Departments concerned or which the official witnesses regarded as even relevant to the discussion. They were one and all satisfied that their regional officers could handle the job successfully in conjunction with local authorities.

The idea that regional planning requires administrative institutions corresponding to the area to be planned and developed has not penetrated the minds of Whitehall. The notion that the planning of the metropolis should be entrusted to a democratic body, responsive and responsible to the wishes of the 9 or 10 million inhabitants of greater London, is entirely foreign to the Departmental outlook. In Whitehall to-day, regionalism means administration by officials of the central Departments stationed in the various regions into which the county is divided for central government purposes. There is no understanding of any other kind of regionalism, and no recognition of the need for decentralisation on democratic lines.

The immense congestion of work in most government Departments at the present time makes this failure to perceive the necessity for democratic regional decentralisation a serious matter; for it is abundantly clear that unless the central departments are prepared to shed a great many of their present functions they will be utterly unable to discharge the heavy tasks required of them during the coming years.[2]

[1] The Board of Trade, for example, has three divisions for the area of the Greater London Plan. The London and South Eastern Division takes in a large part of the metropolis; the Eastern Division includes most of Herts. and Essex, while Bucks., Beds., and Berks. fall into the Southern Division.

[2] For a development of this theme see my pamphlet *Public Administration To-day* (Stevens, 1948).

The attitude of the Board of Trade may be taken as illus-trative of the general tendency. A number of major questions relating to the location of industry are indubitably of national importance, and must be decided by the central government. These include restriction over the entry of new industry into London; diversification of industry; special incentives for the development areas; priorities for factory building, and so forth. It is right and proper that matters of general policy of this kind should be determined by the Board of Trade in consultation with other Departments. But the Board of Trade goes much further than this, and regards it as their business, not merely to decide whether a particular firm shall be allowed to start manufacturing in London, but also to settle the exact district where the factory is to be located. It is true that various regional panels composed wholly or mainly of civil servants are consulted during the process.[1] It is also true that the actual site will be left for the Ministry of Town and Country Planning to determine in conjunction with the local planning authority. But in essence both the major and the minor aspects of the matter remain with the central government.

This is an entirely wrong conception. Restriction on entry into the London Region for industrial purposes is clearly a matter of national policy for the central government. But, if permission to enter is granted, the question of where the factory should be located within the metropolis should be devolved on a regional planning committee acting on behalf of an elected Greater London Regional Council.

The Board of Trade now possesses extensive powers.[2] A local planning authority cannot consent to the erection of an industrial building unless the Board issues a prior certificate. The Board is seeking by means of these powers to maintain full employment; to promote diversification of industry; and to restrict the entry of new industry into London. The full employment objective is pursued by the simple process of taking work to the worker. This means that the labour supply position is the principal factor indicating whether an indus-trialist should be encouraged or discouraged from placing his

[1] *Ante*, pages 454–5.
[2] Town and Country Planning Act, 1947, S. 14. See also S.I. 1948, No. 1309, and Hansard, Commons, for 28 July 1948, Col. 1494.

factory in a particular area. This may be inevitable in present circumstances, but in the long run the general principles laid down by the Barlow Commission will require a considerable amount of movement of workpeople to the places which are suitable for industrial expansion on planning grounds. In consequence the relatively simple criteria which the Board of Trade is using at present must give way, if proper planning principles are to be observed, to much more complex and difficult considerations. Housing, decentralisation, amenities, social services, transport facilities and many other questions will have to be taken into account in determining location of industry, both as between one region and another, and also within each region. The Board of Trade is not properly equipped to obtain the data relating to all these matters; nor is its organisation, either at headquarters or in the regions, suited for this purpose. Direction over the location of industry cannot ultimately be decided solely on industrial grounds, but must depend on a balance of all the factors which come into the planning picture. So far as London is concerned, there should undoubtedly be a considerable degree of decentralisation from the Board of Trade and its various panels to a representative regional body. The radical changes in planning administration, powers and procedure effected by the new legislation, do not affect this question of organisation.

The Greater London Planning Administration Committee was an ineffectual body for several reasons. First, there was a deep cleavage of outlook among the members which sprang from different conceptions of metropolitan government. Second, the Committee was faced with an insoluble problem in trying to deal with a particular function without reforming the whole system of London government, which was clearly outside its terms of reference. Third, in the middle of the Committee's proceedings Mr. Silkin introduced his great Town and Country Planning Bill which entirely altered the local planning organisation in the Greater London area. In consequence all the assumptions on which the Committee had been working were nullified.

After much delay the Committee eventually produced a report which recognised the need for dealing with the region as a whole. The report explained that the Committee was not

The area of the Greater London Plan includes Surrey, Mid
Essex, small but important portions of Bedfordshire and Bucking
across the boundaries of these last-mentioned five counties and th
out the Plan.

A Regional Planning authority covering the whole area is c

SUFFOLK

E S S E X

W. Essex

S. Essex

ssex

N. Kent

Kent

K

Sevenoaks

E   N   T

E   X

GREATER LONDON PLAN BOUNDARY

COUNTY BOUNDARY.

and Hertfordshire in their entirety, substantial parts of Kent and
e and a fragment of Berkshire. The map shows how the Plan cuts
vance of entrusting county councils with responsibility for carrying

eeded.

competent to advise on the reform of London government, and suggested that the Local Government Boundary Commission should be invited to undertake this task. Any such reform would require lengthy investigation and deliberation and would therefore take some time. As a temporary measure a majority of the Committee proposed that a joint advisory committee should be set up by the local planning authorities to act as a co-ordinating body in regard to planning control. This body would have no functions in the field of development. A minority report, to which the author subscribed, advocated a joint planning board as the indispensable minimum for the interim period. This at least would have powers of its own and not be dependent on the willingness of the county councils and county borough councils to carry out its recommendations. Such a board would, in fact, become the local planning authority for the whole metropolis, replacing the county and county borough councils for that purpose. It could, however, delegate powers to the latter councils under a scheme of delegation authorised by Ministerial regulations. Even this proposal was a highly imperfect solution, since it left with the local authorities most of the development functions relating to such matters as housing, highways, drainage, parks and open spaces and other municipal services. Moreover, it had the inevitable disadvantages of introducing yet another *ad hoc* organ in an area where there are already too many single-purpose bodies. I have argued at length against the *ad hoc* authority as a solution of the problems of regional government. It was, however, impossible to set up a more effective body under the Town and Country Planning Act, 1947.

As already mentioned, this far-reaching measure transferred local planning powers from county district councils to county councils. Thus, in the Greater London Region the 23 Joint Planning Committees (into which the 143 separate planning authorities had recently been formed) were replaced by 8 County Councils. These, together with the London County Council (which becomes the planning authority for the City) and the county boroughs of Croydon, East Ham and West Ham comprise the local planning authorities for the whole metropolitan region.

The change in administrative organisation effected by the new town planning legislation contributes little to the solution of the London planning problem. Indeed, in some ways it makes the problem harder to solve. With 143 separate authorities the case for a regional organ was overwhelming. With 23 Joint Committees it remained very strong. With only 12 county and county borough councils possessing planning powers, a regional organ appears superficially to be less necessary.

In fact, however, that is not so. The area of the Greater London Plan includes the whole of Surrey, Middlesex, and Hertfordshire, but it cuts clean through the boundaries of Kent, Essex, Bedfordshire, Buckinghamshire, and Berkshire.[1] The Abercrombie plan covers a small but important part of Buckinghamshire. Most of Bedfordshire is outside, but Luton, the largest town and the chief industrial centre, is included. A trifling fraction of Berkshire is included.

It follows, therefore, that we cannot expect the county councils of Kent, Essex, Buckinghamshire, Bedfordshire, and Berkshire, to be imbued with a regional outlook in their planning activities. Their interest in the Greater London Plan must inevitably be divided and diluted with other conflicting interests.

At present it is scarcely possible for planned development to take place in London, except on paper. Many people will find this a surprising statement because they do not appreciate the gulf which separates a paper plan from its realisation; nor the impossibility of realisation without suitable administrative institutions. Yet it is unfortunately true that, despite the plans for the County and Greater London, post-war development in the metropolis is in danger of proceeding on lines which contravene the principles of good planning owing to the lack of proper administrative machinery suited to the task, apart from the designation of a number of new town corporations. And so it will continue until Parliament and the Government, or the political parties, recognise the simple fact that successful planning control and development cannot take place within an antiquated and incoherent administrative framework such as we now have. To make ambitious plans for the metropolis and expect them

[1] See Map at end of volume.

to be carried out with the present administrative machinery is
as sensible as it would be to expect a modern aircraft to fly
with an old motor-bus engine dating from 1910.

We may now review recent events regarding the reform
of London government. A pamphlet issued in 1943 by the
Labour Party entitled *The Future of Local Government*, contained a pronouncement of the party's post-war policy in this
sphere. A page devoted to local government in Greater
London describes the multiplicity of areas and authorities.
The failure to adjust the structure of local government in
the metropolis to rapidly changing industrial conditions is
diagnosed as a major contributory cause of the sporadic
development and excessive urbanisation of the London region.
The creation of special one-purpose organs for water, police,
the port and other services; the building by the London County
Council of housing estates outside its boundaries; the depen-
dance of outlying authorities on the main drainage system of
the London County Council, are mentioned as indicating
the necessity of carefully reconsidering London's local government structure.

The Labour Party pamphlet recognises that there are far
too many separate authorities in Greater London, and that
their existence produces overlapping and prevents proper
co-ordination and planning. It states that reform must seek
to facilitate effective town planning; unified administration of
education, hospitals and other services; and the preservation
of democratic responsibility and control. The existing structure in Outer London is noted as particularly irrational and
indefensible.

The pamphlet makes no attempt to solve the problem of
London government. The Committee which prepared it
expressed doubt whether a remedy could be found without
treating London and Greater London as one homogeneous
whole—but beyond this daring speculation they did not
venture to go. Instead, the Labour Party was content to
inform a grateful world that "the problem bristles with difficulties, but if there be a general measure of agreement among
the existing local authorities we believe that a satisfactory

solution could be found on the basis of the principles advocated in this Report." These principles include the establishment of regional areas in which there would be major and minor authorities. The democratic tradition in local government must be preserved, which presumably means that both tiers of authorities should be directly elected. The pamphlet refused to grapple with the question of what is a region. It merely advocated the setting up of "suitable machinery" by the central government to survey the country as a whole with a view to determining the areas suitable for regional or major authority, adequate for the efficient performance of large-scale services.

The tone of the pamphlet is vague, nebulous, and full of sonorous commonplace phrases without precise content. There is nowhere any evidence of hard thinking. It is, of course, utterly impossible to reform local government in London on the basis of agreement among the existing authorities. The Labour Party must have been aware of this, yet it did not hesitate to make this futile statement.

With so feeble and half-hearted an approach on the part of the Labour Party, it was unlikely that the Coalition Government would be disposed to treat the London problem seriously. It was therefore not surprising that the White Paper issued in 1945 on Local Government in England and Wales[1] should take elaborate pains to by-pass the exceptional difficulties confronting the metropolis—difficulties which arise partly from its obsolete municipal organisation and partly from the severe damage it suffered in the war.

The White Paper mentions two problems calling for special treatment; they are local government in London and in the county of Middlesex.

As regards the former it discerned two broad issues. First, should the boundaries of the administrative county be extended? Second, are the relations between the metropolitan borough councils and the London County Council satisfactory?

As regards the first of these issues, the White Paper stated that a question of this magnitude, which might well involve the creation of a local government area comprising a fifth of

[1] Cmd. 6579/1945., H.M.S.O.

the entire population of England and Wales, the disappearance of Middlesex as an administrative county and heavy inroads into several other counties, would require special legislation. So large a matter could not be entrusted to a body of Commissioners, however authoritative; and for this reason the Government did not propose that the powers of the Local Government Boundary Commission should extend to London.

Quite apart, however, from the question of the appropriate method of handling the problem, the Coalition White Paper explained why the Government were not satisfied that the present time is opportune for a major reform of this kind. The reason given for this conclusion was that "several county councils and other local authorities in and around London are at present engaged in recasting their educational services and will shortly be confronted with other similar tasks. It would, in the view of the Government, be a mistake to interrupt those tasks by throwing into the melting-pot the whole problem of London and the Home Counties."[1]

This pitiful evasion of a great problem can scarcely be regarded as more than mere face-saving verbiage. It could scarcely convince anyone, since local authorities will always be engaged in some task which is liable to be disturbed by any considerable measure of reform. But a stronger criticism of the argument is that it puts the cart before the horse. The very magnitude of the tasks which confront the metropolis in the era of reconstruction make it essential that its municipal organisation shall be as efficient as possible.

The White Paper emitted another great smoke-screen in order to conceal the truth about the second issue relating to London to which they drew attention. The war, it said, had brought into prominence the civil defence functions of the metropolitan borough councils and the importance of the town hall as a centre of local effort and enthusiasm. The success with which metropolitan borough councils had carried out their civil defence and other wartime functions was deserving of full recognition, but "the very urgency and importance of the work has emphasised the anomaly of dividing London into

Ibid., p. 19.

no less than 29 separate areas differing enormously in size and population. Moreover, the problems not only of civil defence but of reconstruction, have made it clear that a reconsideration of the allocation of functions between the boroughs and the county is overdue."[1] This, again, the White Paper declared, is not a matter which could conveniently be handled by the proposed Boundary Commission. The Government, there-fore, intended to appoint an authoritative body to report to them on local government problems within the county.

The County of Middlesex, the White Paper pointed out, is entirely urban in character. It contains no rural districts. Its two million inhabitants are distributed among 15 non-county boroughs and 11 urban districts. Nine of the former, and one of the latter, exceed 75,000 population. "By aggregating populations now grouped in contiguous non-county boroughs and urban districts it would be possible to create a series of towns, each large enough on any standard for county borough status."[2] In this respect, observed the White Paper, Middlesex is unique—though, one may add, similar proposals have frequently been made for other counties, such as Glamorgan-shire and the West Riding. If once the process of creating county boroughs were started there would be no stopping short of breaking up the entire county and destroying county government. This the Government would not countenance, and hence the Local Government Boundary Commission would not be given power to entertain applications for county borough status from ambitious county districts within the County of Middlesex.

In due course the proposals made in the Coalition White Paper were carried into effect. The Local Government Boundary Com-mission was created by an Act which excluded the adminis-trative County of London from its jurisdiction, and forbade it to constitute any part of Middlesex a county borough.[3]

Thus was the sterile confusion, the impotence and the futility of London government once again sheltered from the cleansing winds of reform which are permitted to sweep freely

---

[1] *Ibid.*, p. 19. [2] *Ibid.*, p. 18.
[3] Local Government (Boundary Commission) Act, 1945, s. 1 (1); and proviso to S. 2 (1).

over the rest of the country. In 1835, some malevolent fate "protected" the City from the recommendations of the Royal Commission on Municipal Corporations and the great local government reforms of 1835. In 1945, political interests have, in turn, "protected" the counties of London and Middlesex from the beneficent activities of the Local Government Boundary Commission.

In April 1945 a Committee was set up under the chairmanship of Lord Reading, to examine and review the number, size and boundaries of the metropolitan boroughs and the distribution of functions between the London County Council, the City Corporation, and the metropolitan boroughs, and to make recommendations.

It was obviously wrong to inquire into the organisation and position of the minor authorities in London before settling the much more important problem of the major organ for the metropolitan region. This was ultimately recognised, for in October, 1946, Mr. Bevan, Minister of Health, announced that the Committee would be abolished, on the ground that "the problem which the Committee was asked to undertake cannot satisfactorily be divorced from that of Greater London and any determination of the areas of metropolitan borough councils and of the distribution of functions between them and the London County Council must await an investigation into the wider problem."[1] Why this obvious truth was not discovered sooner has not been disclosed. When it did at last penetrate the minds of Ministers the path of wisdom would have been either to enlarge the terms of reference of the Reading Committee so as to enable it to deal with the regional problem; or to replace it by another committee; or to authorise the Boundary Commission to consider both London and Middlesex. Unfortunately, none of these courses was adopted, and the Minister contented himself with the evasive statement that an examination of the Greater London problem must await a decision concerning the range of functions, old and new, which are to be entrusted to local authorities.

This is a muddled way of approaching the matter. It may be admitted that some functions are appropriate for local

[1] *The Times*: Parliamentary Report, 25th October, 1946.

government in almost any circumstances; while other services, such as the post office, foreign affairs or the currency, belong inevitably to the central government. But between these extremes is a wide range of functions about which it is impossible to lay down hard and fast rules. There are good reasons why public health, education, police, highways, fire brigades, and many other services should be entrusted to local authorities provided that—and only on that condition—their areas and resources are adequate to the task. But in allocating functions one cannot say, "This is local; that is central," on purely abstract grounds regardless of the constitutional and administrative framework. In short, structure and functions are interrelated. Just as one cannot determine structure without regard to functions, so one cannot settle the functions irrespective of the structure.

I have, for example, shown in this book that a popularly-elected regional council for Greater London could efficiently discharge responsibilities which are beyond the competence of existing local authorities. It could take over the work of the Metropolitan Water Board, the Thames and Lee Conservancy Boards, the Roding Catchment Board, and the Metropolitan Police Commissioner.[1] It could administer on a regional basis the hospital services connected with the National Health Service. It could undertake the distribution of electricity and the gas supply.

I have argued elsewhere[2] that a profound mistake is being made by contemporary legislation in wrenching away from local authorities vast blocks of services for which they have long been responsible, and transferring them to Whitehall departments, public corporations and boards appointed by Ministers. This has already happened in the case of trunk roads, civil airfields, electricity supply, hospitals; it will happen shortly in regard to municipal gas undertakings and probably passenger road transport services. The movement from elected councils to selected boards is highly undemocratic. The trend towards a multiplicity of one-purpose

---

[1] *Ante*, Part III, Chapter III.
[2] See my *Development of Local Government*: Prologue, "Local Government in Crisis," to revised second edition.

statutory bodies is bound to lead to ·confusion of purpose, extravagance, and lack of co-ordination.

These defects, as we have already seen, are clearly visible in the *ad hoc* bodies which were in existence in 1939. Since the end of the World War II, legislation has created many additional regional areas and authorities in the metropolis. Under the new electricity legislation the London Electricity Board will take over the administrative County of London and parts of Essex, Kent, Middlesex, and Surrey; the Southern Electricity Board will cover a large region which cuts through Dorsetshire and Gloucestershire in the west, and through Middlesex, Surrey, and Sussex in the east; the South Eastern Electricity Board comprises parts of Kent, Surrey, and Sussex; while the East Anglian Electricity Board includes in its wide sweep the whole of Hertfordshire and parts of Bedfordshire, Buckinghamshire, Essex, and Middlesex.

Under the National Health Service Act, 1946, London is carved up between four great hospital regions in which Regional Hospital Boards have been appointed. The Bill for nationalising the gas industry envisages an entirely different series of areas for gas supply undertakings in which Regional Gas Boards will be set up.

Since the end of the second World War there has been a substantial transfer of powers from county district councils to county councils. This has occurred in regard to education, fire brigades, police forces, public health, and town and country planning. Broadly speaking, the present tendency is to regard county boroughs and county councils as the primary units of local government. The effect of this movement is to complicate the position in Outer London by thrusting large new responsibilities on county councils whose areas are partly inside and partly outside the metropolis. Despite the fact that the number of local authorities is reduced and their size and resources increased, municipal administration in London is made increasingly incoherent by this transfer of functions from county district to county councils, for it brings into the arena several county councils (such as Kent and Essex), whose main interests lie far outside the metropolitan region. This latest policy is therefore in a sense a movement opposed to regional integration.

In any event, neither Parliament nor the Government can have given the slightest consideration to the needs of the Metropolis in strengthening the county councils in Outer London at the expense of the non-county boroughs, urban and rural districts. London is merely the passive victim of a policy designed to substitute larger units for smaller ones, regardless of the size, shape, or suitability of the former from an administrative or sociological point of view.

This strengthening of the county councils in Outer London is likely to make the creation of a rational system of London government much harder to achieve. For the resistance of the county councils to a Greater London Regional Council will doubtless be intensified in direct relation to the increase of their powers, influence and responsibilities.

It is against this background that London faces the future: battered, bruised and shabby, but unconquered, unbowed, and undaunted. Great plans have been made for her by men who were not only technicians but also poets and dreamers. Great hopes have filled the breasts of soldiers and sailors and airmen returning from the wars, and of their womenfolk and friends who faced the terrors of the skies during the momentous years of war. Whether those plans and hopes receive concrete expression or fade away amid widespread frustration will depend in the last resort on whether our statesmen and legislators possess the imagination and the understanding to see that it is not only India, Burma, Ceylon, Malaya, Palestine, and Hong Kong which require new constitutional machinery but also London. Until London receives as much consideration in the Imperial Parliament as Timbuctoo we shall never have a capital city worthy of the name.

INDEX OF PERSONS AND PLACES

INDEX OF SUBJECTS
(INCLUDING BILLS, ACTS AND PARLIAMENTARY REPORTS)

33

GREATER LONDON

RATES MADE 1935 – 36

Under 9/- in the £.
9/- and under 10/-
10/- ,, ,, 11/-
11/- ,, ,, 12/-
12/- ,, ,, 13/-
13/- and over.

ST ALBANS R D.(Part) 9/9

COUNTY OF HE

BARNET R.D. 9/6

POTTERS BAR U.D. 9/8

WATFORD R.D.(Part) 9/7

BUSHEY U.D. 9/11·8 (av)

BARNET U.D. 9/10·2 (av)

WATFORD M.B.(Part) 11/-

HENDON M.B. 9/4

FINCHLEY M.B. 8/8

RUISLIP NORTHWOOD U.D. 10/2 10/2

HARROW U.D. 8/10

WEMBLEY U.D. 9/3

HAMPSTEAD 10/6

UXBRIDGE U.D.

EALING M.B.

WILLESDEN M.B. (av) 10/5·9

PADDINGTON 11/-

ACTON M.B. 9/9

HAMMERSMITH 11/6

KENSINGTON 10/2·5

YIEWSLEY & WEST DRAYTON U.D. 11/9·7 (av)

HAYES & HARLINGTON U.D. 11/10

SOUTHALL NORWOOD U.D. 10/7

BRENTFORD 10/-

CHISWICK M.B. 10/0·8

FULHAM 11/-

HESTON & ISLEWORTH M.B. 10/6

RICHMOND M.B.

BARNES M.B. 8/9

WANDS 10/8

FELTHAM U.D. 10/4·5 (av)

TWICKENHAM M.B. 10/4

TEDDINGTON U.D. 9/-

9/10

8/8·8

STAINES U.D. 9/0·6 (av)

HAMPTON U.D.

KINGSTON UPON THAMES 8/4

MALDEN & COOMBE U.D. 9/2

WIMBLEDON M.B. 8/6

SUNBURY-ON-THAMES U.D. 10/0·2 (av)

MERTON & MORDEN U.D 9/7

ESHER U.D. 8/9 (Part)

SURBITON U.D. 8/8

SUTTON & CHEAM M.B. 9/-

EPSOM & EWELL U.D.(Part) 9/-

BANSTEAD U.D 9/-(Part)

COUNTY OF

COUNTY OF

RATING MAP

SCALE

½ 0 1 2 3 4 5 Miles

CHESHUNT U.D. 12/6

WALTHAM HOLY CROSS U.D. 12/10

ENFIELD U.D. 12/8

OUTHGATE M.B. 3/6

EDMONTON U.D. 11/8

CHINGFORD U.D. 11/3

CHIGWELL U.D. 10/9

OODGREEN M.B. 10/8

TOTTENHAM M.B. 12/1

SEY B.

WALTHAMSTOW M.B. 13/8

WANSTEAD & WOODFORD U.D. 11/5

ILFORD M.B. 12/4

ISLINGTON 10/2

LEYTON M.B. 13/-

HACKNEY 11/3

DAGENHAM U.D. 14/6

EAST HAM C.B. 16/8

BARKING M.B. 13/6

WEST HAM C.B. 19/4

POPLAR 17/4

STEPNEY 14/3

BERMONDSEY 16/6

DEPTFORD 12/4

GREENWICH 12/5

WOOLWICH 15/4

ERITH U.D. 12/6

CRAYFORD U.D. 10/6

BEXLEY U.D. 10/-

CAMBERWELL 11/8

LAMBETH 11/-

LEWISHAM 10/10

CHISLEHURST & SIDCUP U.D. 10/6

BECKENHAM M.B. (AV) 8/9·7

BROMLEY M.B. 8/11

ORPINGTON U.D. (Part) 10/4

CROYDON C.B. 10/-

REDDINGTON & TON

COULSDON & PURLEY U.D. 8/8

10/-

CATERHAM & WARLINGHAM U.D. (Part)

COUNTY OF ESSEX

COUNTY OF KENT

S U R R E Y

T S

London County Council.

March, 1936.

For Product Safety Concerns and Information please contact our EU
representative GPSR@taylorandfrancis.com
Taylor & Francis Verlag GmbH, Kaufingerstraße 24, 80331 München, Germany

www.ingramcontent.com/pod-product-compliance
Lightning Source LLC
Chambersburg PA
CBHW050328270326
41926CB00016B/3354